Architectures for [illegible]

As greenhouse gas emissions continue to increase, we have embarked on an unprecedented experiment with an uncertain outcome for the future of the planet. The Kyoto Protocol serves as an initial step through 2012 to mitigate the threats posed by global climate change. A second step is needed, and policymakers, scholars, business people, and environmentalists have begun debating the structure of the successor to the Kyoto agreement. Written by a team of leading scholars in economics, law, and international relations, this book contributes to the debate by examining the merits of six alternative international architectures for global climate policy. *Architectures for Agreement* offers the reader a wide-ranging menu of options for post-Kyoto climate policy, with a concern throughout to learn from past experience in order to maximize opportunities for future success in the real, "second-best" world. It is an essential reference for scholars, policymakers, and students interested in climate policy.

JOSEPH E. ALDY is a Fellow at Resources for the Future in Washington, DC. He served on the staff of the President's Council of Economic Advisers, where he was responsible for climate change policy from 1997 to 2000.

ROBERT N. STAVINS is the Albert Pratt Professor of Business and Government at the John F. Kennedy School of Government at Harvard University. He is also Director of the Harvard Environmental Economics Program and Chairman of the Kennedy School's Environment and Natural Resources Faculty Group.

Architectures for Agreement

Addressing Global Climate Change in the Post-Kyoto World

Edited by

JOSEPH E. ALDY AND ROBERT N. STAVINS

CAMBRIDGE UNIVERSITY PRESS
Cambridge, New York, Melbourne, Madrid, Cape Town, Singapore, São Paulo

Cambridge University Press
The Edinburgh Building, Cambridge CB2 8RU, UK

Published in the United States of America by Cambridge University Press, New York

www.cambridge.org
Information on this title: www.cambridge.org/9780521692175

First published 2007

Printed in the United States of America

A catalogue record for this publication is available from the British Library

ISBN 978-0-521-87163-1 hardback
ISBN 978-0-521-69217-5 paperback

To my parents,
Carol and Joe
J.E.A.

To my children,
Daniel and Julia
R.N.S.

Contents

Figures

Tables

Contributors

JOSEPH E. ALDY is a Fellow at Resources for the Future. He served on the staff of the President's Council of Economic Advisers from 1997 to 2000 where he was responsible for climate change policy. He served as lead author of the Clinton Administration's 1998 report *The Kyoto Protocol and the President's Policies to Address Climate Change: Administration Economic Analysis*. His research focuses on climate policy, mortality risk valuation, and energy policy.

SCOTT BARRETT is Professor of Environmental Economics and International Political Economy at the Johns Hopkins University School of Advanced International Studies. He is the author of *Environment and Statecraft: The Strategy of Environmental Treaty-Making* (2003), and *Why Cooperate? The Incentive to Supply Global Public Goods* (2007).

DANIEL BODANSKY is the Emily and Ernest Woodruff Chair in International Law at the University of Georgia School of Law. He was the climate change coordinator in the US Department of State from 1999 to 2001 and has worked with the UN Climate Change Secretariat and the Pew Center on Global Climate Change. His teaching and research interests include public international law, international environmental law, and foreign affairs and the Constitution.

CARLO CARRARO is Professor of Econometrics and Environmental Economics at the Università Ca'Foscari Venezia. He is Director of Research of the Fondazione ENI Enrico Mattei and Co-Director of the EuroMediterranean Center on Climate Change. He was an author of the IPCC's Third Assessment Report and has researched on monetary policy, international economic coalitions, climate policy coalitions, and the dynamics of international environmental agreements.

RICHARD N. COOPER is Maurits C. Boas Professor of International Economics at Harvard University. He is Vice-Chairman of the Global Development Network and has served as Chairman of the National Intelligence Council (1995–1997), Chairman of the Federal Reserve Bank of Boston (1990–1992), Under Secretary of State for Economic Affairs (1977–1981), Deputy Assistant Secretary of State for International Monetary Affairs (1965–1966), and senior staff economist at the Council of Economic Advisers (1961–1963).

DANIEL C. ESTY is the Hillhouse Professor of Environmental Law and Policy at Yale University, where he holds faculty appointments in both the Environment School and the Law School. He is the Director of the Yale Center for Environmental Law and Policy as well as the Yale World Fellows Program. He is the author or editor of eight books and numerous articles on the relationships between the environment and trade, globalization, security, corporate strategy, competitiveness, governance, and development.

JEFFREY FRANKEL is Harpel Professor of Capital Formation and Growth at Harvard University's John F. Kennedy School of Government. He directs the program in International Finance and Macroeconomics at the National Bureau of Economic Research, where he is also a member of the Business Cycle Dating Committee. He served as a member of the Council of Economic Advisers from 1996 through 1999. Until 1999 he was Professor of Economics at the University of California–Berkeley, having joined the faculty in 1979. His books include *American Economic Policy in the 1990s* (2002).

JOYEETA GUPTA is Professor in Climate Change Policy and Law at the Vrije Universiteit Amsterdam and Professor in Water Law at UNESCO-IHE Institute for Water Education, Delft. She is a member of the scientific steering committees of the International Institute of Applied Systems Analysis and of several international research programmes including the Global Water Systems Project. She has published extensively in the area of environmental governance, climate change, and North–South relations. She is an IPCC lead author and her books include *The Climate Change Convention and Developing Countries: From Conflict to Consensus?* (1997).

JAMES K. HAMMITT is Professor of Economics and Decision Sciences at the Harvard School of Public Health and Director of the Harvard Center for Risk Analysis. He is chair of the EPA Science Advisory Board's Advisory Council on Clear Air Compliance Analysis and a member of its Environmental Economics Advisory Committee. His research and teaching concern the development of benefit-cost, decision, and risk analysis and their application to health and environmental policy.

HENRY D. JACOBY is Professor of Management at the MIT Sloan School of Management and Co-Director of the MIT Joint Program on the Science and Policy of Global Change. He was formerly Director of the Harvard Environmental Systems Program, Director of the MIT Center for Energy and Environmental Policy Research, Associate Director of the MIT Energy Laboratory, and Chair of the MIT Faculty. His current research and teaching is focused on economic analysis of climate change and its integration with scientific and policy aspects of the issue.

WARWICK J. MCKIBBIN is Professor of International Economics at the Australian National University. He is also Director of the Centre for Applied Macroeconomic Analysis, a Professional Fellow at the Lowy Institute for International Policy, a Non-Resident Senior Fellow at the Brookings Institution where he co-directs the climate change program, and is a member of the Board of the Reserve Bank of Australia. He is a fellow of the Australian Academy of Social Sciences and was awarded the Centenary Medal for service to Australian society through economic policy and tertiary education in 2003.

AXEL MICHAELOWA is currently doing research on international climate policy at the University of Zurich. He is CEO of the consultancy Perspectives Climate Change that specializes in CDM and emission trading and is an IPCC lead author. He is a member of the CDM Executive Board's Registration Team and on the UNFCCC roster of experts on baseline methodologies. He has written over fifty publications on the Kyoto Mechanisms, including a book on CDM's contribution to development.

JUAN-PABLO MONTERO is an Associate Professor of Economics at the Pontificia Universidad Católica de Chile and a Research Associate at the MIT Center for Energy and Environmental Policy Research. He was Visiting Professor of Applied Economics at the MIT Sloan School of Management during 2001–2002 and Visiting Repsol-YPF Scholar at the John F. Kennedy School of Government in 2005–2006.

RICHARD D. MORGENSTERN is a Senior Fellow at Resources for the Future. He has served as Senior Economic Counselor to the Under Secretary for Global Affairs at the US Department of State; and acting Deputy Administrator, Assistant Administrator for Policy, Planning, and Evaluation, and Director of the Office of Policy Analysis at the US Environmental Protection Agency. His research focuses on the economic analysis of environmental issues with an emphasis on evaluation and design of environmental policies.

SHEILA M. OLMSTEAD is an Assistant Professor of Environmental Economics at the Yale School of Forestry and Environmental Studies. Her research and teaching interests are in the area of environmental and natural resource economics and policy, including both natural resource management and pollution control. Her area of primary research is the economics of water supply and demand, with a focus on urban settings, including the effectiveness of various policy instruments.

JONATHAN PERSHING is Director of the Climate, Energy and Pollution Program at the World Resources Institute. He serves on the Resources Panel and as the facilitator of the Northeast States Regional GHG Initiative, and on the Advisory Boards of the Oregon Climate Trust and the California Market Advisory Committee, as well as on advisory boards to the G8, and to the International Energy Agency (IEA). He has served as the Head of the Energy and Environment Division at the IEA and Deputy Director and Science Adviser for the Office of Global Change at the US Department of State. He is an IPCC lead author and review editor.

WILLIAM A. PIZER is a Senior Fellow at Resources for the Future. He is also Senior Economist at the National Commission on Energy Policy. He was Senior Economist at the Council of Economic Advisers where he worked on environment and climate change issues and was a Visiting Scholar at Stanford University's Center for Environmental

Science and Policy. He won the 2006 Petry Research Prize for the Economics of Climate Change for his research on the effects of long-term uncertainty on discounting.

THOMAS SCHELLING received his PhD in economics from Harvard, served in the US Government 1945–1946 and 1948–1953, was Professor of Economics at Yale 1953–1958, at Harvard 1959–90, and the University of Maryland 1990–2003. He was President of the American Economic Association in 1991. He received the Bank of Sweden Prize in Economic Science in Honor of Alfred Nobel in 2005. His books include *The Strategy of Conflict* (1960), *Micromotives and Macrobehavior* (1978), *Choice and Consequence* (1984), and *Strategies of Commitment* (2006).

ROBERT N. STAVINS is the Albert Pratt Professor of Business and Government at the John F. Kennedy School of Government, and Director of the Harvard Environmental Economics Program. He is a University Fellow of Resources for the Future, the Chair of the Scientific Advisory Board of the Massachusetts Executive Office of Environmental Affairs, and a member of the Board of Directors of Resources for the Future, the Board of Academic Advisors of the AEI-Brookings Joint Center for Regulatory Studies, and the Executive Board of the US Environmental Protection Agency's Science Advisory Board. He has been an IPCC lead author.

LAWRENCE SUMMERS is the Charles W. Eliot University Professor at Harvard, where he served as President from 2001 to 2006. He has taught on the faculty at Harvard and MIT and served in a series of senior public policy positions, including political economist for the President's Council of Economic Advisers, Chief Economist of the World Bank, and Secretary of the Treasury of the United States. In 1993 he received the John Bates Clark Medal, given every two years to the outstanding American economist under the age of forty.

DAVID G. VICTOR is Professor of Law at Stanford Law School and Director of the Program on Energy and Sustainable Development at Stanford University. He previously directed the Science and Technology program at the Council on Foreign Relations, where he remains Adjunct Senior Fellow. His publications include: *The Collapse of the Kyoto Protocol and the Struggle to Slow Global Warming* (2001); *Climate Change: Debating America's Policy Options* (2004); and an edited book of case studies on the implementation of international environmental agreements.

JONATHAN B. WIENER is the William R. and Thomas L. Perkins Professor of Law at Duke Law School, Professor of Environmental Policy at the Nicholas School of the Environment and Earth Sciences, and Professor of Public Policy Studies at the Sanford Institute of Public Policy, all at Duke University. He is also a University Fellow of Resources for the Future, and the President-elect of the Society for Risk Analysis. His publications include the books *Reconstructing Climate Policy* (2003, with Richard B. Stewart) and *Risk vs. Risk* (1995, with John D. Graham).

PETER J. WILCOXEN is an Associate Professor of Economics and Public Administration at the Maxwell School of Syracuse University. He is also the Director of the Maxwell School's Center for Environmental Policy and Administration, and is a Nonresident Senior Fellow at the Brookings Institution. His principal area of research is the effect of environmental and energy policy on economic growth, international trade, and the performance of individual industries.

Foreword

LAWRENCE SUMMERS

THIS is an important book on an important subject. Joseph Aldy and Robert Stavins are to be commended for bringing together, under the auspices of the Harvard Environmental Economics Program, such a distinguished group to think through critical aspects of one of the most important policy problems the world faces.

Many public policy problems are at their root political. In these cases, it is reasonably clear what should be done; the challenge is to get the parties to agree on a plan of action, given the complex political constraints they face. Opinions will differ, but I would place the Israeli–Palestinian conflict, US social security reform, and the completion of the Doha Trade Round in this category. Other problems, such as the repair of the American health-care system or how to address radical Islamic terrorism, are more profound in the sense of needing conceptual work on what the right approach is before it is realistic to aspire to political consensus. Global climate change falls within this latter set of problems, so I think it is particularly appropriate that an institution like Harvard devote its formidable intellectual resources to creative thinking about possible solutions.

As an economist who has served in government, I am particularly pleased to see this volume's emphasis on what might be called policy engineering – the development and implementation of new frameworks to address pressing problems. As traditional engineering rests on but must extend the work of the physical sciences, so also policy engineering rests on but must extend the work of the social sciences. There is a tendency in modern intellectual life to undervalue engineering relative to basic science, and this is no less true with respect to policy engineering and social science. Therefore, volumes like this one make a contribution that goes beyond their particular content. For a number of reasons, I believe that global warming poses uniquely difficult challenges in policy engineering.

First, the essential issues involve a longer time horizon than with respect to any other major public policy problem. Probably the closest alternative is issues associated with entitlements and aging societies where the horizon is a quarter century or so and where success in taking prompt action has not been conspicuous in any democracy. While some costs are already being felt, the greatest costs associated with global warming are unlikely to be felt for half a century or more.

To start, this raises important analytical questions. Traditional approaches to policy analysis rely on what economists refer to as intertemporal discounting. Benefits and costs of a given policy that take place in the future are given less weight to reflect some combination of the human tendency to value benefits today more than tomorrow, the ethical observation that people in the future are likely to be richer than people today, and the observation that through investments it is possible for society to obtain one dollar's worth of goods in the future by spending less than one dollar today.

The Federal government as matter of practice in making investment decisions or doing benefit-cost analyses typically uses a 7 percent discount rate to reflect these considerations.[1] At this rate a $1,000 benefit or cost avoided (after correcting for inflation) fifty years out is worth less than $35 today. If one looks one hundred years out, $1,000 of benefit is worth about a dollar. Even at a 3 percent rate, a $1,000 benefit one hundred years out is worth only about $50 today. Essentially it is on this basis that the Copenhagen consensus group of economists placed global warming at the bottom of a list of development priorities (Lomborg 2004).

While this approach has weight – and ethical force in light of the likelihood that our descendants will be far wealthier than we are – it fails to give adequate weight to the moral intuition that most of us have of our obligation to posterity. The greatest acts of statesmanship – starting with the American Revolution and establishment of our Constitution have been motivated by a concern for posterity not by benefit–cost analyses. How best to recognize this obligation in carrying out policy analysis, while at the same time maintaining some rigor in recognizing that resources are scarce and become increasingly so

[1] US Office of Management and Budget (1992). The September 2003 revision of OMB Circular A-4 on Regulatory Analysis recommends the use of 3 percent and 7 percent real discount rates in government benefit-cost analysis, and consideration of lower discount rates for policies with intergenerational benefits or costs.

over time, is a question that deserves much more analysis than it has received to date.

Equally as important as the analytical questions involved with the long horizons associated with global warming are the issues of structuring decision-making processes. How can democracies of finite-lived citizens electing leaders every several years find their way to taking steps with real costs whose primary benefits will be enjoyed only by subsequent generations? The question is not just one of assuring that commitments are made but also, as the experience in a number of countries with the Kyoto treaty illustrates, that they are honored.

The second respect in which global warming stands out as a public policy problem is our inability to predict with confidence its impacts or even to enumerate with conviction all the possible impacts. We are in the realm of unknown unknowns. My layperson's reading of scientific evidence suggests that there is now little room for debate about the reality of large-scale, anthropogenic climate change. Those who question the reality of human impact on climate look increasingly like those who are still arguing about whether tobacco has harmful health effects. But to say that a human signature on global climate change has been conclusively established is not to say just how large this signature is or what its impact will be. There is a possibility that climate change will prove relatively gradual in its effects and that some impacts will be benign, such as agricultural improvements due to increased carbon dioxide concentrations. On the other hand, there are clearly malign effects such as coastal flooding due to rising sea levels, and a whole range of negative possibilities from increasing storm activity to large-scale melting of polar ice sheets. There are also likely to be important micro-climatic effects that bear on particular regions.

Of particular concern is the possibility of nonlinear effects involving positive feedback effects. If initial climate impacts are self-magnifying, as, for example, the melting of polar ice raises global temperatures, then relatively small impacts can have rapid and profound consequences. But it is very difficult to judge the likelihood or quantify the consequences of such scenarios. It cannot be responsible public policy to ignore risks until it is conclusively established that they will play out. On the other hand, I think economists are right to have great difficulty with the so-called "precautionary principle" favored by many environmentalists, which essentially calls for always assuming the worst. What is needed is an approach that recognizes our inherent ignorance and seeks to preserve as much flexibility for future generations – by neither allowing greenhouse gases past a point of unreasonable risk nor preventing the energy use that is essential if standards of living globally are going to rise. In this regard, it would be useful to imagine how some of the approaches described in this volume could evolve with different possible scenarios for the impact of global climate change so as to evaluate their robustness.

The third respect in which global warming stands out as a public policy problem is that for any successful solution it requires international cooperation at a scale to which we are not accustomed. Although developed countries still account for the majority of carbon emissions, the balance will soon tilt towards developing countries. Today, most of the growth in emissions of greenhouse gases is coming from developing countries, and within a decade, if current projections prove accurate, developing countries will account for more emissions than all OECD member countries combined (US Energy Information Administration 2006). Moreover, the areas of the world that are likely to experience the greatest impacts from climate change, such as Bangladesh, are not especially important sources. Conversely some important source countries such as Russia are likely to be much less burdened and may even benefit from global climate change.

There is not much precedent for international cooperation involving the combination of the breadth of countries and the magnitude of the policy commitments that will be necessary to address global warming. Tom Schelling in his chapter in this volume rightly holds out NATO as perhaps the most significant historical example of countries making major commitments to one another. I cannot help but wonder whether NATO would have been formed in the late 1940s if the Soviet threat was fifty years off. Nor, I suspect, could it have been formed without the singular role of the United States whose security commitment to Europe made participation an easy choice for European countries. In a world where there is no dominant actor in global climate change, reaching agreement will be more difficult.

An alternative and perhaps more encouraging example of international cooperation is the European Union in which countries have made very significant economic commitments to one another, sacrificed substantial sovereignty, and indeed engaged in significant resource transfers. Yet the Common Market and now the European Union arose out of particular historical circumstances that seem difficult to replicate quickly on a global scale.

The difficulty of reaching and enforcing international agreements with the necessary scope is reinforced by particular aspects of the global climate challenge. The countries where the opportunities to reduce the emission of greenhouse gases most efficiently and with the greatest leverage, such as India and China, currently have very low emissions per capita and have contributed less cumulative emissions to the atmosphere than some developed countries, and so assert a strong moral claim to being left alone to catch up with the industrial world. Moreover, unlike with CFCs or lead in gasoline, there is no qualitative principle like zero emissions that can serve as a focal point for agreement. Any targets or commitments that result from negotiation will inevitably appear somewhat arbitrary.

New approaches to international cooperation will be required if strong steps are to be taken with respect to global climate change. It has been said that in democracies fear does the work of reason. Perhaps as concern increases about the consequences of global warming, the willingness of nations to enter into truly binding agreements will increase. But I suspect considerable imagination will be required as to how agreements can be made attractive to the major developing countries or made to be effective without their participation.

Policy observations

During the 1990s, I was an active participant in the Clinton–Gore administration's policy process leading up to Kyoto. It gave me a clear sense of the difficulties involved and broadened my perspective beyond the economists' narrower view I had previously taken. While I continue to believe that careful, rigorous, and thorough economic analyses of alternative proposals is a sine qua non for effective policymaking in this area, in the remainder of this foreword, I want to focus on some of the noneconomic considerations and perspectives I believe, based on my experience in government and at Harvard, need careful attention in any informed debate on global climate change policy.

First, there is the constraint imposed by politics that governments are unlikely to write substantial checks to each other pursuant to international treaties in the not too distant future. It is revealing in this regard that the Maastricht Treaty, which called for relatively limited financial penalties within the European Union for the violation of fiscal rules to which everyone was deeply committed, has not succeeded in constraining behavior or led to the successful imposition of fines.

Estimates differ as to the size of the potential transfers connected with global climate change, but it seems likely that achieving full international efficiency could require transfers in the range of tens of billions of dollars. As one who has sought, with mixed success, to induce the US Congress to support transfers in low hundreds of millions of dollars to international financial organizations at a time when the US economy was imperiled by international financial instability, I am skeptical that US policy would ever contemplate transfers in the billions of dollars. I fear this kind of political constraint may be every bit as real as the various natural constraints imposed by the laws of chemistry and physics.

This is not a counsel of despair. It is rather a call for incorporating these constraints into the design of optimal frameworks. One potentially fruitful approach involves assigning emission rights not to nations but to corporate entities, and allowing these rights to be traded even across international borders. Clearly we need some national allowances. We could construct national aggregates from total transfers, just as there is a focus on current account balances today. Other important questions raised by this approach concern the treatment of new enterprises and whether they are differentially handled in developed and developing countries. These issues are taken up in chapters 2 and 3, where approaches characterized by "targets and timetables" are examined.

The second approach – the focus of chapters 4 through 7 – involves seeking to coordinate domestic policies alternatively categorized as "harmonized domestic policies" and "coordinated and unilateral policies." The idea here parallels what is done in international trade agreements where nations accept commitments made by other nations. While subject to the difficulties noted above, this approach has, I believe, considerable potential. It is reinforced by international emulation effects of a kind not factored into standard "realist" analyses in political economy. Because of emulation effects, the scope for coordination may actually be greater than it seems. I am struck by the example of the Basel Banking Accords, where countries have adjusted their domestic banking systems even when they are not of great systemic importance, simply because they want to be part of the international club. It would be a serious mistake for the United States not to recognize the considerable moral impact on the international community of any meaningful actions it takes to reduce greenhouse gas emissions.

A second broad observation goes to the need to be catholic in pursuing multiple areas of policy research. I do not think it is jaundiced to recognize that there is a possibility that climate change will prove to have been much more seriously consequential than today's scientific consensus suggests and at the same time that we will have done too little to prevent the problem. I very much hope on both counts that such a situation never materializes. It does, however, raise the question of geoengineering, a topic which is an anathema to environmentalists. Geoengineering involves the kind of intrusion into a highly complex ecosystem that we should all fear most. The prospect of trying to balance one man-made global effect with another is certainly disconcerting, and it may well turn out that there are no viable geoengineering solutions. Nonetheless, there is a prospect and a possibility that viable solutions could be identified. Such solutions could be crucial in a time of what otherwise could be moments of environmental catastrophe. They could also contribute to reducing the adverse effects of climate change at a fraction of the cost of more direct approaches.

Without necessarily being as enthusiastic as Tom Schelling is in Chapter 8 or Scott Barrett in Chapter 6, I think it would be a shame if this was not a more active area of research in the future than in the past. Much as it is prudent to invest in carbon dioxide emissions abatement because of the potential for enormous unforeseen costs, it is also prudent to invest in geoengineering research for the prospect of its enormous and as-of-yet unforeseen benefits. If successful strategies were to be identified, carefully tested and implemented, and the difference between benefits and costs proved to be sufficiently large, then even a relatively small probability of success would justify carrying on an active research program.

In the same way that many who were rightly concerned about nuclear war abhorred any consideration of a post-nuclear environment for fear that it would lower the nuclear threshold, many environmentalists fear that legitimating discussions of geoengineering will undermine efforts at more direct approaches to the climate change policy challenge. I sympathize with this concern, and this is why I have treated the geoengineering issue gingerly in these observations. I suspect, however, that there is at least a small chance we would live to regret treating the whole subject as taboo.

The third broad observation concerns a policy aspect that is probably very important in practice, but which is very uncomfortable for economists. It is the area that environmentalists refer to as "win-win" situations. These are situations where reducing energy use or carbon dioxide emissions makes a company more profitable or lets a nonprofit organization free up valuable resources. Economists tend to be skeptical of such opportunities, because it is an article of faith for them that profit-maximizing firms do not systematically leave hundred dollar bills on the table. Others seem to see such opportunities as pervasive. It is not uncommon to hear claims that increasing energy efficiency would yield double-dividends. On the one hand, it would decrease emission of greenhouse gases; on the other hand, such improvements would pay for themselves.

My experiences as a kind of CEO at Harvard have frankly made me more sympathetic to these claims than my previous experience as an academic economist. I have come to see that what economists would call agency problems often lead to inefficiency. At Harvard, for example, building architects were judged on what their designs cost per square foot to build with no attention paid to subsequent operating costs. Similarly, those with construction responsibility were judged on whether they brought a building in under or over projected cost, not on what happened to energy outcomes. It is little wonder campus buildings are often less energy-efficient than optimal, and that opportunities to make marginal investments in energy efficiency with an internal rate of return far greater than any sensible hurdle rate – of 15 or even 20 or 30 percent – are forgone. Establishing the so-called "green campus fund" during my tenure had the effect of changing budgeting procedures to recognize some of these impacts. The gains were substantial.

I doubt that Harvard is unique in not providing all the right internal incentives for energy efficiency. The appropriate policy to address this reality is less clear. I worry about trying to overcome these internal problems of corporations with regulation that by its nature is likely to be imperfectly targeted. Universal government standards cannot be the answer. Voluntary efforts associated with ideas like the "triple bottom line" seem more fruitful to me, but it is hard for them to be the basis for binding national commitments. We now have a rich theory of why and how firms exhibit what Harvey Leibenstein called "x-inefficiency," based on the problems of harmonizing internal incentives (Leibenstein

1966). These kinds of problems are probably even more important in the case of public enterprises and firms in complex environments where maximizing profits may not be the most important objective. Applying some of these ideas to the environmental area would seem to be very fruitful.

There is an additional aspect related to efforts to force technological improvements. A great deal of experience with public policies ranging from new weapon systems to infrastructure projects such as Boston's "Big Dig," to programs such as Medicare prescription drugs suggests that the costs of new public initiatives that represent a step into the unknown typically exceed initial estimates by a wide margin. On the other hand, in the environmental area, the tendency has been the opposite, most conspicuously in the case of sulfur dioxide control programs introduced to control acid rain. Once commitments were made, the costs were much less than initially predicted.

One plausible hypothesis as to why the experience is so different in the environmental area comes from considering the sources of forecasts. Weapon systems, infrastructure projects, and entitlement programs all have cost estimates produced by or at least relying on information from those who strongly support them. The major source of information on the costs of environmental policies is likely to be businesses who would rather avoid new regulation and who may underestimate the new technologies that will be created if sufficient incentive is provided. It may be right to extrapolate somewhat from previous environmental initiatives and conclude that the cost of policy commitments may actually prove over time to be less than now appears.

Conclusion

In the physical and life sciences, it is taken for granted that basic research pays off remarkably over time. No one doubts that research on the inner working of cells unconnected to any particular disease will pay off over the long run, or fails to recognize that fundamental research on quantum theory ultimately pays off in new materials or even that sophisticated research in pure mathematics pays dividends in the encryption algorithms that protect our privacy. It often seems less intuitive that basic research pays off in the social scientific realm. Yet over time the impacts can be very large. I remember back in the 1970s, when the idea of market-based solutions to environmental problems was first put forward and was being dismissed as "licensing pollution." Auction theory was seen as a mathematical economist's plaything for many years but has come to shape public policy in areas as disparate as the leasing of government lands, the realizing of spectrum rights, and indeed the creation of markets for sulfur dioxide. Congestion tolls were seen as an unworkable and impracticable economic idea when first put forward and yet have come to be highly successful in managing traffic and pollution in central London. There are many more such examples.

If the world is to grapple successfully with a challenge like global climate change, basic social scientific research in economics, international relations, organizational theory, and many other branches of the social sciences can make an important contribution. I commend the editors and the authors who have contributed to this volume, and I look forward to the widespread discussion it will undoubtedly generate.

References

Leibenstein, H. (1966). "Allocative Efficiency versus 'X-Efficiency,'" *American Economic Review* 56(3): 392–415.

Lomborg, B. ed. (2004). *Global Crises, Global Solutions*, Cambridge: Cambridge University Press.

US Energy Information Administration (2006). *International Energy Outlook 2006*, Report DOE/EIA-0484(2006). Washington, DC: Department of Energy.

US Office of Management and Budget (1992). "OMB Circular A-94, Revised," October 29, Washington, DC: Executive Office of the President.

1 Introduction: International policy architecture for global climate change

JOSEPH E. ALDY AND ROBERT N. STAVINS

EXCEPTIONALLY diverse aspects of human activity result in greenhouse gas emissions that are the source of anthropogenically induced global climate change. Such emissions occur in every part of the world – a coal-fired power plant in the United States, a diesel bus in Europe, a rice paddy in Asia, and the burning of tropical forest in South America. Today's emissions will affect the global climate beyond our lifetimes – most greenhouse gases reside in the atmosphere for decades to centuries. The impacts of global climate change pose serious, long-term risks. Global climate change is the ultimate global-commons problem, with the relevant greenhouse gases mixing uniformly in the upper atmosphere, so that damages are independent of the location of emissions. Because of this, a multinational response is required. To combat the risks posed by climate change, efforts that draw in most if not all countries over the long term will need to be undertaken. The challenge lies in designing an *international policy architecture* that can guide such efforts.

This focus on climate policy architecture reflects the need to establish a foundation of policy principles and institutional infrastructure to inform and frame multilateral and national actions. Richard Schmalensee highlighted the need for a long-term policy architecture in his 1998 review of the Intergovernmental Panel on Climate Change's assessment of policy instruments. He called for the "establishment of effective institutions for policymaking, as well as a policy architecture that permits efficient transitions between particular policies. When time is measured in centuries, the creation of durable institutions and frameworks seems both logically prior to and more important than choice of a particular policy program that will almost surely be viewed as too strong or too weak within a decade" (Schmalensee 1998: 141).

The current climate policy architecture has evolved since 1992 under the United Nations Framework Convention on Climate Change

(UNFCCC) and the Kyoto Protocol. These agreements have provided several near-term emission goals and commitments for some countries but have failed to set long-term quantitative goals for the entire international community. The Kyoto Protocol is merely a first step, stipulating emission targets for industrialized countries through 2012. Significant interest has focused on designing post-Kyoto policies, either building on the Kyoto framework or transitioning to a different policy approach.

This book addresses this need to design a post-2012 international climate change policy architecture. Building on a May 2006 workshop at Harvard University that brought together the world's leading scholars on climate policy architecture, this volume presents six proposals for successors to the Kyoto Protocol. Some of these proposals clearly build on the foundation established by the Kyoto agreement, while others focus on the need for developing an entirely new policy infrastructure. A dozen commentaries provide critical reviews of the policy designs and political questions raised by the proposals. The book closes with an epilogue by Thomas Schelling plus our synthesis chapter that provide insights both for the international policy community regarding the design of climate policy architecture and for the academic community as it considers how to address unresolved research questions.

The next section briefly surveys the state of climate science to establish the need for policies to mitigate climate change risks. We then describe the evolution of global climate policy architecture from the UNFCCC through the Kyoto Protocol. After describing the current policy architecture, we elaborate on the strengths and weaknesses of the UNFCCC/Kyoto Protocol and look beyond the Kyoto Protocol, with a discussion of post-2012 policy processes. The final part of the chapter provides an overview of the six proposals and their associated commentaries for post-2012 climate policy architectures that serve as the core of this book.

Human activities and the global climate

Over the past several decades, progress in global climate change research has found with increasing levels of confidence that human activities are affecting and will continue to affect the global climate. Over the last century, global anthropogenic emissions of carbon dioxide (CO_2) from fossil fuel combustion increased from 0.5 billion metric tons of carbon to 6.7 billion metric tons annually. Over this same time, the atmospheric concentration of carbon dioxide increased from 295 parts per million (ppm) to 369 ppm. From 1900 to 2000, global average temperatures increased by 0.6°C (1°F), with the 1990s the warmest decade on the instrumental record (dating to 1861), and likely the warmest decade in 1,000 years in the Northern Hemisphere. During the twentieth century, sea levels rose on average 10 to 20 centimeters (Marland, Boden, and Andres 2006; Neftel *et al.* 1994; Keeling and Whorf 2005; IPCC 2001).

The state of knowledge has improved with respect to detection and attribution of the human impact on climate as summarized periodically by the Intergovernmental Panel on Climate Change (IPCC). Established by the World Meteorological Organization and the United Nations Environment Programme in 1988, the IPCC convenes thousands of natural and social scientists periodically to review and synthesize the state of scholarly research on global climate change for the policy community. The IPCC has published four major assessments of the climate change literature, and with each review the IPCC has found stronger evidence of human impacts on the global climate.

In its first assessment, the IPCC (1990) reported that greenhouse gas emissions from human activities were increasing atmospheric concentrations of these gases. Reflecting on the quickly expanding academic literature on climate science, the IPCC concluded in its second assessment report that "the balance of evidence suggests a discernible human influence on global climate" (IPCC 1996: 4). In its third assessment report, the IPCC stated "[M]ost of the observed warming over the last 50 years is likely to have been due to the increase in greenhouse gas concentrations. Furthermore, it is very likely that the twentieth-century warming has contributed significantly to the observed sea level rise, through thermal expansion of seawater and widespread loss of land ice" (IPCC 2001: 10).

In response to a request in the United States by the George W. Bush Administration to review the IPCC's conclusions on climate science, a committee of the National Academy of Sciences agreed with the IPCC's general findings. The National Academy committee opened its 2001 report by stating: "Greenhouse gases are accumulating in the Earth's atmosphere as a result of human activities, causing surface air temperatures and subsurface ocean temperatures to rise. Temperatures are, in fact, rising" (National Research Council 2001: 1).

The IPCC stated in its fourth assessment report that "warming of the climate system is unequivocal, as is now evident from observations of global average air and ocean temperatures, widespread melting of snow and ice, and rising global mean sea level" (IPCC 2007: 4). Having established that the global climate is warming, the IPCC concluded that "understanding of anthropogenic warming and cooling influences on climate has improved since the Third Assessment Report, leading to *very high confidence* that the globally averaged net effect of human activities since 1750 has been one of warming" (IPCC 2007: 3; emphasis in original). With 90 percent confidence, the IPCC attributed most of the warming over the past half-century to the increase in anthropogenic greenhouse gas concentrations.

The IPCC (2007) forecasts accelerated warming under a variety of scenarios. Even if atmospheric concentrations of greenhouse gases could be held constant at the year 2000 levels through the twenty-first century, the global climate would still warm 0.6°C (+/− 0.3°C). Under a variety of long-term emission scenarios, temperature increases could range from 1.1 to 6.4°C by 2100.

The changing climate will result in a myriad of impacts. The sea level will rise, on average globally, about 20 to 60 centimeters through 2100. The severity and frequency of hurricanes, floods, droughts, and other extreme weather events may increase. Heat waves will become more common. Agricultural, fishery, and forest productivity will change, with adverse impacts more likely with higher levels of warming. The ranges of vector-borne diseases such as malaria will expand. Some species of plants and animals, especially those inhabiting unique ecosystems, may be at risk as the climate changes, especially since the rate of change may exceed their capacity to migrate and adapt. The capacity to adapt to such impacts, as evident by endowments of human capital and technology as well as effective government institutions, varies substantially around the world, and such heterogeneity in capacity may persist for some time.

The potential risks from increasing atmospheric greenhouse gas concentrations include potential catastrophic events. A warmer world may weaken the Atlantic Ocean's thermohaline circulation – the Gulf Stream that currently carries warm water from the Caribbean north and east to Europe – resulting in colder temperatures and different precipitation patterns for Europe as the rest of the world warms. A changing climate could also result in relatively rapid and large increases in sea level – on the order of ten or more meters – if Greenland or the West Antarctic ice sheet effectively melts. A warmer climate may induce strong, positive feedbacks, such as through the release of large amounts of methane from thawing of permafrost.

The global, long-term impacts of increasing emissions of greenhouse gases provide some insights about how to design policies to address climate change. A ton of carbon dioxide has the same effect on the global climate regardless of whether it is emitted in Shanghai, Stockholm, or San Francisco. This ton of carbon dioxide could remain in the atmosphere for more than a hundred years. Moreover, most other greenhouse gases are both more potent than and longer lasting in the atmosphere than carbon dioxide. Uncertainty about the effect of a ton of carbon dioxide on the climate and the small probability of major events does not suggest inaction. Rather, such uncertainty commends policy action as a hedging strategy or insurance policy as more information about climate science is developed (Manne and Richels 1992).

These characteristics of the climate change problem, as noted by Lawrence Summers of Harvard University in his Foreword to this book, also provide formidable challenges for effective policy responses. The long-term effects of climate change raise questions about how we weight the welfare of today's generation versus the welfare of future generations in making decisions. The decade- to century-long time frame of climate change does not square well with the shorter political time horizon that most elected officials operate under. The uncertainty about the impacts of climate change requires the pursuit of flexible policy approaches that are robust under a variety of possible climate change scenarios. The global nature of the problem calls for international cooperation, and, as Summers suggests, considerable imagination will be required to design climate policy architectures that can effectively address these challenges.

The evolution of climate change policy architecture

The United Nations Framework Convention on Climate Change (UNFCCC)

The global-commons nature of the climate change problem motivated several international conferences in the 1980s to consider coordinated goals and policies. In 1990, the United Nations General Assembly, based in part on the IPCC's first assessment report, initiated negotiations for a multilateral framework to address the risks posed by global climate change. This negotiation process resulted in the United Nations Framework Convention on Climate Change, signed at the United Nations Conference on Environment and Development in Rio de Janeiro, Brazil in 1992. The US Senate voted unanimously to ratify the agreement later that year, and the treaty entered into force in 1994. With 190 countries as parties to the UNFCCC, this treaty enjoys broader participation than nearly any other multilateral agreement.

The UNFCCC created a global policy architecture with four key elements: a general long-term environmental goal; a near-term environmental goal with specific quantitative targets; concerns about equity; and preference for cost-effective implementation.[1] These elements are important because they have largely defined international policy architecture to address climate change since 1992.

The UNFCCC recognized the very long-term impacts of greenhouse gas emissions by setting a long-term environmental goal. The UNFCCC established as the primary objective of climate change policy the "stabilization of greenhouse gas concentrations in the atmosphere at a level that would prevent dangerous anthropogenic interference with the climate system" (Article 2). Although the world community agreed that climate change policy should strive to maintain atmospheric greenhouse gas concentrations at a "safe level," they did not articulate what this meant. Some suggested quantifying this objective with a long-term greenhouse gas concentration stabilization goal (e.g., 550 parts per million – about double preindustrial carbon dioxide concentrations) or a temperature increase stabilization goal (e.g., 2°C above current levels). However, in the absence of consensus within the scientific, economic, and political communities, negotiators at subsequent Conferences of the Parties (COPs) to the UNFCCC did not agree to specific quantitative expressions of the ultimate objective.

The UNFCCC set a near-term goal for industrialized countries, consisting of most members of the Organisation of Economic Co-operation

[1] The FCCC also initiated, *inter alia*, processes for monitoring greenhouse gas emissions, communicating countries' climate policies, reporting on how climate change may affect parties to the Convention, and financing technology transfer via the Global Environment Facility. Although all play important roles in climate change policy, they are secondary to the four key elements identified in the text above.

and Development (OECD) and most countries with economies in transition (together, forming the so-called Annex I countries in the treaty). These countries agreed to a nonbinding aim to stabilize their greenhouse gas emissions at 1990 levels starting in 2000. This focus on a country's "output" instead of its actions or "inputs" created the precedent that policy commitments would take the form of quantitative emission targets. This approach also created the precedent for country-level discretion about how to implement policies to meet those targets. Compliance with these voluntary goals was not impressive – most countries with year 2000 emissions below their 1990 levels met their goal through substantial economic decline and transformation (e.g., Russia and Germany, the latter because of reunification) or non-climate-related energy sector reforms (e.g., the United Kingdom).

Reflecting concerns about equitable burden sharing, the UNFCCC declared that the principle of "common but differentiated responsibilities" (Articles 3 and 4) should guide climate change policy. This translated into a very clear policy dichotomy between the industrialized countries and the developing countries. Industrialized countries took on emission targets, but developing countries had no quantitative emission goals or any other policy obligations. OECD member countries also had financial and technology transfer obligations to developing countries. The "obligations" of developing countries include occasional reporting on their climate vulnerabilities and climate change policies, monitoring and reporting of greenhouse gas emissions, and accepting financial and technology transfers from OECD countries.

To provide some experience with cost-effective implementation policies, the UNFCCC established a pilot program for so-called "Joint Implementation" (JI). This would allow an industrialized country to invest in an emission-reducing project in a developing country and use the emission reductions toward its 2000 emission goal. It was thought that allowing industrialized and developing countries to implement such projects jointly would exhibit some of the potential for emission trading to lower costs of achieving emission goals. This project-based emission trading draws on the fundamental characteristics of greenhouse gases – they mix globally and reside in the atmosphere for very long periods of time – so the climatic impact of an emission reduction does not vary with location. A broad and efficient emission trading program would have the potential to reduce emissions at lowest possible cost, by providing market-based incentives for emission sources to seek out the least-cost emission abatement opportunities. But the pilot program resulted only in a modest number of jointly implemented emission reduction projects.

The Kyoto Protocol

At the UNFCCC's first Conference of the Parties in Berlin, Germany, in 1995, the international community decided to begin a new round of negotiations for a second set of commitments by industrialized countries. The "Berlin Mandate" called for commitments by industrialized countries after 2000 and reiterated the UNFCCC's "common but differentiated responsibilities" language in effectively exempting developing countries from emission commitments. The following year at the second COP in Geneva, Switzerland, the United States advocated in favor of binding quantitative emission commitments. These two years of negotiations set the stage for the third COP in Kyoto, Japan, in December, 1997.

On the eleventh day of the ten-day Kyoto conference, the parties to the UNFCCC agreed on the terms of what came to be known as the Kyoto Protocol. This agreement built on the foundation laid by the Framework Convention on Climate Change, by serving as the first step toward the UNFCCC's ultimate objective through ambitious, near-term quantitative targets for industrialized countries with policy mechanisms for cost-effective implementation.

The Kyoto Protocol established emission commitments for industrialized countries for the 2008–2012 time frame, the so-called first commitment period. As such, it was intended to be a first step toward a long-term, but still unspecified objective. Several European countries initially advocated much longer-term emissions, concentrations, and temperature goals, but these received limited attention at the Kyoto conference. Instead, industrialized countries agreed to ambitious, binding quantitative emissions targets for the 2008–2012 commitment period. At the time, the agreement was expected to result in industrialized countries' emissions declining in aggregate by 5.2 percent below 1990 levels.

A grand bargain to secure acceptance by countries with concerns about the economic burden of these targets included an array of market-based mechanisms to promote cost-effective implementation. The Kyoto Protocol created tradable emission allowances for industrialized countries with quantitative targets that would serve as the basis for an international emissions market. This same set of countries could also engage in JI projects among each other. The agreement also established the Clean Development Mechanism (CDM), a framework for JI projects to generate emission reductions in developing countries that would be financed and used as credits by industrialized countries to satisfy (partially offset) their targets.

The agreement included other elements of flexibility to promote cost-effectiveness. The five-year commitment period allowed for implicit trading over time – short-term banking and borrowing. A country's annual emissions could fluctuate between 2008 and 2012, for example, because of business-cycle effects or weather variations, as long as that country's aggregate, five-year quantitative emissions did not exceed its five-year emissions budget under the agreement. In addition, creating commitments based on a basket of all six types of greenhouse gases would allow for implicit inter-gas trading. For example, if a country with a 1990 target could abate methane at lower cost than carbon dioxide, then it would have the flexibility to lower its total cost of compliance by reducing methane more than carbon dioxide so long as the carbon equivalent for all greenhouse gases equaled the 1990 level.

The Kyoto Protocol stipulates that industrialized countries' quantitative emission commitments are legally binding. If a country's emissions exceed its target for the 2008–2012 period, then it is obligated to "repay" those tons in the second commitment period plus a 30 percent penalty. For example, if a country had 10 million tons of carbon equivalent in excess of its target over 2008–2012, then it would have to reduce its emissions 13 million tons below its second commitment target. The Protocol, as in most international treaties, also includes a provision for a country to withdraw from the agreement simply by stating its intent to do so and waiting one year after notification of withdrawal.

As in the UNFCCC, the Kyoto Protocol calls only on industrialized countries to limit their emissions, requiring no emission restrictions or other greenhouse gas policies of any kind for developing countries. Developing countries can participate in the global effort to address climate change by cooperating in CDM projects, submitting reports to the United Nations, and benefiting from technology transfer.

The Kyoto Protocol did not settle all climate policy issues; negotiations at the next four COPs addressed a variety of implementation details in the Kyoto agreement. After the 2001 COP in Marrakech, Morocco, industrialized countries began to ratify the Kyoto Protocol. By that time, however, the George W. Bush Administration in the United States had declared that the United States would not ratify the Kyoto Protocol. The Government of Australia soon thereafter echoed its lack of support. Despite the withdrawal of these two countries, the Kyoto Protocol entered into force in 2005, having met the dual requirements that 55 countries had ratified the agreement and jointly accounted for 55 percent of 1990 Annex I emissions.

Strengths and weaknesses of the existing international policy architecture

The international climate policy architecture embodied in the Kyoto Protocol, building on the foundation provided by the UNFCCC, has been both lauded and criticized. To provide context for the most frequently identified strengths and weaknesses of the Kyoto Protocol, we identified six important criteria for evaluating climate policy architectures in previous work with our co-author (and contributor to this volume) Scott Barrett. We employed the following six criteria to frame our assessments of the Kyoto Protocol and alternative climate policy architectures: (1) environmental outcome; (2) dynamic efficiency; (3) dynamic cost-effectiveness; (4) distributional equity; (5) flexibility in the presence of new information; and (6) participation and compliance.

Environmental outcome refers to a policy's time path of emissions or concentrations of greenhouse gases, or the impacts of climate change. A dynamically efficient policy maximizes the aggregate present value of net benefits of taking actions to mitigate climate change impacts. The criterion of dynamic cost-effectiveness refers to the identification of the least costly way to achieve a given environmental outcome. Distributional equity refers to the distribution of both benefits and costs across populations within a generation and across generations, and can account for responsibility for climate change, ability to pay to reduce climate change risks, and other notions of equity. Given the significant uncertainties that characterize climate science, economics, and technology, and the potential for learning in the future, a flexible policy infrastructure built on a sequential decision-making approach that incorporates new information may be preferred to more rigid policy designs. Finally, incentives for participation and compliance are important, since a climate policy architecture that cannot promote participation and compliance will not satisfactorily address the climate change problem (refer to Aldy, Barrett, and Stavins 2003 for more details on these criteria).

Strengths of the Kyoto Protocol

To some degree, the Kyoto Protocol fares well on the cost-effectiveness criterion in terms of creating market-oriented institutions and rules, including international emission trading, broad coverage of emission sources and sinks, and some temporal flexibility in complying with emission commitments. These market-based approaches, by lowering marginal and total costs of a climate change policy, can deliver environmental benefits at lower cost than without such provisions. The European Union, formerly a critic of emission trading, launched the world's largest emission trading market in 2005. The lessons learned by countries participating in the Kyoto Protocol will benefit future policymakers as they endeavor to revise and improve the climate change policy architecture. On the other hand, the cost-effectiveness of the agreement is severely limited by its virtual exclusion of developing countries and its abrupt, short-term targets.

This same focus on industrialized countries, however, comports well with several notions of distributional equity. First, the Kyoto Protocol calls on those countries responsible for most of the anthropogenic greenhouse gases in the atmosphere to adopt the first binding emission commitments. Second, these countries have a much greater ability to pay for emission mitigation than the poorer, developing countries without commitments.

Weaknesses of the Kyoto Protocol

The most glaring weakness in the Kyoto Protocol is obvious: three of the five largest greenhouse gas emitters in the world do not face constraints on their emissions. China and India do not have quantitative emission targets, and Russia's Kyoto Protocol commitment is so generous that it is unlikely to bind before 2012. In addition, the largest greenhouse gas contributor, the United States, has not ratified the agreement. These omissions raise questions about the environmental outcome, dynamic efficiency, cost-effectiveness, and the incentives for participation in the Kyoto agreement.

The environmental benefits from the Kyoto Protocol are likely to be quite modest, particularly without US participation. Moreover, the agreement does not stipulate medium- or long-term emission goals. The absence of a long-term goal, although at an apparent detriment in terms of the environmental outcome, may be helpful in light of the policy flexibility criterion. Setting quantitative commitments only through 2012 allows for the international policy community to respond to future information and adapt policy tools and goals accordingly. The Kyoto approach of simply focusing on the near-term may be inferior, however, to one that sets long-term goals and allows for policy flexibility in deciding on short-term goals and means of implementation.

The Kyoto Protocol imposes costs on some sources, primarily those in countries with emission commitments, but no costs on sources outside these industrialized nations. This discrepancy in costs lowers the cost-effectiveness of the policy. The difference in costs across countries can also cause emission leakage – carbon-intensive firms in countries with emission commitments and high costs may relocate some operations to countries without commitments, where costs are therefore less. This can further reduce the efficiency and the environmental benefits of the agreement.

In theory, establishing market-based institutions could promote substantial cost-savings, but the failure to include the United States, China, and India eliminates much of the potential gains from trade. This illustrates the trade-off between approaches that promote efficiency and cost-effectiveness and concerns about equity embodied by the notion of "common but differentiated responsibilities." As an emission reduction credit program operating outside the Kyoto Protocol's cap-and-trade system, the Clean Development Mechanism (CDM) suffers from the problem of unobservable, hypothetical baselines for all credits. In addition, the cumbersome bureaucracy associated with project review under the CDM may serve as an example of how not to design an emission trading program.

The Kyoto Protocol clearly does not provide sufficient incentives for participation (take, for example, the United States' withdrawal, and the lack of developing country commitments) and may provide inadequate incentives for compliance. Even worse, the agreement effectively prohibits developing countries from voluntarily joining the set of industrialized countries with emission commitments. Argentina challenged this idea with a proposal for an emission commitment in 1999, but it received little support for modifying the Kyoto Protocol to allow for voluntary accession. Finally, the provision in the Kyoto Protocol for withdrawal from the agreement suggests that "legally binding commitments" may not be so binding, thereby weakening incentives for compliance.

Despite the various shortcomings of the Kyoto Protocol and its rejection by the United States and Australia, the Protocol entered into force in February of 2005. Leading up to the Kyoto Protocol's first commitment period (2008–2012), some industrialized countries have begun to consider or implement policies to abate their greenhouse gas emissions. The European Union (EU) launched its Emission Trading Scheme in 2005 to cover approximately half of the carbon dioxide emissions of EU member countries, primarily from power plants and large industrial boiler facilities. Since 1997, Japan has promoted emission abatement through implementation of the Keidanren Voluntary Action Plans on the Environment, which aim to limit carbon dioxide emissions in thirty-four industries to their 1990 levels by 2010 (Government of Japan 2006). As of the end of 2006, industrialized countries had financed nearly 500 approved and registered CDM projects in developing countries. The Kyoto Protocol and the UNFCCC will serve as the starting point, either as a foundation or a point of departure for future climate change policy architectures.

Looking beyond Kyoto

As the industrialized countries prepared for the Kyoto Protocol's first commitment period (2008–2012), the international policy community began to consider policies beyond Kyoto at the 2005 COP in Montreal, Canada. One process, based on a provision under the Kyoto Protocol, initiated negotiations on emission commitments for the industrialized countries in the post-2012 period. This process would not provide an opportunity to incorporate countries currently without emission commitments under the Protocol – including the United States, China, and India. A second process established a "dialogue" among all countries under the UNFCCC to discuss various "approaches for long-term cooperative action to address climate change" (Decision CP.11, 2005 Montreal COP to the UNFCCC). This second process was designed to be much more open-ended, and neither limited the consideration of any policy options nor mandated a new negotiating process.

In addition to these ongoing UN-sponsored negotiations and conferences on climate change policy, world leaders have also pursued several alternative venues for considering policy responses to the climate change problem. In 2004, Paul Martin, then Prime Minister of Canada, proposed an L20 group of leading nations to address an array of transnational issues, including global climate change. This group, composed of the leaders of twenty major developed and developing countries, would have included all of the largest emitters of greenhouse gases and representatives from every region of the world. By focusing on a much smaller number of countries in discussing climate change policy, Prime Minister Martin envisioned a more efficient process for developing new ideas and policies. In 2006 and early 2007, a series of preparatory meetings were held to discuss and advance the establishment of a leader's forum on climate change and energy security issues.

In a similar vein, the United Kingdom, as host of the 2005 G8 Summit in Gleneagles, Scotland, launched a G8+5 Climate Change Dialogue. This dialogue included the Group of Eight member countries (Canada, France, Germany, Italy, Japan, Russia, the United Kingdom, and the United States) plus five key developing countries (Brazil, China, India, Mexico, and South Africa). At the Gleneagles Summit, four policy-oriented working groups were launched on adaptation, energy efficiency, market-based mechanisms, and technology transfer. The Dialogue established a series of meetings for representatives of the thirteen countries in the four working groups through at least the 2008 G8 Summit in Japan. The Dialogue did not, however, reach agreement on any major policy architecture proposals at the Gleneagles Summit.

In 2005, the United States engaged five developed and developing countries to initiate climate-related technology cooperation through the Asia-Pacific Partnership on Clean Development and Climate. Australia, China, India, Japan, South Korea, and the United States established a series of task forces comprising representatives of their public and private sectors to address climate change and energy issues in several energy-intensive and energy-producing industries. The partnership charges these task forces with the goal of accelerating the development and deployment of clean energy technologies. This effort does not, however, provide a broader framework to guide climate policy and the founding agreement explicitly notes that this partnership serves as a complement to the Kyoto Protocol.

In civil society, other efforts have been undertaken to provide venues for the discussion of post-Kyoto climate change policy architectures. For example, the Pew Center on Global Climate Change, based in the United States, convened a "Climate Dialogue" at the Pocantico conference center in Tarrytown, New York, in 2004 and 2005. This Dialogue promoted informal discussion of various climate policy ideas by policymakers and stakeholders from fifteen developed and developing countries. In November of 2005, the participants agreed to a set of policy elements that they believed could strengthen the multilateral approach to climate change and merit consideration by the international policy community (Climate Dialogue at Pocantico 2005).

Proposals for the Post-Kyoto world

Some policymakers and scholars have been critical of the Kyoto Protocol, noting that because of specific deficiencies it will be ineffective for the problem, and relatively costly for the little it accomplishes. Others have been more supportive of the Protocol, noting that it is essentially the "only game in town." Both sides agree, however, that whether this first step is good or bad, a second step will be required. Given the global-commons nature of the climate change problem, a central element of the second step will most likely be an international agreement. The basic shape and structure of that international agreement – its architecture – is the focus of this book.

In May 2006, the Harvard Environmental Economics Program, with the support of the Harvard University Center for the Environment and the Belfer Center for Science and International Affairs, hosted a two-day workshop at the John F. Kennedy School of Government in Cambridge, Massachusetts. The workshop brought together twenty-seven leading thinkers from around the world who approach the climate policy architecture question from a variety of perspectives, including economics, law, political science, business, international relations, and the natural sciences (refer to Table 1.1 for a list of the workshop participants and their affiliations).

For the workshop and this book, the Harvard Environmental Economics Program commissioned six papers that propose specific international policy architectures to succeed the Kyoto Protocol. Accompanying each paper are two commentaries, some of which provide their own visions for future policy architectures. Both the

Table 1.1 *Participants at May 2006 workshop on architectures for agreement: addressing global climate change in the post-Kyoto world*

Joseph Aldy	Resources for the Future
Scott Barrett	Johns Hopkins University
Daniel Bodansky	University of Georgia
Carlo Carraro	Università Ca'Foscari Venezia
Richard Cooper	Harvard University
Daniel Esty	Yale University
Brian Flannery	Exxon-Mobil
Jeffrey Frankel	Harvard University
Joyeeta Gupta	Vrije Universiteit Amsterdam
James Hammitt	Harvard University
Chris Harrison	Cambridge University Press
Thomas Heller	Stanford University
Henry Jacoby	Massachusetts Institute of Technology
Axel Michaelowa	Hamburg Institute
Juan-Pablo Montero	Pontificia Universidad Católica de Chile
Richard Morgenstern	Resources for the Future
Christopher Mottershead	BP
Sheila Olmstead	Yale University
Jonathan Pershing	World Resources Institute
William Pizer	Resources for the Future
Thomas Schelling	University of Maryland
Daniel Schrag	Harvard University
Robert Stavins	Harvard University
Lawrence Summers	Harvard University
David Victor	Stanford University
Jonathan Wiener	Duke University
Peter Wilcoxen	Syracuse University

proposals and the commentaries were presented at the two-day workshop and received substantial discussion and comments. The authors incorporated these comments in revising and refining their contributions for this volume.

We categorize these six proposals and their associated commentaries into three principal types of architectures: (1) targets and timetables; (2) harmonized domestic actions; and (3) coordinated and unilateral policies. In the remainder of this introductory chapter, we provide an overview of the chapters that follow, including two final chapters that synthesize the key insights from the proposals.

Targets and timetables

The UNFCCC through its voluntary aims and the Kyoto Protocol through its emission commitments provide illustrations of the targets-and-timetables approach – quantitative goals over a specified timeframe. Proposals for targets and timetables for the post-2012 world expand and modify the current policy architecture. Proposals by Jeffrey Frankel of Harvard University and Axel Michaelowa of the Hamburg Institute focus on expanding participation to include developing countries and the United States, while maintaining the international emission trading institutions that can address cost concerns and serve as a means for eliciting participation by reluctant countries.

Frankel presents his ideas for a climate policy architecture along these lines in chapter 2, "Formulas for Quantitative Emission Targets." Frankel proposes the use of a sequence of formulas that combine historic greenhouse gas emissions, population, income, and other variables to set country-level decade-by-decade quantitative targets. These formulas would be specific to different classes of countries, primarily based on income, and evolve over time as more countries adopt emission commitments. In the long term, all countries would have equal per capita emission targets. Formulas for near-term emission targets should be progressive, in Frankel's view, resulting in more-stringent targets for countries with higher per capita income. To promote developing country participation, Frankel endorses international emission trading and formulas that produce generous near-term targets for developing countries such that they would benefit from gains from trade in emission permits. To respond to concerns about uncertainty in target stringency, Frankel also recommends indexing emission targets to economic growth.

Daniel Bodansky of the University of Georgia and Jonathan Wiener of Duke University comment on Frankel's proposal. Bodansky, in his commentary "Targets and Timetables: Good Policy but Bad Politics?", recognizes the virtues of a cap-and-trade approach but raises questions about the political viability of a formula-based system of setting targets. In the near term, Bodansky argues that developing countries have rejected and would continue to reject proposals like Frankel's, and questions whether the United States would embrace a Kyoto-like framework after withdrawing from the Kyoto Protocol. Bodansky then considers whether setting a long-term concentration target (e.g., 450 ppm) could enhance salience and public interest in addressing climate change and thereby catalyze action. He concludes with questions about the need for a top-down architecture versus an evaluation of what can be achieved through a decentralized, bottom-up approach – an issue addressed in several chapters and commentaries in this volume.

Wiener, in his commentary "Incentives and Meta-Architecture," discusses Frankel's ideas within the context of climate policy architecture and the larger suite of multilateral institutions and structure of international relations. He challenges the idea that countries would allow for a quantitative formula to substitute for negotiated allocation of targets. Instead, he suggests that the sequence of formulas Frankel recommends could serve as a "conversation-starter" to frame future negotiations. Wiener is not persuaded that setting near-term aggregate emission goals more stringently than benefit-cost analysis indicates is efficient would promote long-term credibility, and advocates a long-term emission pathway based on benefit-cost analysis. His focus on meta-architecture results from the need to elicit national consent for multilateral policies given the absence of a supranational institution that can coerce action. Wiener suggests that a successful and robust climate policy architecture will need side payments or an explicit consideration of development objectives to secure developing country participation.

Michaelowa also builds on the Kyoto Protocol framework with his proposal in chapter 3, "Graduation and Deepening." He rejects benefit-cost analysis in setting long-term goals and instead embraces an indicative greenhouse gas atmospheric concentration target of 550 ppm. This long-term goal motivates his proposal for emission commitments more stringent over 2013–2017 than during the Kyoto commitment period – on average, 23 percent below 1990 levels during this second period. To expand the set of countries with commitments, Michaelowa advocates a graduation index, based on a country's per capita income and per capita emissions, which would determine when a developing country would "graduate" to emission targets. The stringency of targets would vary across countries, in part a function of their graduation index. This proposal includes near-term commitments for the United States and Australia, and a number of developing countries, including Mexico and Brazil, but not China and India. Michaelowa also supports international emission trading and a more flexible Clean Development Mechanism.

Richard Cooper of Harvard University and Joyeeta Gupta of Vrije Universiteit Amsterdam comment on Michaelowa's proposal. Cooper criticizes the Kyoto-style targets and timetables in "Graduation and Deepening," and advances the idea of a harmonized carbon tax in his commentary "Alternatives to Kyoto: The Case for a Carbon Tax." Cooper disagrees with Michaelowa's decision to reject benefit-cost analysis and argues that the high costs of a stringent long-term atmospheric stabilization goal would be unacceptable to the general public. He also expresses concerns that allocating emission permits and allowing for international emission trading may suffer from favoritism and corruption in many countries. Instead, Cooper recommends a carbon tax levied domestically by individual national governments and harmonized across all countries. He argues that such a tax could result in the same cost-effective mitigation of emissions as an efficient trading program, but without the large income transfers that could support corruption in some developing and transition countries. Cooper suggests setting the tax initially at $50 per ton of carbon and updating it regularly to account for improved knowledge over time about abatement costs and technologies and the climatic impact of greenhouse gas emissions.

Gupta provides a more supportive assessment in her commentary "Beyond Graduation and Deepening: Toward Cosmopolitan Scholarship." She embraces Michaelowa's long-term atmospheric concentration goal and recommends the precautionary principle as a tool for guiding policy. The more stringent near-term targets and the differentiation among countries – based on the polluter pays (emissions per capita) and ability to pay (income per capita) principles – are also in line with Gupta's own proposal for a climate policy architecture. Like Michaelowa, Gupta's approach places countries in various categories of commitments based on their incomes and emissions but also focuses on "mainstreaming" or linking climate policy to other international policy issues, including trade, investment, and development. Gupta describes the potential for litigation and local and state actions to advance efforts to address climate change, and calls for ways to promote good governance and rule of law at the international level.

Harmonized domestic actions

Instead of the UNFCCC/Kyoto approach that attempts to create a new international institutional infrastructure, some have advocated international policy architectures that would essentially accept and adapt to the limits of international institutions. National governments do not cede their sovereignty when they establish global institutions, and some have argued that because of this the design of policy architectures should focus on national and, in some cases, regional institutions. David Victor of Stanford in chapter 4 and Warwick McKibbin of Australian National University and Peter Wilcoxen of Syracuse University in chapter 5 propose policy architectures that attempt to circumvent the disadvantages of global institutions by harmonizing actions across national and regional institutions.

Victor elaborates on the weaknesses of the Kyoto-style architecture and advocates a decentralized approach in "Fragmented Carbon Markets and Reluctant Nations: Implications for the Design of Effective Architectures." His critique of the status quo approach begins with the fundamental difficulty of securing participation in an international agreement that must be perceived by all countries as in their individual interest for them to join. Victor then challenges the conventional wisdom about the key elements in the Kyoto Protocol, including binding targets, international emission trading, and compensation for developing country participation through the Clean Development Mechanism. These criticisms motivate his alternative approach to climate change policy that allows for variable geometry of participation. On process, Victor suggests that a small group of the most important countries in terms of climate change should engage in negotiations on these variable forms of participation. This would involve some country-level pledges of action and could also follow, in part, from fragmented carbon markets, such as the EU Emission Trading Scheme, in order to harmonize the price of carbon across participating countries. Victor's proposal concludes on the topic of developing country participation with the goal of refocusing climate change actions in terms of issues of interest to these countries, such as energy development, local air pollution, and other issues.

Carlo Carraro of the Università Ca'Foscari Venezia and Sheila Olmstead of Yale University comment on Victor's proposal. Carraro's commentary "Incentives and Institutions: A Bottom-Up Approach to Climate Policy" applauds Victor's emphasis on participation incentives as the starting point for designing a climate policy architecture. Carraro compares Victor's political science-oriented assessment of existing policy institutions and carbon markets with the game-theoretic literature in economics that arrives at the same conclusion: the incentives to free ride are strong without a coercive supranational institution. This economic analysis also supports the idea of an evolution of "climate blocs" of countries through a bottom-up process, in a similar fashion as regional trade pacts evolve and complement global trade agreements. Carraro raises several outstanding questions about Victor's proposal, including whether this approach can generate sufficient effort to address climate change, whether it can do so in a measurably equitable manner, and how the focus on mitigation should interact with necessary adaptation investments and actions.

Olmstead endorses some of Victor's critique of the status quo but takes exception with some elements of his proposal in her commentary "The Whole and the Sum of Its Parts." She notes that a climate policy architecture must address "who," "how," and "when." On the "who" question, Olmstead agrees that participation does not need to be universal and likes the focus on a small group of important countries. She disagrees, however, about "how" and questions the dynamics of Victor's proposal. Olmstead supports the use of international emission trading because it provides incentives for developing country participation (through generous emission allocations) and incentives for industrialized countries to invest in developing countries, in contrast with Victor's proposal. She concludes that Victor is silent on "when" and suggests the need for further consideration of the long-term evolution of bottom-up approaches.

McKibbin and Wilcoxen propose an alternative architecture relying on the strength of domestic institutions in "A Credible Foundation for Long-Term International Cooperation on Climate Change." Their proposal relies on an internationally coordinated set of national emission trading regimes. In a participating country, the government allocates to the private sector a set of long-term permits that grant the owner the right to emit a specified amount of carbon equivalent every year for the life of the permit. In addition, each government offers to sell an unrestricted number of annual permits at a predetermined price. Firms could buy and sell permits within the country, but there would be no international emission trading. The policy architecture would require harmonization and coordination of the price of annual permits and guidance on setting the number of long-term permits. The free allocation of long-term permits to the private sector, effectively a large wealth transfer, provides the incentive for these firms to lobby to maintain the emission trading program over the long term.

Richard Morgenstern of Resources for the Future and Jonathan Pershing of the World Resources Institute comment on McKibbin and Wilcoxen's proposal. Morgenstern questions the political economy of their idea and compares it with his work on a safety valve in "The Case for Greater Flexibility in an International Climate Change Agreement." Although he appreciates the elegance of McKibbin and Wilcoxen's proposal, Morgenstern expects difficulty in securing broad international support for a one-size-fits-all domestic architecture. The current variation in domestic emission mitigation policies among industrialized countries with Kyoto commitments does not provide much evidence that a homogenous policy implementation would be well received. Among other concerns, he also wonders how such an approach could promote developing country participation without potential gains from trade through international emission trading. Morgenstern describes how the McKibbin and Wilcoxen proposal differs from his own proposal on a domestic trading program with a safety valve that would auction permits and allow for firms to buy additional permits from the government at the pre-set (safety-valve) price.

Pershing raises questions on the political viability of the McKibbin and Wilcoxen proposal, and offers ideas about using sustainable development policies to build support in developing countries for climate change policy efforts in his commentary, "Using the Development Agenda to Build Climate Mitigation Support." He is concerned about the potential need for the government to buy back the freely allocated long-term permits if climate science deems that necessary. The credibility of long-term permits could be undermined by future governments who decide to renege on previous commitments and take back the permits. Pershing expresses concerns about the use of a price ceiling via the annual permit sales and forgoing potential cost-savings through international emission trading. He recommends incorporating climate change mitigation actions into development priorities that could promote political buy-in among developing countries.

Coordinated and unilateral policies

Very different from the top-down architecture of the UNFCCC/Kyoto Protocol are the third set of proposals, which feature bottom-up climate change policies. Such approaches allow for countries and regions to experiment and learn about the efficacy and effectiveness of various forms and stringencies of alternative policies. As opposed to an international agreement dictating the goals and/or policies countries should pursue, these approaches rely on countries to coordinate or implement unilateral policies. Proponents of bottom-up architectures believe that this can eventually evolve into a more cohesive international architecture, given appropriate experience. Scott Barrett of Johns Hopkins University and William Pizer of Resources for the Future propose climate policy architectures that include a mix of coordinated and unilateral country-level policies.

Barrett identifies the need to take action in four primary areas in chapter 6, "A Multitrack Climate Treaty System." First, he states that countries should take "appropriate measures" to address climate change risks, refocusing attention on actions instead of outcomes. Second, Barrett recommends coordination on a multilateral research and development (R & D) program. To increase diffusion of new climate-friendly technologies, he calls for setting international technology standards and argues that as enough countries adopt these standards, the technologies will become the de facto global standard. Third, Barrett advocates adaptation assistance, with the focus on promoting development in low-income countries with the least capacity and resilience to climate change impacts. Finally, he suggests that the international community should invest in R & D for possible geoengineering responses to climate change. Barrett believes that this suite of unilateral and coordinated policy responses could effectively address the multifaceted characteristics of global climate change.

Daniel Esty of Yale University and Henry Jacoby of the Massachusetts Institute of Technology comment on Barrett's proposal. Esty compliments Barrett's assessment of the current challenges in designing a climate policy architecture but suggests that previous multilateral efforts to address stratospheric ozone depletion provide a better model for moving forward in his commentary "Beyond Kyoto: Learning from the Montreal Protocol." Esty identifies several elements of the Montreal Protocol that resulted in its success and should be incorporated in a global climate policy architecture. He notes that the agreement required that all countries phase out ozone-depleting substances, in contrast with the Kyoto approach of focusing only on industrialized countries. Second, the Montreal Protocol included carrots and sticks to promote compliance, in the forms of subsidies to developing countries to transition to ozone-friendly products, and trade sanctions for both nonparticipation and noncompliance. Third, greenhouse gas emission targets, like the CFC phase-out, should occur over decades, not years, and be achievable at reasonable cost.

Jacoby argues that the international climate change policy regime is moving away from the dream of an elegant, comprehensive top-down architecture, akin to a gothic cathedral, and towards a bottom-up, decentralized architecture that he compares to the ramshackle neighborhoods of Rio de Janeiro in his commentary "Climate Favela." He notes that the UNFCCC, even if it cannot support a universal, comprehensive architecture, does provide a number of useful activities that will support any future climate policy architecture, including information collection and capacity building. Jacoby expresses concerns about the lack of attention given to the institutional context in Barrett's four-pronged policy approach. He raises several questions about the feasibility of implementing agreements on technology standards and adaptation. Jacoby also calls for more analysis of the issue of geoengineering, especially with respect to the issue of coordinating decisions about "engineering" remedies for global climate change.

Pizer addresses what a global climate policy architecture proposal *can* be, not what it *should* be, in chapter 7, "Practical Global Climate Policy." His evaluation of the current behavior by countries within the UNFCCC/Kyoto framework motivates his conclusion that national action, not international coordination, should be the centerpiece of the international climate policy effort. Pizer calls on country-level actions that reflect their domestic interests and influences through a nonbinding pledge and review process. An agreement to coordinate these unilateral actions would provide a venue both for countries to lobby other countries about their commitments and for periodic review. Countries could agree to link their domestic actions, including the integration of emission trading programs and credit-based approaches, such as a streamlined Clean Development Mechanism. The periodic reviews would focus on three types of actions: emission mitigation, climate-friendly technology innovation and deployment, and engaging developing countries, especially through actions that address climate and other development goals.

James Hammitt of Harvard University and Juan-Pablo Montero of the Pontificia Universidad Catolica de Chile comment on Pizer's proposal. Hammitt agrees that Pizer's policy and measures framework may enjoy more support than Kyoto-style targets but raises questions about whether this approach can satisfactorily mitigate climate change in his commentary "Is 'Practical Global Climate Policy' Sufficient?" In response to Pizer's observation that environmental advocates prefer a "safe-level" approach to setting goals as reflected by the ultimate objective of the UNFCCC, Hammitt explains that pursuing policies based on a marginal-damage approach (i.e., setting policy goals based on benefit-cost analysis) may be more protective of the climate. He also notes that a safe-level goal may suffer from more credibility problems than a goal motivated by marginal analysis. Hammitt questions Pizer's conclusion that unilateral policies and measures would not be much less efficient than a functioning international emission trading system because of implicit and explicit harmonization of the marginal costs of abatement.

Juan-Pablo Montero identifies several reservations he has with Pizer's proposal and suggests an alternative approach in his commentary "An Auction Mechanism in a Climate Policy Architecture." Montero believes that Pizer has reached his conclusions about the ineffectiveness of the current policy architecture in haste, and that more time is necessary to evaluate climate policies. He also has concerns about unilateral approaches in terms of their effect on climate change and challenges the notion that coordinated multilateral policies cannot work. Montero advocates a global auction of emission permits by countries, where the auction quantity reflects the demands set by the participating countries. In addition, a portion of the auction revenues would be returned to bidders in a manner that provides the incentive for countries to bid truthfully in the auction. Developing countries could be incorporated into the auction as sellers of emission permits, much in line with David Bradford's 2002 climate policy architecture proposal.

Synthesis and conclusions

This volume closes with two chapters that draw out the key conclusions from the proposals and commentaries and provides guidance for the international policy community. In the Epilogue, Thomas Schelling, the 2005 Nobel Laureate in economics and professor at the University of Maryland, reflects on the range of architectures from the top-down to the bottom-up frameworks. He provides an assessment of the viability of these various approaches. Schelling notes, based on his experiences, the prospect for a pledge and review system to mobilize action, such as with the Marshall Plan and the founding of the North Atlantic Treaty Organization. He raises questions about the incentives for participation and compliance in some of the more-holistic, top-down architectures. Schelling also discusses the potential interaction between the structured international institutions envisioned in some proposals and domestic policy regimes. He asks whether a climate policy architecture is possible or necessary in order to make progress.

We close the book with our chapter: "Architectures for an International Global Climate Change Agreement: Lessons for the Policy Community." We review and synthesize the six proposals and their associated commentaries. We draw conclusions about some issues that reflect a general consensus and highlight others for which these contributors, and the broader academic community, have yet to come to agreement. The lessons provided in this chapter can serve to inform the international policy community, from decision-makers and diplomats to those in the private sector and civil society who actively are engaged in the design of climate change policies.

References

Aldy, Joseph E., Scott Barrett, and Robert N. Stavins (2003). "Thirteen Plus One: A Comparison of Global Climate Policy Architectures," *Climate Policy* 3(4): 373–397.

Bodansky, Daniel (2004). *International Climate Efforts Beyond 2012: A Survey of Approaches*, Arlington, VA: Pew Center on Global Climate Change.

Bradford, David (2002). "Improving on Kyoto: Greenhouse Gas Control as the Purchase of a Global Public Good," Dept. of Economics Working Paper, Princeton University.

Government of Japan (2006). *Japan's Fourth National Communication under the United Nation's Framework Convention on Climate Change*, Tokyo: GOJ.

IPCC (1996). *Climate Change 1995: The Science of Climate Change*, Cambridge: Cambridge University Press.

(2001). *Climate Change 2001: The Scientific Basis*, Cambridge: Cambridge University Press.

(2007). *Climate Change 2007: The Physical Science Basis – Summary for Policymakers*, Published online by IPCC on February 2, 2007 at www.ipcc.ch.

Keeling, C. D., and T. P. Whorf (2005). "Atmospheric CO_2 records from sites in the SIO air sampling network," in *Trends: A Compendium of Data on Global Change*, Carbon Dioxide Information Analysis Center, Oak Ridge National Laboratory, US Department of Energy, Oak Ridge, TN, USA.

Manne, Alan, and Richard Richels (1992). *Buying Greenhouse Insurance: The Economic Costs of CO_2 Emissions Limits,* Cambridge, MA: MIT Press.

Marland, G., T. A. Boden, and R. J. Andres (2006). "Global, Regional, and National CO_2 Emissions," in *Trends: A Compendium of Data on Global Change*, Carbon Dioxide Information Analysis Center, Oak Ridge National Laboratory, US Department of Energy, Oak Ridge, TN, USA.

National Research Council (2001). *Climate Change Science: An Analysis of Some Key Questions*, Washington, DC: National Academy Press.

Neftel, A., H. Friedli, E. Moor, H. Lötscher, H. Oeschger, U. Siegenthaler, and B. Stauffer (1994). "Historical CO_2 record from the Siple Station ice core," in *Trends: A Compendium of Data on Global Change*, Carbon Dioxide Information Analysis Center, Oak Ridge National Laboratory, US Department of Energy, Oak Ridge, TN, USA.

Pew Center on Global Climate Change (2005). *International Efforts Beyond 2012: Report of the Climate Dialogue at Pocantico*, Arlington, VA: Pew Center on Global Climate Change.

Schmalensee, Richard (1998). "Greenhouse Policy Architectures and Institutions," in W. D. Nordhaus (ed.), *Economics and Policy Issues in Climate Change*, Washington, DC: Resources for the Future Press, pp. 137–158.

PART I

Targets and timetables

2 Formulas for quantitative emission targets

JEFFREY FRANKEL

We are sorely in need of ideas as to how to proceed to address the emission of greenhouse gases (GHGs):

(1) Global climate change (GCC) is a huge and genuine problem, as is now more widely recognized than even a few years ago.
(2) The Kyoto Protocol, and the United Nations Framework Convention on Climate Change (UNFCCC) within which it sits, constitute the only multilateral framework we have to address the problem.
(3) The Protocol, as actually negotiated in 1997 or as it went into force in 2005, is inadequate in three important ways: its goals could be costly to achieve if interpreted literally, neither the largest nor the fastest-growing emitters have signed up, and it would have made only the tiniest dent in global GHG concentrations even if it had entered into force with good prospects for compliance and even if all countries had participated.

Few American economists support the Kyoto Protocol.[1] I have spoken and written in support, at least, of the Clinton–Gore version of it, perhaps because I was (one of many) involved in its design during 1996–1999. My claim is that – given the combination of political, economic, and scientific realities as they are – Kyoto is a good foundation, a good first stepping stone on the most practical path if we are to address the global warming problem more seriously, as we should. Nobody would say that the text negotiated in Kyoto is ideal. But, as I phrased it in the title of a recent article, "You're Getting Warmer: The Most Feasible Path for Addressing Global Climate Change Does Run through Kyoto."[2]

The author would like to thank Joe Aldy, Robert Stavins, and Jonathan Weiner for useful comments.

[1] Goulder and Pizer (2006) survey recent economists' research.

[2] Frankel (2005). For another defense of Kyoto as a good foundation, see Sandalow and Bowles (2001).

In this chapter I will try to take a constructive approach, by asking what are the desiderata, the requirements, for a second step in the process, a successor to the Kyoto regime of 2008–2012, one that would build on what is good about it and fix what is most lacking.[3]

What are necessary requirements for the next multilateral treaty?

I see a list of requirements that any new agreement must meet. They are: *More comprehensive participation, Efficiency, Dynamic consistency, Robustness under uncertainty, Equity, and Compliance.*[4]

The first big desideratum: ***More comprehensive participation***, specifically getting the United States, China, and as many other developing countries as possible to join the list of those submitting to binding quantitative commitments.[5] The absence of the developing countries is probably the most serious and most intractable shortcoming of the Kyoto Protocol, except perhaps for the absence of the United States – which is to some extent a function of the competitiveness fears generated by the absence of the developing countries. We need to get developing countries into the system for three reasons:

(1) The developing countries will be the source of the big increases in emissions in coming years according to the business-as-usual path (BAU), that is, the path along which technical experts forecast that countries' emissions would increase in the absence of a climate change agreement. China, India, and other developing countries will represent up to two-thirds of global carbon dioxide emissions over the course of this century, vastly exceeding the expected contribution of the Organisation of Economic Co-operation and Development (OECD) of roughly one-quarter of global emissions. Without the participation of major developing countries, emissions abatement by industrialized countries will not do much to mitigate global climate change.

(2) If a quantitative international regime is implemented without the developing countries, their emissions are likely to rise even faster than

[3] This extends the ideas in Aldy and Frankel (2004) and Frankel (1999, 2005).

[4] Stewart and Weiner (2003) have a similar list.

[5] As pointed out by Barrett (2006), Nordhaus (2006), Olmstead and Stavins (2006), Victor (2004), and just about all other Americans who have written on the subject. Even the Clinton–Gore Administration would not have submitted the treaty to US Senate ratification without meaningful participation by developing countries.

the BAU path, due to the problem of *leakage*. Leakage of emissions could come about by relocation of carbon-intensive industries from countries with emissions commitments under the Kyoto Protocol to nonparticipating countries, or by increased consumption of fossil fuels by nonparticipating countries in response to declines in world oil and coal prices. Estimates vary regarding the damage in tons of increased emissions from developing countries for every ton abated in an industrialized country. But an authoritative survey concludes "Leakage rates in the range 5 to 20 percent are common" (IPCC 2001: 536–544).

And (3) the opportunity for the United States and other industrialized countries to buy relatively low-cost emissions abatement from developing countries is crucial to keep the economic cost low. This would increase the probability that industrialized countries comply with the system of international emissions commitments.

The second big category of concern is ***Efficiency***, which I would argue in this particular context must come down to cost-effectiveness: an attempt to minimize the economic costs of achieving an environmental goal. The critics of Kyoto sometimes fail to give it credit for the market mechanisms and flexibility that were built into the treaty. They are the primary reason why I think it a positive first step on which to build. Yet they could just as easily not have been in the treaty. Most obviously, there is the international trading of emissions permits ("where flexibility"), which is critical. The Clinton Administration Council of Economic Advisers (CEA) reported to Congress in its 1998 Administration Economic Analysis[6] estimates of sharp cost reductions from trading (80 percent, if key developing countries are in the system). These estimates were similar to the independent estimates of Edmonds *et al.*[7] But there is also some scope for smoothing over time ("when flexibility") in the form of averaging over a five-year period, and the possibility of banking credits. And insufficiently appreciated is "*what* flexibility" – scope for substituting across

[6] *The Kyoto Protocol and the President's Policies to Address Climate Change: Administration Economic Analysis*. Summaries included Yellen (1998) and Frankel (1998).

[7] Edmonds, Pitcher, Barns, Baron, and Wise (1992) and Edmonds, Kim, McCracken, Sands, and Wise (1997). We at the CEA made adjustments for the effect of the six-gas objective, as the existing economic models focused on carbon dioxide alone. Joe Aldy was the CEA staff economist who implemented the model.

the six GHGs, not just carbon dioxide, or between the GHGs and sequestration through forestation ("sinks"). These kinds of flexibility not only reduce the expected value of the economic cost but also ease response to fluctuations in the price of carbon.

My definition of efficiency was cost-minimization for a given environmental goal. But that is not the same as the full benefit–cost optimization that is favored by economists. The cost-minimization approach takes the quantitative goal as given, rather than determining the optimal plan by weighing up environmental benefits of more or less stringent paths against the economic costs. Economists argue that the optimal path involves shallower cuts in the earlier years and deeper cuts late in the century. This is the result that usually emerges from the integrated assessment models that Nordhaus (1994) inaugurated and others have emulated and extended.[8] For example, it is the strategy suggested by the Manne and Richels (1997) research.[9]

The Kyoto Protocol itself failed to address future periods. Because the problem is the stock of cumulative concentrations of GHGs in the atmosphere, not the flow as in most other kinds of pollution, an agreement that does not specify long-term measures seems to accomplish little. Most of those participating in the Kyoto negotiations (though not all) understood that the cuts under discussion would make only minute contributions to reducing greenhouse gas concentrations, temperatures, and sea-level rises later in the century. We estimated that if participation were limited to the countries undertaking commitments at Kyoto, then the effect of the agreement would be to reduce the temperature in 2050 by only roughly one-tenth of one degree Celsius (relative to a baseline increase of 1.15 degrees). Even so, the longest journey begins with a single step.

It is true that it would be optimal to save the deep cuts for later in the century, within the terms of those models. Scrapping coal-fired

[8] For example, Nordhaus (2006) wants the numbers to be derived from "ultimate economic [and] environmental policy objectives." Olmstead and Stavins (2006) also argue for "moderate targets in the short term to avoid rendering large parts of the capital stock prematurely obsolete, and more stringent targets for the long term."

[9] CEA found the Manne–Richels model to be the most useful of all that existed as of early 1997, for the specific purposes of planning for Kyoto. Subsequently, Hammitt (1999) estimated that the least-cost emissions path for stabilizing at 550 (parts per million) lies below the fully optimizing path of benefit–cost analysis, until 2024, and then crosses above it.

power plants today is costly, while credibly announcing that the goals will be stringent fifty years from now would not be as costly, because it would give businessmen and consumers, and scientists and engineers time to plan. But I believe that benefit–cost maximization, though obviously right in theory, is in this case the wrong logic in practice for two reasons. First, the uncertainties are so great – not only regarding unknown future technologies, but even more so regarding the probabilities and nature of tremendous global catastrophes and what is the appropriate discount rate for events centuries into the future – as to make benefit–cost estimates worthless. It is impossible to put probabilities on the catastrophe scenarios (an end to the Atlantic circulation pattern, melting of the Antarctic ice shelf, an unstable feedback loop through release of methane from thawing permafrost, and so on). Just the range of uncertainty about the appropriate discount rate can give every answer, from large immediate cuts to a path that begins with no cuts.[10]

Of course economic costs have substantial uncertainties too, but nothing to compare with the uncertainties on the side of the benefit of mitigation. Surely, you will say, the existence of uncertainty does not change the logic that we should "take our best shot" at estimates of the optimal path of emission reductions. Here comes the second reason why I believe the benefit–cost approach is not practical in this context.

Dynamic consistency. Governments cannot bind their successors. Call it institutional or political myopia, if you will. But nobody knows what the objective scientific, technical, economic, and political realities will look like in the future. Thus it is rational and sensible that the system is unable to commit to specific targets for the rest of the century. If the designers of a treaty specify a path for steep future reductions, their successor governments will not likely be willing to pay the high economic costs necessary to follow through when the time comes. Those making investment decisions today – from power plant construction, to housing, to technological research – know that tomorrow's policies will not be consistent with today's promises and will act accordingly. This is the definition of dynamic inconsistency, a term I am borrowing from

[10] E.g., Cline (1992) and Dasgupta (2001). For a general review of the timing issues, see Aldy, Orszag, and Stiglitz (2001).

monetary economics where it refers to the problem of noncredible anti-inflation announcements by central banks.[11]

In considering the desiderata for a successor to Kyoto, another aspect of dynamic consistency belongs on the list: the importance of standing by precedents so as to avoid the incentive for countries to stay out of an agreement as long as possible or to ramp up emissions in the meantime, so as to be able to negotiate from a higher base. This is the principle well recognized in legislating tax-law changes, even in so dysfunctional an institution as the US Congress, that the date on which (e.g.) investment tax credits take effect is (retroactively) the date that the legislation is first proposed, not the date it passes. This principle was recognized in the 1997 treaty when 1990 emission levels were chosen as the overall baseline, even while concessions were made to subsequent facts of life (e.g., allowing the US target to cut less deeply below the baseline than the European target, because US emissions even by then had been growing more rapidly). I think whatever framework is established now to negotiate targets for 2013–2017 and beyond, there should not be 100 percent forgiveness of the rapid increases that countries like the United States and China have incurred in the meantime, i.e., the Europeans should get some credit for the cuts they have made. Otherwise, whatever countries are the last holdouts at any particular stage have even less incentive to join.

Others, to be sure, have come to recognize the importance of dynamic consistency. The answer, from Olmstead and Stavins (2006), is that "once national governments have ratified the agreement, implementing legislation within respective nations would translate the agreed long-term targets into domestic policy commitments . . . This represents a logical and ultimately feasible chain of credible commitment." I am afraid I just do not agree that this is politically feasible. Bringing up politics is problematic, because every analyst can simply pronounce the proposals of others politically infeasible, and there is no way of verifying which of them are in fact more or less infeasible than others. But none of the country delegations at Kyoto proposed specific

[11] I want to be clear that it still makes sense for academic models often to work in terms of long-term paths. But it is also important to realize that no given policymaker gets to choose a long-term path. The job of each, at most, is to choose the size of one link in the chain. The economic adviser who insists on talking about 100-year paths is making the common mistake of refusing to answer the question that the policymaker has hired him to answer.

policy measures that went beyond 2012, and it was not because many of us were not acutely aware of the unpleasant reality that the cuts for the first budget period were but a drop in the bucket in terms of effect on climate. Perhaps there is a reason for this empirical fact, a reason that governments cannot in this context bind their successors, that is not readily apparent to academics.[12]

Let me try a new analogy: the regular fiscal budget process. My guess is that almost all policymakers in the United States or Europe or other countries would agree that the budget ought to be roughly balanced over the business cycle, or in countries where debt is too high that it ought to run surplus on average to undo steadily over time the excess accumulation of debt in the past. And yet budgets are almost always passed by politicians at a one-year horizon. Occasionally a country sets a "medium-term fiscal path" of budget consolidation over ten years, or a "social security fix," or adopts an institutional constraint like Europe's Stability and Growth Pact or the Gramm–Rudman legislation of the 1980s. Usually these are failures. Even where they succeed, the regime at the very most lasts a decade. Has anyone proposed to legislate a 75-year path of budget deficits or surpluses to achieve long-run fiscal sustainability? The idea is not practical. There is too much genuine uncertainty and too many imperfections in the political process to make such a thing feasible.

President Clinton would have been laughed out of town if he had announced in 1997 that his administration's climate change proposal was to do almost nothing until 2050 and then institute tough emission cuts in the second half of the twenty-first century. Certainly there would have been no rush by the engineers to develop new technologies to be ready for the big day in 2050 when the price of carbon would abruptly rise, because nobody would have believed it.

In climate policy, as in fiscal policy or monetary policy, an announcement of future plans is not in itself credible. Nor can lock-in institutions take us as far as our models sometimes make it sound. Problems of genuine unpredictability and political legitimacy (under democracies, elected legislative bodies have to decide how money is spent), even leaving aside the possibility of myopia or excessively high discount rates, mean that the quantitative stringency of targets will have to be

[12] McKibbin and Wilcoxen (this volume) also seek to address the long-term credibility problem.

set one political term at a time. The bottom line is that there is a good reason for policymakers to try to begin with a "down payment," a more aggressive first step than would be implied by an "optimally" calculated path as might be found in a standard forward-looking integrated assessment model. To pursue the analogy of monetary policy, one way central bankers can address the problem of dynamic inconsistency and attain credibility is by establishing reputations for monetary discipline. Both Paul Volcker and Alan Greenspan did this by raising interest rates *at the beginning* of their terms as Chairman of the Board of Governors of the Federal Reserve System.

If Americans were collectively to decide, hypothetically in 2008, that they wanted to get serious about addressing global climate change, it would be far too late to sign on to the quantitative targets agreed to at Kyoto, which pertain to the budget period 2008–2012. These targets are no longer attainable. Thus it is a matter of setting emission targets for the subsequent budget period, which begins in 2013. Given that most of the rest of the world has gone ahead without us, we might have to begin by establishing some credibility, by taking some dramatic short-term action – if not agreeing to substantive short-term cuts, then passing a big national gas tax or carbon tax.[13]

The conventional wisdom has always been that big energy taxes are as politically unacceptable to the American public as Kyoto itself.[14] After the fiasco of the proposed BTU (British Thermal Unit) tax and gas tax in the first year of the Clinton Administration, one could not even mention the word "tax" out loud in a discussion of GCC options in the late 1990s. But that should not stop us academic economists from talking about it. Political realities change. If we had had true leadership

[13] The spirit is the same as Pizer's (2006, p. 9) "Recognizing the evolutionary nature of a global climate policy. . . useful near-term steps should be the priority and should be viewed in the context of ongoing negotiations to refine the international framework."

[14] Some take encouragement from results such as a *New York Times* / CBS News Poll conducted February 22–26, 2006, with 1,018 respondents. Whereas 85 percent oppose an increased federal gas tax when the question is posed unconditionally and only 12 percent favor it, the number in favor rises to 55 percent (with 37 percent opposed) if the question is preceded by "What if the increased tax on gasoline would reduce the United States' dependence on foreign oil. . ." and to 59 percent (with 34 percent opposed) if preceded by "What if the increased tax on gasoline would cut down on energy consumption and reduce global warming. . ." But it is not clear that the latter formulations are necessarily the more politically relevant.

in the White House on September 11, 2001, we could have had such taxes adopted, in a bid to reduce dependence on oil from the Middle East, with the proceeds going either to cut other taxes or to fight a war on terrorism or both. I suggest we should be ready with such a plan on the shelf for the next shock (whether it is a 9/11 with unconventional weapons, a loss of oil from the Gulf, or the replacement of governments in Pakistan or Saudi Arabia by radical anti-Western regimes, etc.).

Robustness under uncertainty. Two defining aspects of Kyoto merit particular support. The first is the desirable market mechanisms.[15] Second is the adoption of quantitative targets for countries.

Cooper (1998), Nordhaus (2006), and other economists prefer price targets to quantity targets. More specifically they prefer a global carbon tax. Their arguments are as valid as they are familiar. One of the most important is robustness under uncertainty. But let me pick one major problem with the beloved global harmonized carbon tax (e.g., Nordhaus, 2006): what is to determine the baseline? Lots of countries already have lots of government policy measures governing energy and the environment. Europe has long had very high taxes on petroleum products. Britain phased out coal in the 1980s. Others have phased out subsidies. Many signatories are currently undertaking major further steps. It is neither "fair" nor intertemporally consistent to say that they must add a new carbon tax at an internationally unified rate. What if their existing array of policies already comes very close to a carbon tax? Are they then exempt? It's not that these problems are totally insoluble; the point, rather, is that they are on the same order of magnitude as the practical problems with agreeing quantitative targets that the critics like to point out. With one difference: the signatories at Kyoto have already demonstrated (surprisingly) that they can in fact agree on differentiated national quantitative targets.

The main point is that harmonizing taxes internationally seems to me infeasible while agreeing quotas internationally is feasible, however

[15] I am not, however, a big fan of the Clean Development Mechanism (CDM), Joint Implementation (JI), or other project-based credits. I think the problems of baselines and "additionality" are nearly insurmountable, and the use and abuse of such provisions will probably only undermine the respectability of international trading of emission permits – for countries that have agreed to baselines – where there is at least a hope of compliance because there is at least something to comply with (as opposed to deals to buy pieces of paper with no property rights).

countries choose to implement their commitments domestically. A carbon tax would still be my first choice for US domestic implementation, if it were politically acceptable.

I agree with many others that a good way to increase robustness with respect to uncertainty is a hybrid system such as an escape clause, also called a safety valve. This would be a tradable emissions permit scheme plus a promise to sell additional permits at a pre-specified price threshold, thus easing the quantitative limit, if the price of an emissions permit threatens to rise above that cap. In effect it would be a sort of insurance policy against unexpectedly high costs in a tradable permit policy.[16]

Equity. The developing countries point out that it was the industrialized countries, not they, who created the problem of global climate change, and they should not be asked to limit their economic development to pay for it. The developing countries are said to have contributed only about 20 percent of the carbon dioxide that has accumulated in the atmosphere from industrial activity over the past 150 years.[17] Moreover, in contrast to richer countries, they do not have the ability to pay for emissions abatement. Developing country governments consider the raising of their people's economic standard of living the number one priority. Achieving this objective requires raising market-measured income as well as improving the local environment, particularly reducing air and water pollution.

In their more unrealistic moments, spokesmen for developing countries argue that equity requires setting quantitative targets at equal amounts per capita. (This has long been India's position.) It is true that equity in itself suggests moving in this direction. In fact this proposal would not even take into account that the industrialized countries have done most of the emitting to date while the environmental damage falls disproportionately on the already-hot, largely agrarian, poor countries. But the rich countries would never accept the huge effective transfer of wealth from them to the poor that is implicit in the per capita formulation. The status quo of high emissions from rich countries

[16] E.g., Pizer (1997) or Kopp, Morgenstern, Pizer, and Toman (1999).

[17] It has been estimated that if one accounts for the contribution of land use change and deforestation to the atmospheric build-up of CO_2, developing countries are in fact responsible for about 43 percent of all CO_2 in the atmosphere now. Austin, Goldemberg, and Parker (1998).

cannot be ignored, because the status quo is the fall-back position when international negotiations fail (the "threat point" in the language of game theory).

Compliance. No country will agree to a treaty that can be expected to impose disproportionate economic costs on it. If a country somehow agrees to a plan that turns out to have this consequence *ex post*, it will subsequently back out. Whatever other carrots or sticks can be dreamed up – trade sanctions, moral suasion, or penalties in the form of steeper cuts in the future (this last being almost comically impractical, in my view) – minimizing the chances of substantial economic loss is the sine qua non of assuring compliance.

Ensuring countries are not hurt *ex post* by having signed up is harder than ensuring that they are not hurt *ex ante*. It brings us back to the question of robustness under uncertainty. Even before we get to an escape clause, we can tinker with the Kyoto framework in future rounds to make it more robust with respect to unknown future developments in the desired degree of stringency, arising from growth surprises, technology surprises (e.g., will fuel cells work?), climate surprises (whether their importance lies in the changing state of scientific knowledge, popular priority on the question or both), and so forth. We now turn to specific proposals for the architecture, which include provisions to maximize the probability that no country pays a disproportionately high cost, either *ex ante* or *ex post*.

Proposed architecture for quantitative emissions targets

So what is the answer? What could be the next step after Kyoto, an architecture that balances the needs for comprehensive participation, efficiency, dynamic consistency, robustness under uncertainty, equity, and enforceability?[18] I offer my own proposed architecture. It would be built upon the principles of quantitative emission limits and international trading, the foundations laid at Kyoto. Unlike Kyoto, it seeks realistically to bring in all countries and to look far into the future. But to look into the future does not mean to pretend that we can see with as fine a degree of resolution at the century-long horizon as at the five-year horizon. We consider different horizons in turn.

[18] Aldy, Barrett, and Stavins (2003) and Victor (2004) review alternative proposals.

The century-long horizon: building confidence within a long-range framework

The entire path of emission targets for the twenty-first century cannot be chosen at once, because of the problems of uncertainty and dynamic inconsistency discussed above. It would be too constraining to write down a complete set of numerical targets for each country from now to 2100, and expect them to hold.

Rather we need a sequence of negotiations, all of which fit within a common framework. The framework, like the US Constitution, would ideally be flexible enough to last a century or more. An example of such a framework in another policy area is the postwar General Agreement on Tariffs and Trade, which gave us fifty years of successful rounds negotiating trade liberalization, culminating in the foundation of the World Trade Organization, whose rules are more binding. Nobody at the beginning could have predicted the precise magnitude or sequence of the cuts in various trade barriers. But the early stages of negotiation worked, and so confidence in the process built, more and more countries joined the club, and progressively more ambitious rounds of liberalization were achieved.

Another analogy would be with the process of European economic integration, culminating in the European Economic and Monetary Union.[19] Despite ambitions for more comprehensive integration, nobody at the time of the founding of the European Coal and Steel Community, or the subsequent European Economic Community, could have forecast the speed, scope, magnitude, or country membership that this path would eventually take. The aim should be to do the same with the UNFCCC.

When the United States (and Australia) are politically ready to join, it will also be time to set a standard for participation of developing countries. The threshold for expected participation in any multilateral agreement could be something like two tons of CO_2 emissions per person from fossil fuel use.[20] (Not an income threshold. Why penalize

[19] Tom Heller raised this analogy at the May 2006 Architectures for Agreement Workshop hosted by the Environmental Economics Program at Harvard University.

[20] This threshold would include Brazil and China, but not yet, for example, India, Pakistan, Indonesia or most African countries (Baumert, Herzog, and Pershing 2005: Figure 4.1, p. 22). I have deliberately chosen the metric of carbon dioxide

economic growth, if it is clean?) But declaring a threshold will not be sufficient to persuade countries to join if its economic incentives are strongly against accepting the targets demanded.

How should the sequence of quantitative targets be set? Since fixing targets a century ahead is impractical, I would propose doing it a decade at a time. I have little to say about how to set the decade targets for the global or aggregate level of emissions. For the reasons I have said, I don't think an optimal trade-off of the economic costs and the environmental benefits is practical. How ambitious national leaders turn out to be will be determined by the extent of popular support, which will in turn be influenced by such unpredictable vicissitudes as weather disasters (which may or may not be scientifically attributable to global climate change) and any new 9/11 tragedies or instability in the Persian Gulf (which alone would generate the political will in the United States to reduce oil consumption). The key attempted contribution of this chapter is, rather, to propose how the framework would allocate the relative targets *across* countries. Table 2.1 attempts to adumbrate how the sequence of formulas would produce decade-by-decade allocations across countries of emission reductions.

The decade horizon: formulas for setting targets

The proposal is to allocate quotas across countries in any given budget period according to a nested sequence of formulas for emissions. The formula would be very general for the distant future, and become increasingly specific as the budget period in question draws close. Viewed at the horizon of one decade, the formula for emission limits would be phrased as cuts from the expected business-as-usual path (BAU). The BAU baseline entails, for example, rapid increases in emissions for such countries as China and India. Thus, even with cuts relative to the baseline, we are talking about "growth targets" for such countries, not cuts in the absolute levels of emissions as were agreed by the industrialized countries at Kyoto. The formula for targeted reductions would include among its determinants the following variables:

- 1990 emissions
- emissions in the year of the negotiation

rather than including land use for two reasons: measurability, and to make sure that we are capturing only countries that have begun the industrialization process.

Table 2.1 *Proposed framework for emission targets and timetables*

A sequence of formulas setting country targets as functions of income, population, and lagged emissions.

Proposed time table	*Europe and Japan*	*US and Australia*	*China and other developing countries*
2008–2012	Cuts below 1990 levels, agreed at Kyoto (as illustrated in Fig. 2.1 relative to BAU)	BAU, as estimated at the date of joining	No targets
2013–2020	Further cuts (relative to BAU estimated in the year of negotiation, say 2008). Bigger cuts for richer countries, analogously to pattern in Fig. 2.1	Targets put less weight on 2008 levels and more on 1990 levels, to move in direction of Europeans	Targets = BAU as estimated at the date of joining, say 2008
	Targets partially indexed to income within the decade (though perhaps numbers to be fixed in 2015, to facilitate trading as part of settling up during 2016–2020)		Partially indexed to income (though perhaps intensities are fixed in 2015)
2021–2030	Weights on 1990 emissions vs. 2008 emissions begin to converge (so that USA is not unduly rewarded for staying out). Bigger cuts for richer countries, analogously to pattern in Fig. 2.1		First cuts, relative to BAU, steeper for those having attained higher income, analogously to pattern in Fig. 2.1
	Targets are again partially indexed to income within the decade		
2031–2040	Weight on population starts to increase, relative to weights on lagged emissions and income Partial indexation within the decade		

A sequence of formulas setting country targets as functions of income, population, and lagged emissions.

Proposed time table	*Europe and Japan*	*US and Australia*	*China and other developing countries*
2041–2050	Weight on population continues to increase, weights on lagged emissions and income continue to diminish, with weight on 1990 emissions vanishing altogether. Partial indexation within the decade		
. . .			
2091–2100	No weight on lagged emissions. Countries with higher income still get higher targets, but much less so than early in the century		
Infinitely long run	Equal per capita emission targets		

- population
- income, and
- perhaps a few other special variables like whether the country in question has resources like coal or hydroelectric power.

A uniform formula would in a sense apply to all countries. But the straitjacket must not be too rigid if it is to be applied uniformly to all. The presence of lagged emissions on the list recognizes the reality that no country, neither the United States nor developing countries, will agree to a costly sharp reduction relative to the status quo. Other special circumstances would also be recognized. For example, at Kyoto the eastern European transition economies were in effect given an extra adjustment, the target for Iceland recognized its abundant hydroelectric and geothermal resources and related plans for aluminum smelting, and the target for Australia (would have) recognized its abundant coal deposits.

One property of the proposed sequence of formulas would be that a country joining the commitment regime for the first time would agree to a target for the subsequent budget window given by its estimated BAU path. In the first year of membership, that level is very close to what the country had been emitting in the previous year. (In "year zero," the year of negotiation, the weights on "emissions in the year of negotiation" is 1.0 and the weights on the other factors enumerated above are zero.) This assures that the country is not being asked to take on a large economic cost the moment it joins, in which case it will not agree. Rather, the joiner can only *gain* in the short run, by the right to sell permits in the first budget period, thereby giving it an economic incentive to join. Existing members gain economically from the right to buy cheap reductions from the joining developing countries. The world environment benefits as well, by pre-empting leakage even within the first budget period, aside from the benefit of having the country in the system for the longer term. For the United States, 2008–2012 would (retroactively) be considered the first budget period, i.e., the period in which the United States is asked for no more than to limit its emissions to the BAU path. This recognizes the reality that large cuts relative to what in 1997 appeared to be the US BAU path – as are implicit in the Kyoto numbers for 2008–2012 that the United States has since rejected – would by now be far too costly to make. In this sense, the United States would in the near term be treated more as a developing country than on a par with Europe and Japan.

In the second budget period of participation, countries are asked to make more serious cuts: their target cuts relative to BAU assign less weight to emissions in the year of negotiation, and more to 1990 levels. The reason for giving at least some weight to 1990 levels in the medium term is to assure the Europeans that the United States is not being unduly rewarded for having dropped out of Kyoto and "ramping up" its emissions over the period 1990–2008 to a much higher level from which to negotiate any future reductions. The weight on 1990 in the medium-term targets could be relatively larger or smaller, depending on the relative bargaining position of the United States versus Europe.

Persuading developing countries to join a system of quantitative targets

To entice developing countries, who fear being made to ask larger cuts in the long run, the framework of a nested sequence of formulas could include the possibility, at least at a rhetorical level, that the target for emissions in the limit as the year under consideration approaches infinity puts zero weight on income or past levels of emissions, and complete weight on population. In plainer English, in the very long run, the developing countries would notionally achieve their equity-based demand for equal levels of emissions per capita. Lest this sound outlandish, it is worth considering that by the twenty-second century, China could well have caught up with Western countries in income per capita (other Asian countries like Singapore have already done so), in which case the proposal that the emission targets should put all weight on population gives an answer similar to putting all weight on any combination of population or income.

Rhetorical promises regarding the twenty-second century will not go far to convince poor countries today. But even in the shorter term, if quantitative emission commitments are set for developing countries in a very careful way, they can address their concerns, at the same time as addressing the concerns of the rich countries. Three important principles should guide the formulation of such targets:

- gains from trade
- progressivity, and
- protection against inadvertent stringency.

The remainder of this chapter elaborates on how an agreement on targets under such principles can bring economic and environmental benefits for developing countries as well as for rich countries. Thus everyone should be able to agree that these targets represent an improvement, relative to the alternative of not having developing countries in the system. This is true regardless of how much weight one wants to put on the economic interests of poor countries versus rich, and regardless of how much weight one wants to put on environmental goals versus economic goals.

The gains from trade

If developing countries were to join a Kyoto-like system of targets-with-trading, it would not only have environmental and economic advantages for the rest of the world; it would also have important environmental and economic advantages for the developing countries themselves. For the sake of concreteness, consider a plan under which developing countries do no more in their first budget period of participation than commit to their BAU emission paths and then join the trading system.

A climate regime with a BAU emission target would not hurt developing countries. These countries would have the right to emit whatever amount they would have emitted anyway in the absence of an international climate change policy. They need not undertake emission abatement unless a foreign government or foreign corporation offers to pay them enough to persuade them voluntarily to do so. Importantly, however, by constraining their emissions to business as usual, these commitments would forestall emissions leakage and improve the environmental effectiveness of emissions abatement efforts in industrialized countries.

Developed countries' governments and private firms would likely offer to pay developing countries enough to persuade them voluntarily to reduce emissions below their BAU paths. How do we know this? It would be expensive for the United States, Europe, and Japan to reduce emissions below 1990 levels if the reductions are made only at home. The cost of emission abatement would be far lower in many developing countries, and so rich countries could offer terms that make emission reductions economically attractive to them. The economic theory behind the gains from trading emission rights is analogous to the economic theory behind the gains from trading commodities. By doing what they do most efficiently, both sides win. Table 2.2 illustrates.

Table 2.2 ***If developing countries accept targets and trade, everyone wins***

Gains if poor countries join and trade	*Economic gains*	*Environmental gains*
Gains for developing countries	Price received for emission cuts > cost to poor country	Auxiliary benefits: less air pollution
Gains for industrialized countries	Price paid for emission permits < cost to cut only at home	Precludes leakage of emissions to non-members

Why is it so much cheaper to make reductions in China or India than in the United States? One major reason is that, in industrialized countries, one would have to prematurely scrap coal-fired power plants in order to replace them with natural gas facilities or other cleaner technologies. This would be expensive to do, because it would mean wasting a lot of existing capital stock. In rapidly growing countries, by contrast, it is more a matter of choosing to build cleaner or more-efficient power-generating plants to begin with. When contemplating large increases in future demand for energy, it is good to be able to plan ahead. The benefits include learning from the mistakes of others that have gone before, and taking advantage of their technological advances.

An extreme example of how measures to reduce carbon emissions can have low costs in developing countries is the case of subsidies to fossil fuels. Energy subsidies run as high as 58 percent for Venezuela and 80 percent for Iran. Even oil importers subsidize energy: 11 percent for China and 14 percent for India.[21] Eliminating such subsidies would create substantial immediate benefits – fiscal, economic, and environmental – even before counting any benefits under a Kyoto agreement. Subsidy cuts within a target-and-trade system would pay developing countries twice over – once in the form of the money that is saved by eliminating wasteful expenditure, and then again in the form of the money that is paid by a developed country for the claim to the resulting emission abatement.

Progressive emissions commitments

Developing countries fear that they will be asked to accept emission targets that are more stringent than BAU, and perhaps lower than current or past emissions (such as what the industrialized countries accepted in the Kyoto negotiations). It would not be reasonable for the rich countries, however, to insist that the poor accept targets that fail to allow for their future economic growth. It is useful to begin by expressing all possible emission targets as relative to BAU, as already suggested. Any proposed emission abatement is relative to BAU, not relative to the past.

A reasonable lower bound for developing country emission targets would be the "break-even" level. This is the level that leaves them neither better off nor worse off economically than if there had been no treaty at all. In other words, it is a level where they have to make some

[21] These statistics are from International Energy Agency (1999: Table 6, p. 64), which estimates that elimination of the subsidies among a sample of eight developing countries would yield annual economic efficiency gains of 0.73 percent of GDP and would reduce CO_2 emissions by 16.0 percent.

low-cost reductions from the start, but where sales of emission permits at an intermediate price are sufficient to compensate them for their marginal reduction. The aim should be to fall somewhere in the range that is bounded above by BAU and bounded below by the break-even level. As long as the target is above the break-even lower bound, the developing countries benefit economically from the arrangement. Developed countries still enjoy the opportunity to invest in low-cost emissions abatement in developing countries and the lower global emissions of such an arrangement. Everybody gains from having taken the first step to fight global climate change.

There is probably a moderately large range between business as usual and the break-even point. What would constitute a "fair" emission commitment within this range? To give some reasonable degree of progressivity, targets for emission per capita would also assign some weight to income per capita. A fair target for developing countries might be one that fits the pattern that in retrospect holds among the existing targets agreed at Kyoto, which are illustrated in Figure 2.1. Even though the emission targets agreed at Kyoto reflected the outcome of political negotiations, rather than economists' calculations of some definition of optimality, retroactively it is possible statistically to discern systematic patterns in the targets. This approach turns out to allow some *progressivity*, with richer countries committing to larger emissions abatement efforts than poor ones. Yet it does not go nearly so far as the massive redistribution of wealth that some poor-country representatives unrealistically ask for, based on a tabula rasa notion of equity.

Out of thirty industrialized countries' targets agreed at Kyoto (those with adequate data, including some that have not subsequently ratified the agreement), the average reduction from BAU was 16 percent. For the less-rich half of the countries, the average reduction was 5 percent below BAU, which shows the progressivity in a very simple way. The progressivity of the Kyoto system was also revealed within the EU's "bubble" allocation: wealthier countries such as Germany and the United Kingdom accepted emission targets much more stringent than the EU-wide commitment (1990 minus 8 percent), to allow less wealthy countries such as Portugal and (at the time) Ireland to have less stringent targets.

Further statistical analysis can help illustrate the progressivity of the targets. Controlling for countries' projected BAU emission growth,

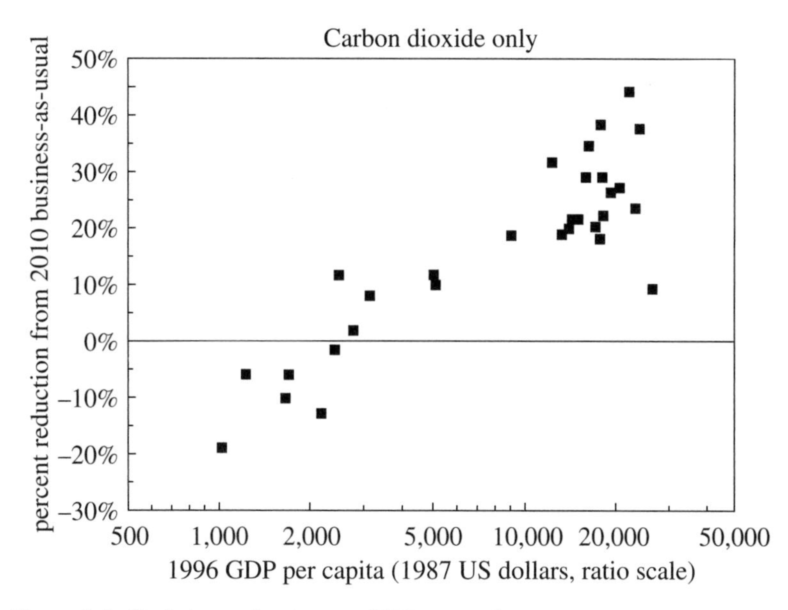

Figure 2.1. Emission reduction vs. GDP per capita

Sources: The World Bank, US Energy Information Administration, national communications to the UNFCCC

their coal intensities, and whether they are beginning the transition from central planning, we estimated that a 1 percent increase in per capita income implied a target of 0.11 to 0.17 percent greater emissions abatement from BAU.[22]

As an illustrative example, when this pattern was extrapolated to China, looking forward from 1998, the projected target was about 5 percent below BAU. This emissions level happened to lie inside the desirable range: below BAU but above the break-even point (based on a number of global energy-economic models). In other words, if China accepted such a target, economic benefits would accrue both to it and to the rich countries that would pay China to reduce emissions further. As a rough guideline, 5 percent is not an unreasonable benchmark for other developing countries as well. To repeat, it still allows for growth.

[22] The graph and estimates are based on the data that was publicly available as of the end of 1997, when the Conference of Parties met at Kyoto. The regression line fits almost as well, but not quite, when subsequently revised data are used – Frankel (1999). But it is more appropriate to use the original data if the attempt is to discern the degree of progressivity that the negotiators implicitly decided was "fair."

Resolving concerns about unintended target stringency

One important objection concerns uncertainty regarding how stringent given numerical targets would turn out to be during the course of a given decade. Calculations regarding the BAU path or the cost of deviations from it are subject to great imprecision and unpredictability. Poor countries worry that uncertainty surrounding their forecasted economic performance is so great that they cannot now risk adopting an emission target that would be binding five or ten years in the future. Even if a particular numerical target appears beneficial *ex ante*, it might turn out to be something different *ex post*. If the country turns out to achieve unexpectedly rapid growth, the last thing it wants is to have to put a stop to it because the accompanying emissions threaten to overrun its target. A response to this concern would be to structure international agreements regarding these countries' targets so as to reduce the risk of being inadvertently stringent.

Symmetrically, environmentalists have also expressed a concern on the other side that a target may turn out *ex post* to be too lax. They fear that such a target might fail to result in environmental benefits in terms of actual emission reductions relative to what would have happened in the absence of a treaty. If, for example, Korea or Thailand had accepted targets at Kyoto in 1997, the sharp slowdowns that began in East Asia in the same year would have turned out to imply that they might have been paid for emission abatement (relative to *ex ante* BAU) that would have happened anyway. This is the so-called "hot air" problem. Thus it is desirable to mitigate the risk of inadvertent laxity while also mitigating the risk of inadvertent stringency – to narrow the variability of the effective stringency of the target without relaxing or tightening the intended target itself.[23]

One solution is indexation of the emissions target. The general notion is to agree today on a contract under which the numerical target depends in a specified way on future variables whose values are as yet undetermined.[24] Future economic growth rates are probably the biggest source of uncertainty, especially in developing countries. A simple approach would index a country's aggregate emissions to future

[23] See also Lutter (2000).

[24] An analogy is a cost-of-living adjustment clause in a labor contract. It specifies a given increase in the wage for every dollar increase in the Consumer Price Index – thus reducing uncertainty over *real* wages.

income alone. Other possible proposals could allow the targets to vary with other variables such as population or temperature.

More specifically, for every percentage point in GDP growth that is higher or lower than forecast, the emission target is raised or lowered by a corresponding amount.

A special case of indexation is "intensity targets," which the Bush Administration has talked about. An intensity target is a target for the ratio of energy use or emissions to GDP. Not only are emissions indexed to GDP, but the indexation is full, i.e., the relationship is fully proportionate. This has the advantage of simplicity over my formulas.[25] I would still argue that from the year when a country negotiates entry into the system until the year before the budget window starts (whether it is a five-year or ten-year budget window), the emission target should be indexed to GDP in a *less-than*-proportionate way. For example, every 1 percent of extra growth might call for an automatic 0.7 percent increase in the target. Or the coefficient could be 0.5, which would make the formula into a simpler "square root" rule.[26] The proposal would require countries that are doing a bit better to contribute more than those that are not, maintaining principles of progressivity and insurance, without penalizing them unduly for their success.

A China that turns out to grow at 10 percent a year for six years has raised its standard of living 77 percent. It should not be unduly penalized for its growth by failing to allow an increase in the numerical target (i.e., there should be some indexation). But neither should it be allowed a fully proportionate increase over this six-year period. Rather, it is capable of making a larger reduction relative to GDP than had been anticipated six years earlier, and it should do so (i.e., the indexation should be less than proportionate).

Indeed a shortcoming of the proposal to set intensity targets, i.e., to have fully proportionate index, is that it actually asks *less* sacrifice of countries that have experienced the good fortune of an unexpected economic boom, and more sacrifice for those that have suffered an

[25] Was simplicity the Bush Administration's motive for preferring emissions efficiency targets? An alternative possible motive is rhetorical: a plan that involves "gradual reductions in emissions per GDP" sounds better – even though that also describes the BAU path – than a path that allows rising emissions.

[26] The Argentine government proposed an emission target indexed to the square root of its economic growth at the 1999 Conference of Parties in Bonn, Germany.

unexpected slowdown. The reason is that even in the absence of policy measures, energy use and emissions tend to rise less rapidly than GDP.[27]

There is a separate question as to whether the target levels should be finally fixed numerically at the beginning of the budget period, or whether there should also be indexation *within* the budget period. An argument for continuing to index is that it would make it more feasible for a government to hit its emission targets in the face of economic fluctuations. An argument for setting the numerical targets at the beginning of the budget period is that it would facilitate international trading of emission permits because each country would know exactly what its targets were. The important point is that approaches that explicitly account for at least some of the uncertainty that characterizes emission abatement would make it more likely that the target will turn out to fall within the range intended, where it brings benefits – economic as well as environmental – to developing countries and industrialized countries alike.

References

Aldy, Joseph E. (2004). "Saving the Planet Cost-Effectively: The Role of Economic Analysis in Climate Change Mitigation Policy," in R. Lutter and J. F. Shogren (eds), *Painting the White House Green: Rationalizing Environmental Policy Inside the Executive Office of the President*, Washington, DC: Resources for the Future Press, pp. 89–118.

Aldy, Joseph E., and Jeffrey Frankel (2004). "Designing a Regime of Emission Commitments for Developing Countries that is Cost-Effective and Equitable," written for *G20 Leaders and Climate Change*, Council on Foreign Relations, September 20–21.

Aldy, Joseph E., Peter R. Orszag, and Joseph E. Stiglitz (2001). "Climate Change: An Agenda for Global Collective Action," paper presented at workshop on the Timing of Climate Change Policies, Pew Center on Global Climate Change, Washington, DC, October 2001, AEI-Brookings Joint Center for Regulatory Studies.

[27] Aldy (2004) analyzes the Bush Administration intensity goal. He shows that if the US economy grows faster than currently forecast (the high economic growth case of the Energy Information Administration [EIA] in its Annual Energy Outlook), then the required emission sacrifice to meet the intensity goal would be substantially lower in absolute number of tons than under the expected growth case. If the economy grows more slowly than expected (EIA's low economic growth case), then the required emission abatement to meet the goal would he higher.

Aldy, Joseph, Scott Barrett, and Robert Stavins (2003). "Thirteen Plus One: A Comparison of Global Climate Policy Architectures," *Climate Policy* 3(4): 373–397.

Austin, D., J. Goldemberg, and G. Parker (1998). *Contributions to Climate Change: Are Conventional Metrics Misleading the Debate?* World Resources Institute Climate Notes, October, Washington, DC: World Resources Institute.

Barrett, Scott (2006). "Climate Treaties and 'Breakthrough' Technologies," *American Economic Review, Papers and Proceedings* 96(2): 22–25.

Baumert, Kevin, Timothy Herzog, and Jonathan Pershing (2005). *Navigating the Numbers: Greenhouse Gas Data and International Climate Policy*, Washington, DC: World Resources Institute.

Cline, William (1992). *The Economics of Global Warming*, Washington, DC: Institute for International Economics.

Clinton Administration (1998). *The Kyoto Protocol and the President's Policies to Address Climate Change: Administration Economic Analysis*, Washington, DC: White House, July 1998.

Cooper, Richard (1998). "Why Kyoto Won't Work," *Foreign Affairs* (March/April).

Dasgupta, Partha (2001). *Human Well-Being and the Natural Environment*, Oxford: Oxford University Press.

Edmonds, J. A., H. M. Pitcher, D. Barns, R. Baron, and M. A. Wise (1992). "Modeling Future Greenhouse Gas Emissions: The Second Generation Model," in Lawrence Klein and Fu-chen Lo (eds.), *Modeling Global Climate Change*, Tokyo: United Nations University Press, pp. 295–340.

Edmonds, J. A., S. H. Kim, C. N. McCracken, R. D. Sands, and M. A. Wise (1997). "Return to 1990: The Cost of Mitigating United States Carbon Emission in the Post-2000 Period," Working Paper PNNL-11819, Washington, DC.

Frankel, Jeffrey (1998). "Economic Analysis of the Kyoto Protocol," *After Kyoto: Are There Rational Pathways to a Sustainable Global Energy System?*, Aspen Energy Forum, Aspen, CO, July 6.

(1999). "Greenhouse Gas Emissions," *Brookings Policy Brief*, No. 52, Brookings Institution, Washington, DC. June.

(2005). "You're Getting Warmer: The Most Feasible Path for Addressing Global Climate Change Does Run Through Kyoto," in John Maxwell and Rafael Reuveny (eds.), *Trade and Environment: Theory and Policy in the Context of EU Enlargement and Transition Economies*, Cheltenham: Edward Elgar, pp. 37–55.

Goulder, Lawrence, and William Pizer (2006). "The Economics of Climate Change," National Bureau of Economic Research Working Paper No. 11923, January.

Hammitt, James (1999). "Evaluation Endpoints and Climate Policy: Atmospheric Stabilization, Benefit-Cost Analysis, and Near-Term Greenhouse Gas Emissions," *Climatic Change* 41: 447–468.

International Energy Agency (1999). *World Energy Outlook 1999 Insights – Looking at Energy Subsidies: Getting the Prices Right*, Paris: IEA.

IPCC (2001). *Climate Change 2001: Mitigation*, Cambridge: Cambridge University Press.

Kopp, Raymond J., Richard D. Morgenstern, William A. Pizer, and Michael Toman (1999). "A Proposal for Credible Early Action in US Climate Change Policy," *Weathervane*, Resources for the Future, Washington, DC.

Lutter, Randy (2000). "Developing Countries' Greenhouse Emissions: Uncertainty and Implications for Participation in the Kyoto Protocol," *Energy Journal* 21(4): 93–120.

Manne, Alan, and Richard Richels (1997). "On Stabilizing CO_2 Concentrations – Cost-Effective Emission Reduction Strategies," *Environmental Modeling and Assessment* 2(4): 251–265.

Nordhaus, William (1994). *Managing the Global Commons: The Economics of Climate Change*, Cambridge, MA: MIT Press.

(2006). "After Kyoto: Alternative Mechanisms to Control Global Warming," *American Economic Review, Papers and Proceedings* 96(2): 31–34.

Olmstead, Sheila, and Robert Stavins (2006). "An International Policy Architecture for the Post-Kyoto Era," *American Economic Review, Papers and Proceedings* 96(2): 35–38.

Pizer, William (1997). "Prices vs. Quantities Revisited: The Case of Climate Change," Resources for the Future Discussion Paper 98-02, October, Washington, DC.

(2006). "The Evolution of a Global Climate Change Agreement," *American Economic Review, Papers and Proceedings* 96(2): 26–30.

Sandalow, David, and Ian Bowles (2001). "Fundamentals of Treaty-Making on Climate Change," *Science* 292: 1839–1840.

Stewart, Richard, and Jonathan Weiner (2003). *Reconstructing Climate Policy: Beyond Kyoto*, Washington, DC: American Enterprise Institute Press.

Victor, David (2004). *Climate Change: Debating America's Policy Options*, New York: Council on Foreign Relations.

Yellen, Janet (1998). "The Economics of the Kyoto Protocol," Testimony before the Subcommittees on Energy and Power, Committee on Commerce, US House of Representatives, March 4.

Commentaries on Frankel

2.1 *Targets and timetables: good policy but bad politics?*

DANIEL BODANSKY

From a policy perspective, cap-and-trade systems have many virtues. They give countries flexibility to determine how to reduce emissions – whether through domestic trading schemes, taxes, subsidies, efficiency standards, voluntary agreements with industry, or some other policy instrument. They harness the marketplace to encourage climate mitigation efforts by whoever can reduce emissions at the lowest cost. They can easily be scaled up or down, in response to new information about the risks of climate change. They can be differentiated among countries, to address equity concerns. And they can be set a level that, while requiring a country to do better environmentally than business-as usual, still allows it to profit by trading, thereby serving as a type of side payment to encourage participation by otherwise reluctant countries such as China and India – a virtue that, as Jonathan Wiener (1999) emphasizes, is particularly important in an international system that relies on voluntary assent rather than coercion.

In his chapter, "Formulas for Quantitative Emission Targets," Jeffrey Frankel presents a particularly appealing version of the targets-and-timetables approach, with a formula for setting targets intended to address the equity concerns of developing countries, as well as design features such as indexation and a safety valve that respond to one of the principal criticisms of Kyoto-style emission targets, namely that they represent an economic straitjacket and could impose unacceptably high costs on countries. Of course, even Frankel's architecture fails to solve the compliance issue, which some see as the Achilles' heel of the targets-and-timetables approach (Barrett 2003). His only discussion of compliance focuses on one potential source of noncompliance, namely high costs, without addressing the more general problem of free riding that characterizes collective action problems such as climate change. Nevertheless, while Frankel's proposal is vulnerable to criticism on this score, I think that there are plausible stories of how a state might comply, even in the absence of coercion (see Chayes and Chayes 1995; Koh 1997), so Frankel's failure to solve the compliance problem is not, to my mind, a fatal flaw.

But even if we are willing to concede that targets and timetables represent good climate *policy*, I think it is far more questionable whether they represent good climate *politics*. Frankel does not specifically say which perspective he is adopting. At the beginning of his article, he sets forth a list of the requirements that, in his view, "any new agreement *must* meet" (Frankel, this volume, p. 32, emphasis added). But the sense in which he uses the term "must" is not altogether clear. Does he mean the requirements that any agreement must meet in order to be good climate policy, or to be politically realistic? Quite conveniently, he appears to think that there is a significant overlap between the two and that the architecture he proposes not only satisfies his policy desiderata, but is also politically realistic, even in the near term, as the "next step after Kyoto" (Frankel, this volume, p. 41).

The distinction between policy and politics is, of course, fuzzy. Even pure policy design is not a strictly formal exercise; it must take into account how people and governments behave. For example, in arguing that benefit-cost maximization is "right in theory" but "wrong in practice," because governments cannot effectively bind their successors, Frankel is making what usually we would consider a policy argument, but one based on political considerations. Politics and policy fall along a continuum. At one end of the spectrum, we can take present political realities as a given; at the other, we can regard them as contingent and transitory. But even at the ideal-policy end of the spectrum, where current political realities do not represent a constraint, analysis must be consistent with immutable, intrinsic features of the political system – it cannot assume, for example, that states will cease being motivated by self-interest. Otherwise it will become purely utopian.

Where along this spectrum does Frankel's climate architecture fall? Critics of the targets-and-timetables approach, such as David Victor (2001) and Scott Barrett (2003), see it as essentially utopian – Victor because states will simply not be able to negotiate a series of progressively stricter emission targets, Barrett because, in the absence of an effective sanctioning mechanism, states will be unlikely to participate and comply. For the reasons mentioned earlier, I am willing to concede that targets and timetables may represent good policy, at least for the longer term – one could envision a world in which successfully stricter targets and timetables were adopted and achieved the policy desiderata that Frankel articulates. But, for reasons that I will explain in a moment, I am skeptical about the political viability of this approach in the nearer term. Frankel, in contrast, clearly thinks that his proposed climate architecture represents both good policy and good politics – it represents a sensible political strategy for the post-2012 period, when Kyoto's first commitment period expires, as well as for the longer term.

As an academic economist with significant government experience, it is perhaps unsurprising that Frankel manifests an ambivalent attitude toward the world of politics. On the one hand, he appears to regard statements about politics as unscientific, because they are infinitely malleable. As he observes, "bringing up politics is problematic, because every analyst can simply pronounce the proposals of others politically infeasible, and there is no way of verifying which of them are in fact more or less infeasible than others" (this volume, p. 36). And he does not always feel fully constrained by judgments of political viability, since "political realities change" (p. 38). On the other hand, he is not above making such judgments himself. For example, he rejects a carbon tax as "infeasible," even though it would be his first choice if it were politically acceptable. Similarly, he rejects proposals to allocate emission targets to achieve equal per capita emissions on the political grounds that "rich countries would never accept the huge transfer of wealth from them to the poor that is implicit in the per capita formulation" (p. 40). Since Frankel himself raises the test of political feasibility,[1] I think it is fair to assess his architecture not simply in terms of whether it represents good policy, but also in terms of whether it represents good politics. It is to that question that I now turn.

Politics is, of course, the art of the possible. And, in thinking about what is politically possible, we need to begin by acknowledging the modest results achieved in recent negotiating sessions. At the Eleventh Conference of the Parties to the United Nations Framework Convention on Climate Change (UNFCCC), held in Montreal in December 2005, countries agreed to begin two processes – one a dialogue under the UNFCCC on long-term cooperative action to address climate change, the other under the Kyoto Protocol to consider further commitments for developed country ("Annex I") parties for the post-2012 period. Neither process appears likely to produce significant results, however.

[1] In describing his proposed architecture, he specifically states that it "seeks *realistically* to bring in all countries" (this volume, p. 41, emphasis added).

In recent years, the international climate negotiations have been bogged down, which makes it critical to ask: What have been obstacles to progress? And what can we do to overcome these obstacles?

Jeffrey Frankel's paper does not specifically ask these questions, but the policy he prescribes makes some implicit assumptions about the nature of the recent political impasse. Interestingly, his paper focuses much more on developing countries than on the United States. Indeed, the absence of developing countries is, in his view, "the most serious and intractable shortcoming of the Kyoto Protocol." Although he adds the caveat, "except perhaps for the absence of the United States," he sees the US absence largely as a function of the absence of developing countries. In his view, the United States is reluctant to join Kyoto because it does not include developing country targets. If we can solve the developing country issue, then the issue of US nonparticipation will take care of itself.

Frankel's proposed solution is to give developing countries indexed growth targets, which will alleviate their concern that greenhouse gas emission targets could adversely impact their economic growth. If developing countries are given targets that, while below business-as-usual, are achievable at a lower cost than the international carbon price, then the potential gains from emissions trading should provide developing countries with an upside incentive to participate. The trick is to set developing country targets at a level that will allow them to make more from the sale of surplus emission allowances than it costs to produce those surplus allowances by reducing emissions. Moreover, if emission targets are not fixed but are tied to a country's GDP, then this will protect developing countries against the downside risk that rapid economic growth will make it costly for them to achieve their targets, since as their economies grow, their permitted emissions will rise as well.[2] Frankel believes that by setting indexed emission targets at an appropriate level, developing countries can be enticed to participate. And so long as developing countries participate, and the US target is set at a relatively moderate level initially, the United States will be willing to join as well.

Now, this analysis may prove correct, but the history of the climate negotiations counsels us to be cautious. The unfortunate reality is that

[2] From an environmental standpoint, indexed targets also help protect against the danger of hot air, since if a country's economy declines, its emissions quotas will shrink.

developing countries have been turning down proposals along essentially similar lines as Frankel's for the last seven years, first, in the late 1990s when they were being pushed by the United States during the Clinton Administration; then, after the Bush Administration rejected the Kyoto Protocol in 2001, when the European Union tried a similar approach at the Conference of the Parties (COP-8) in Delhi.

Earlier I noted that Frankel tries to take into account issues of political feasibility. But he is far more sensitive to feasibility vis-à-vis developed countries, and in particular the United States, than developing countries. He assumes that an international carbon tax is off the table, despite its policy advantages, because it would be unacceptable to the United States. And he assumes that target allocations based on equal per capita emissions are a non-starter, because they would be unacceptable to developed countries. But he fails to take seriously developing country opposition to emission targets. He seems to assume that their opposition is transitory and changeable. Perhaps it is just a negotiating posture. Or perhaps it simply results from ignorance, which could be overcome through a better sales job regarding the benefits of targets. Whatever the explanation, Frankel seems to think that it is just a matter of time until developing countries come to their senses and realize that indexed growth targets are in their interest – just as, after Kyoto, the European Union eventually gave up its benighted opposition to emissions trading and accepted trading as a desirable policy instrument.

Perhaps . . . but I have my doubts. I think we need to at least consider the possibility that developing countries such as China and India mean what they say, namely that economy-wide, binding emission targets are unacceptable because they would unduly restrict their national sovereignty. True, emission targets give countries flexibility as to the choice of national implementing measures. States can implement their targets through a domestic trading scheme, taxes, efficiency standards, and so forth. But because virtually every aspect of a country's economy contributes to climate change – not only energy production, but also transportation, manufacturing, and even agriculture – an economy-wide target represents, both symbolically and in practice, a constraint on a country's economy as a whole.

Although Frankel is more attuned to US political considerations than to those of developing countries, his assessment of the US political situation seems somewhat tone deaf as well. It is true, of course, that concern about developing country participation in a targets-based approach has been a theme of US political debate for many years, highlighted by the Senate's adoption of the Byrd–Hagel Resolution in the run-up to Kyoto.[3] But US opposition to a Kyoto-like solution now runs much deeper. In part, it stems from the Bush Administration's ideological opposition to any mandatory regulation of carbon emissions. But, in part, it reflects a deeper opposition to international constraints on US sovereignty, which has made it difficult for the United States to join multilateral treaties generally, even when they otherwise enjoy very widespread support (as is the case, for example, with the 1982 UN Convention on the Law of the Sea, which is currently tied up in the Senate, despite support from the Bush Administration, the US military, and the business community). It is conceivable that US opposition to mandatory carbon regulation could change as a result of the 2008 elections; but the difficulty of joining multilateral treaties such as Kyoto is a structural problem, which is unlikely to change anytime soon. Thus, even if developing countries were to change their stripes, it is by no means certain that this would induce the United States to reverse course and accept binding targets and timetables. Indeed, the current US position points exactly the opposite way. Rather than seeking developing country acceptance of emission targets, the Bush Administration has sided with developing countries in opposing a targets-based approach, both for itself and others.

In essence, Frankel's paper starts from the assumption that, from a political standpoint, there is nothing fundamentally wrong with targets and timetables – they were negotiated in Kyoto, and they could be negotiated again. It is just a matter of fine-tuning the formula for allocating targets in order to win over reluctant countries. But, I think, the trajectory of the climate negotiations over the past decade raises questions about whether this diagnosis of the situation is correct.

[3] However, more recent Senate action suggests that the developing country issue is evolving. In 2005, the Senate adopted by a vote of 53–44 the Bingaman–Domenici resolution (State Amendment 866), which calls for the enactment of a "a comprehensive and effective national program of mandatory, market-based limits and incentives on emissions of greenhouse gases," without any precondition of developing country action. Instead, the Bingaman–Domenici resolution flips the issue around, viewing US action as a means of encouraging comparable actions by other nations. Other proposals to develop a domestic cap-and-trade system also view developing country action not as a precondition for the United States taking a first step, but rather as relevant to the adoption of more-ambitious emission targets in the future.

Like most policy analyses of the climate change issue, Frankel implicitly sees it as a collective action problem. According to this view, individual actors will be unwilling to take action, unless they can be assured that their actions will be reciprocated by others; otherwise, they will simply incur costs without getting any real environmental benefit. The task of the international climate change regime, on this view, is to ensure some reciprocity of effort, by defining appropriate commitments for each party and providing some assurance of compliance. That is what Frankel's proposed target formula is intended to do.

But, although the collective action analysis of climate change mitigation seems correct in theory, it has not done very well in practice in accounting for the behavior of key actors. On the one hand, some actors are pushing forward to address climate change, even though their efforts are not being reciprocated by others. In the United States, for example, states and cities are developing their own climate policies, rather than waiting for action by the Bush Administration.[4] On the other hand, the Bush Administration has opposed binding emission targets not just for itself, but for others as well. It is not trying to free ride, as game theory would predict – it is trying to stop the bus altogether.

Let me suggest an alternative way of conceptualizing the climate change problem, which starts not by defining the policy desiderata for a collective response to climate change, but from the efforts that are already emerging from the bottom up, and asks, how might we make some incremental progress? If we start from this end of the telescope, so to speak, there are two questions to consider. First, are we doing all that we can to exploit the level of political will that currently exists? Are we getting the most bang for our buck? Second, what can we do to increase the level of political will, in order to build on the efforts that are already under way?

Starting with the second issue first, how might we increase the level of political will? Now I recognize that the whole idea of political will is fuzzy. What do we mean by it? How do we measure it? These questions do not have easy answers. And I think that Jeffrey Frankel is

[4] The Senate's adoption of the Bingaman–Domenici resolution in 2005, combined with its defeat the same year of a resolution focusing on international climate change policy, support the view that bottom-up, domestic approaches may be more promising in the short term than international treaty negotiations.

correct that many of the factors that might increase the political salience of climate change, such as extreme weather events, are exogenous to the climate change regime and, as such, we have little control over them (except perhaps in the conspiratorial world of Michael Crichton).

Nevertheless, to some extent, we may be able to influence the level of political will by the way we design the climate regime; in that sense, political salience may be partly endogenous to the system. This is one of the rationales behind the framework convention/protocol approach, namely, that a framework convention helps generate political will by focusing public attention on a problem, thereby building concern (Haas, Keohane, and Levy 1994; Bodansky 1999). It is also one of the rationales for periodic scientific assessments, and for establishing a long-term target for greenhouse gas concentrations or temperature increase, toward which the international climate change effort should aim. On this view, setting a long-term concentration target of, say, 450 ppm, could help serve as a catalyst for greater political action, the way that John F. Kennedy's pledge to go to the moon in the 1960s helped galvanize public opinion.

Of course, thus far the UNFCCC has not been terribly successful in building public concern. The IPCC, by contrast, has had a much greater impact, particularly its fourth assessment report issued in 2007. What more might the international negotiating process do to build political will to address climate change? One option would be to link the climate change agenda more closely with the development agenda, in order to piggyback climate change on other issues that developing countries care about (see Pershing, this volume). Similarly, in Western countries, many suggest linking climate change with issues of greater public concern, such as energy security.

Turning to the first issue, are we wringing out as much progress as possible from what little political will currently exists? In my view, the answer is no. The existing international climate change regime unnecessarily limits what can be accomplished internationally. First, it includes virtually every country in the world, a factor that makes the negotiations unnecessarily complex and difficult. Although it is often said that climate change is a global problem requiring a global solution, in fact just twenty-five countries account for more than 80 percent of global greenhouse gas emissions. If the climate change negotiations were limited to a smaller group of countries – the so-called "big emitters," for example, or the big economies, or like-minded states, or perhaps regional groups – this would simplify the negotiations considerably (Victor, this volume).

Second, although the Kyoto Protocol sets individualized targets for each Annex B country, it contains only one form of commitment: absolute targets and timetables, tied to historical emissions. Given the wide range of differences in national perspectives and preferences regarding climate change, the climate change regime needs a more flexible approach, which allows different countries to assume different types of international commitments – not only absolute targets, but also indexed targets, taxes, efficiency standards, and so forth. This is one of the important conclusions of the Pew Climate Dialogue at Pocantico, in which senior policymakers, business leaders, and NGO representatives participated.[5] A more flexible, bottom-up approach would attempt to build on what countries (and their subdivisions and businesses) are already doing, rather than imposing a solution from the top down (Pizer, this volume).

Would a more flexible, bottom-up approach provide a long-term solution to the climate change problem? Probably not, at least if climate change mitigation proves expensive. The more costly climate change mitigation is, the more states will want to ensure that their efforts are being reciprocated by other states, as the collective model action predicts – and the more policy architectures like Frankel's will become crucial. But a bottom-up approach might, at least, help break the current impasse and get the ball rolling. It reflects, not ideal policy, but rather less than ideal politics.

References

Barrett, Scott (2003). *Environment and Statecraft: The Strategy of Environmental Treaty-Making*, Oxford: Oxford University Press.

Bodansky, Daniel (1999). *The Framework Convention/Protocol Approach*, WHO Technical Briefing Series, Framework Convention on Tobacco Control, Coc. WHO/NCD/TFI/99.1.

[5] Pew Center on Global Climate Change (2005), The Pocantico dialogue met four times in 2004 and 2005 and involved policymakers and stakeholders from fifteen countries, including the United States, the United Kingdom, Germany, China, India, Japan, Australia, Canada, South Africa, Brazil, Argentina, and Mexico. The discussions focused on options for advancing the international climate change effort in the post-2012 period.

Chayes, Abram, and Antonia Chayes (1995). *The New Sovereignty: Compliance with International Regulatory Agreements*, Cambridge, MA: Harvard University Press.

Haas, Peter M., Robert O. Keohane, and Marc Levy, eds. (1994). *Institutions for the Earth: Sources of Effective International Environmental Protection*, Cambridge, MA: MIT Press.

Koh, Harold (1997). "Why do Nations Obey International Law," *Yale Law Journal* 106: 2599–2659.

Pew Center on Global Climate Change (2005). *International Efforts Beyond 2012: Report of the Climate Dialogue at Pocantico*, Arlington, VA: Pew Center on Global Climate Change.

Victor, David G. (2001). *The Collapse of the Kyoto Protocol and the Struggle to Slow Global Warming*, Princeton, NJ: Princeton University Press.

Wiener, Jonathan B. (1999). "Global Environmental Regulation: Instrument Choice in Legal Context," *Yale Law Journal* 108: 677–800.

2.2 *Incentives and meta-architecture*

JONATHAN B. WIENER

Jeffrey Frankel's paper in this volume helps advance thinking about two architectures: the architecture of a climate treaty regime, and the larger surrounding structure or "meta-architecture" of the institutions and decision framework within which the climate regime must be constructed. The former usage of "architecture" denotes the design of a climate policy regime (see Schmalensee 1998). The latter usage is much broader, embracing the structure of international relations and the evolution of institutions over time (following Keohane 1982 and North 1990) – the foundation on which the climate policy architecture is built or the constitutional framework within which climate regulation is adopted (see Wiener 1999a). Here I offer comments on each of these two architectures.

Climate policy architecture

Comprehensive, cap-and-trade

Frankel advocates the climate policy architecture in the Kyoto Protocol, with improvements. It is important to recognize that the basic climate policy architecture in both the 1992 United Nations Framework Convention on Climate Change (UNFCCC) and the 1997 Kyoto Protocol arose from several years of expert analysis yielding a bipartisan insistence on these elements by US negotiators. This climate policy design was developed in both the Bush (father) and Clinton administrations over the decade from 1989 to 1999: a "cap-and-trade" regime that is "comprehensive," covering multiple gases, sources and sinks (see Stewart and Wiener 1990, 1992, 2003; Wiener 2001, 2002).

The author thanks Joe Aldy, James Hammitt, and Robert Stavins for comments on a prior draft, and The Eugene T. Bost, Jr., Research Professorship of the Charles A. Cannon Charitable Trust No. 3 at Duke Law School for support.

As Frankel states, the "where" and "what" flexibility in this comprehensive cap-and-trade architecture would reduce the costs of global greenhouse gas emissions limitations by roughly 70–90 percent, compared to a CO_2-only regime with no sinks and no trading. In addition, there are environmental advantages which Frankel does not mention: the comprehensive approach would prevent the perverse cross-gas shifts (e.g., from CO_2 to methane) that would occur under a narrow CO_2-only regime, and it would foster conservation of biodiversity in forest sinks (Wiener 1995, 2002; Stewart and Wiener 1990, 1992, 2003).

Some critics assert that such a regime is unworkable because the task of allocating allowances is too difficult for it to get off the ground. This criticism is misplaced. First, all international climate regimes confront problems of burden-sharing negotiations and compliance assurance. A cap-and-trade regime arguably eases rather than worsens these problems – by sharply reducing costs, and by fostering post-allocation transferability (which, as Coase (1960) showed, makes the initial allocation less binding and hence easier to negotiate). Second, empirically such systems do work: this approach is essentially the same policy architecture that was used successfully to phase out ozone-depleting substances such as CFCs (though with less trading), to reduce lead (Pb) and SO_2 in the United States (though not internationally and not multi-gas), and to manage fisheries in New Zealand. It is also the same architecture currently being employed in the European Union's Emission Trading Scheme (ETS) (though not yet multi-gas) – a remarkable irony, because the EU had earlier opposed the comprehensive cap-and-trade policy design that the United States insisted be included in the 1992 UNFCCC and the 1997 Kyoto Protocol.

Fixing the flaws in Kyoto: participation and long-run credibility

At the same time, however, two key problems remained inadequately addressed by this policy architecture: engaging participation by major developing countries, and setting targets over the long run.[1] Frankel tackles both. In his advocacy of the comprehensive cap-and-trade

[1] The Kyoto Protocol and its follow-on agreements at Bonn and Marrakech also contained some undue constraints on the use of sinks and emissions trading and lacked clear rules for compliance. See Stewart and Wiener 2003.

regime, and his strategies for addressing these two remaining problems, Frankel is in close agreement with Olmstead and Stavins (2006) and Stewart and Wiener (2003), all of whom advocate the use of favorable allowance allocations ("headroom" or "growth targets") to deliver the side payments needed to engage the participation of developing countries, and the use of long-run emissions pathways derived from benefit-cost analysis. Frankel also adds some new proposals.

Participation

Although the 1992 UNFCCC called for action by all parties, the 1995 Berlin Mandate decided that quantitative emission targets would apply first only to wealthy countries, with no initial obligations for developing countries. The result was that the Kyoto Protocol targets covered only the industrialized countries listed in its Annex B. But without covering the rapidly growing emissions of countries like China, India, and Brazil, the regime cannot be environmentally effective. Developing country GHG emissions are forecast to surpass industrialized country emissions around 2020; China is set to surpass the United States by 2009. Even with US ratification, the Kyoto Protocol would only achieve a small reduction in global emissions. Worse, this omission could accelerate the emission growth in developing countries via "leakage" of emissions-intensive activities from regulated Annex B to unregulated Non-Annex B countries. Such leakage undermines the efficacy of the treaty even further and also makes developing country economies more carbon-intensive and hence less eager to join the treaty regime in the future (Schmalensee 1998). The omission of major developing countries also cost Kyoto the participation of the United States and Australia. The Clinton Administration signed the Kyoto Protocol but declined to submit it to the US Senate for ratification until "meaningful participation" by major developing countries was attained; and the Bush (son) Administration withdrew the United States from Kyoto for essentially the same reason. Engaging participation by major developing countries would accomplish four major advantages: addressing their rapidly growing emissions, preventing leakage from Annex B to Non-Annex B countries, reducing the cost of abatement via low-cost opportunities in developing countries, and attracting the United States and Australia to participate as well.

To engage full participation, Frankel proposes a formula for allowance allocations that seeks to satisfy efficiency and equity criteria by incorporating income, population, and recent emissions (see his Table 2.1). My reading is that Frankel intends this formula as a conversation-starter, to get countries talking about phasing in more fully global participation. But it is not a substitute for negotiated allocations. Countries are unlikely to agree to a set of variables in principle and then wait to see whether their allocations turn out well or not; they will bargain directly over the allocations. Formulas for targets seem to perform at best an anchoring role in guiding negotiations. Country governments will decide what allocations to accept based on their perceived net benefits (Wiener 1999a), as mediated through the complex mix of their foreign policy interests, domestic institutions and interest groups (Wiener 1999b). Indeed, three of Frankel's factors seem aimed at ensuring national net benefits. Progressivity of allocation burdens is important if the willingness to pay for climate protection rises with wealth. Indexation of targets (e.g., to economic output) offers insurance against escalating abatement costs when the economy turns out to grow faster than expected but also yields tighter targets when the economy turns out to grow more slowly (SueWing, Ellerman, and Song 2006). And a safety valve (price cap on allowances) is a way to avoid cost escalation from uncertainty about eventual true costs. (Frankel does not say at what price the safety valve should be set.)

Frankel's formula seeks convergence – at least in the very long term – to "equal per capita emissions" worldwide. But this seems unnecessary to attract participation (which, as just noted, depends on net benefits and should be addressed directly). Fairness does not seem to require equal emissions per capita, just as fairness does not require equal land area per capita or equal water use per capita. It is not clear why, say, Rawlsian justice or even Marxian "to each according to his needs" would require rights to strictly equal amounts of physical resources. If countries or individuals producing more-useful goods and services have an argument for being able to use more physical resources, then a better long-run goal might be equal emissions per unit of economic output. (Frankel suggests that over the long term, if per capita incomes converge, per capita emissions will also converge, but empirical research suggests that this need not be so [Aldy forthcoming].)

Long-term targets and credibility

The Kyoto Protocol targets cover only a "first commitment period" from 2008 to 2012, with subsequent targets left to future talks. For many of the Annex B countries, especially for the United States, these Kyoto targets posed costly cuts from business-as-usual (BAU) emissions forecasts (though Russia received extra allowances above its BAU forecasts for 2010).

Academic studies at the time suggested that a least-cost path to stabilizing atmospheric concentrations (e.g., at 550 ppm of CO_2-equivalent) would involve almost no abatement below BAU in the near term (i.e., much less stringent than Kyoto), and then very sharp reductions later (say, after 2040), when new energy technologies would presumably become available (Wigley, Richels, and Edmonds 1996). But this approach neglects the interim benefits of reducing climate change damages from emissions along the way. A full benefit–cost optimizing approach, incorporating even a low estimate of gradual interim climate damages (with no account taken of potential abrupt damages), suggested that the optimal abatement path would begin to reduce emissions below BAU somewhat earlier, around 2010, that is, more aggressively than the least-cost path to stabilizing concentrations although still less sharply than Kyoto (Hammitt 1999).

Frankel would set long-run targets based on benefit-cost analysis, as would Stewart and Wiener (2003). But in defense of the more stringent, near-term Kyoto targets, he contends that governments could not credibly announce targets that would take effect only forty years after they left office, because political officials can only act one political term at a time and are unable to bind future governments. To be credible, he argues, a policy requires some extra near-term bite even if it is more aggressive than an optimal long-run path.

There are problems with this argument. First, it is not clear why a credible long-term target demands more stringent as opposed to less stringent initial targets than the optimal path. A government could seek credibility for limiting emissions, or credibility for limiting abatement costs. The choice depends not on credibility per se but on which kind of error (rising emissions and climate change damages, or rising abatement costs) one would rather avoid.

Second, even if a tighter, near-term target is needed to be credible, benefit–cost analysis is still needed to decide how much tighter than the optimal long-term target pathway would be *truly* optimal to remedy the credibility problem. Frankel offers no other method for deciding this question. He agrees that uncertainties are no obstacle to benefit–cost analysis because they should still be handled explicitly through benefit–cost analysis by "taking the best shot" in light of the uncertainties. Thus, rather than counseling against using benefit–cost analysis in setting optimal targets, the task of setting a credible target actually necessitates using it, if credibility requires some optimal adjustment from the ordinary optimal path.

Third, a tighter target does not solve the credibility problem, because it can still be undone by future governments. For climate policy, setting targets for a later period than the present political term is unavoidable. The Kyoto targets, for example, were set for a date (2008–2012) that was long after the political term of the leaders in office in 1997. Frankel lampoons as non-credible a target set on least-cost criteria to require no action initially and then a sudden "big day in 2050" when emissions would be slashed; but a target set for the "big day" in 2012 (or even 2008) shares the same criticism, from the vantage of 1997. Perhaps Frankel would have favored an earlier target period in Kyoto, say for 2000, but of course targets set in 1997 for the years *during* or just after that presidential term would have had to be very minor to be at all achievable.

In short, the credibility problem – derived from the inability to bind future governments – is distinct from the duration and optimality of the policy pathway. Even if a current government must demonstrate credibility with a near-term constraint, it need not stop at that near-term constraint; it must and should still adopt a long-term phase-down schedule that would have to be implemented by future administrations. That pathway should still be guided by an optimal benefit–cost analysis. In contrast to the Kyoto target period of 2008–2012 set in 1997, a long-term pathway could have been set for, say, the four or five decades from 2000 to 2040 or 2050. This would have been similar to the multi-year emissions reduction schedules used in the cap-and-trade programs to phase down lead (Pb) in gasoline (petrol) and to phase down SO_2 emissions. Such a longer-term pathway gives industry more guidance on investment decisions (and a better means for banking and borrowing allowances over time) and gives regulators a better basis for tracking actual progress than does a single five-year target period. Announcing a long-term pathway shifts the default option, from no constraint after the end of current target period (2012) to the designated pathway, and may therefore make it more costly to change course; but so does the present default option of no constraint (after 2012) make it difficult to adopt a new second target. The question, inescapably, remains the optimality of the pathway in light of the twin costs of erring through excessive emissions or excessive abatement costs.

Then this longer-term pathway should be (and would inevitably be) *adjusted*, sequentially, by future negotiations among governments (as Frankel proposes). At each point in time, our best benefit–cost analysis charts a plausible optimal pathway (derived from a portfolio of scenarios), and we must both act now based on that optimal pathway and also update that policy over time. As noted above, benefit–cost analysis recommends doing more than nothing about greenhouse gas emissions in the near term. Adjustability allows learning and adaptive management in light of new information. In a dynamic problem characterized by uncertainty and learning over time, it makes sense to set targets through an iterative process of sequential decisions: policy, research, evaluation, and adaptive policy revision. We will learn as we go, both about the climate system and about economic responses and technology options. Thus, even if a current government did adopt a fifty-year pathway for emissions reductions, while it knew it could not bind future governments, it would still make sense to set and adjust that pathway over time.

Meta-architecture

Any specific policy regime must be formulated within the overarching decision framework for government action. That framework is a meta-architecture, a set of structures and rules that are fixed at least in the short term. As James Buchanan has emphasized, one must understand these decision-making institutions before designing policy options; there is no "benevolent despot" to adopt by fiat the economist's efficient policy design (Buchanan 1987). Yet the climate policy literature has repeatedly spoken as if there were a benevolent despot to impose a global carbon tax or a long-run abatement path (see Wiener 1999a). In reality, at the international level, apart from coercion through military force or trade sanctions, the basic legal structure is that policy regimes are binding only on those countries who agree to join. As Frankel says, countries will not join a regime that imposes disproportionate costs. And the costs and benefits of joining a regime to prevent climate change vary widely across countries.

Thus the problem is how to produce a global public good, via national consent, among heterogeneous countries. The direct implication is that the successful policy architecture must be not only collectively rational

(Kaldor–Hicks efficiency), but also individually rational for those joining the regime (actually Pareto improving) – joining must offer net benefits to each participating country over not joining (Wiener 1999a, Gruber 2000). Those benefits can include climate protection as well as national co-benefits, such as in public health (e.g., from reducing other pollutants), biodiversity conservation, energy security, and reputation. Where national net benefits are not positive, some form of side payment will be needed to attract participation.

In other words, although a substantially global regime is needed to protect the global environment, countries have diverse interests, so some mechanism is needed to make cooperation attractive to the diverse countries significantly influencing the outcome. For major developing countries whose priority is development and who view climate change as of low importance or even as benign, this means that climate policy must support their development goals, not impose obstacles or costs to their development.

Further, countries are not monoliths but have contending domestic factions and internal institutions (Wiener 1999b). To take one example, the US Constitution requires a two-thirds majority to ratify a treaty in the Senate, which accords greater representation to voters in sparsely populated states (often western resource-rich states) Business groups and environmental groups lobby Senators. Meanwhile, these internal institutions and factions are often linked across countries in transnational networks (Wiener 2001).

The spatial challenge

This meta-architecture poses a spatial challenge and a temporal challenge. The spatial challenge is of engaging cooperation among countries. How to engage major developing countries such as China, India, and Brazil? How to engage the United States? Frankel, as well as Olmstead and Stavins 2006 and Stewart and Wiener 2003, opts for cap-and-trade with BAU or "growth" allowances to attract developing countries. Frankel explains why this would work well but does not compare it to alternatives, such as a tax. The comparison is complex; what follows is but a brief overview:

- Decentralized unilateral action would not adequately respond to the problem of a global public good. Global benefits of abatement will exceed national benefits, so uncoordinated national action will be inadequate. See Olmstead and Stavins (2006). In certain situations under favorable conditions – generally involving small close-knit groups – some tragedies of the commons can be successfully addressed through informal development of shared norms among the players. Perhaps energy subsidies could be reduced through coordinated disarmament, as Frankel suggests. But where measures are costly, and at larger scales where reciprocity and monitoring are more elusive, the tragedies persist unless stronger institutions provide incentives for action in the common good.

- A global carbon tax, even if collectively efficient, will not attract participation by those countries who see little benefit (or a loss) from preventing climate change. To them the tax is pure cost. Yet such countries are precisely the ones needed to address global greenhouse gas emissions. See Wiener (1999a); Endres and Finus (2002). (Perhaps a set of coordinated national GHG taxes, with offsetting reductions in other taxes to yield a revenue-neutral double-dividend, would be attractive in many countries; but unless the net costs of such taxes would be negative [in which case one wonders why such taxes are not being adopted unilaterally already], there will need to be side payments to cover the remaining net cost or else countries will not adopt such taxes. The question remains how best to make those side payments. Further, countries can be expected to play "fiscal cushioning" games that will undermine the effect of taxes more seriously than they would affect quantity-based emission limits (Wiener 1999a).
- Developing new technology through government-funded research and development (R & D) is desirable where private firms fail to invest in basic research with long lead times. But it is insufficient, because we also need some demand-side incentives to generate adoption and diffusion. And taxes or tradable allowances would also stimulate technology R & D by private firms.
- Hence, to engage participation, we need a policy instrument combined with some system of side payments across countries. Direct payments (project finance, such as via the Clean Development Mechanism [CDM]) could involve roughly the same level of international transfers as under cap-and-trade. But without emissions caps, they risk within-country leakage (see Frankel, footnote 14). They also risk the perverse aggregate effects of a subsidy for abatement that reduces emissions at each firm but expands the emitting sector (as pointed out by Oates and Kohn; see Wiener 1999a).

If direct payments are handled by government aid agencies, they also risk the political distortions that market-based allowance trading investments would avoid. And if aid payments are close to the price of formal allowances, they risk discouraging recipient countries from joining a formal cap-and-trade regime.

Alternatively, a global tax could be coupled with compensatory side payments to indemnify reluctant countries against the costs of the tax. But these side payments must cover countries' net costs at the margin (to achieve actual Pareto-improvement) and this would undermine the incentive effect of the tax.

The best method of delivering these side payments is a cap-and-trade regime with "headroom" allowance allocations (or "growth targets"). Headroom allowances can be resold to deliver the side payment; the cap prevents the perverse incentives that other side payments could generate (Wiener 1999a, 2002; Stewart and Wiener 2003). Frankel suggests that emission targets for developing countries should be set below their BAU and above their "break-even point" so that they face no net cost from joining (i.e., so that their revenues from the sale of extra permits would just exceed their costs of abatement). But the net costs of joining would also reflect the benefits of climate protection (and other co-benefits). Thus developing countries that get some benefit from climate protection could accept a target that is tighter (fewer allowances) than Frankel's "break-even" point, whereas developing countries that perceive benefits from a warmer world would need extra allowances to meet a no-net-cost criterion.

The temporal challenge

The temporal challenge is to engage present and future governments in action on a sensible path with credible signals to motivate private sector investments.

First, as Frankel asks, can current governments bind future governments to continue limiting emissions?[2] Put another way, can the

[2] Current governments can in theory partially bind future governments, such as by adopting policies that are especially difficult to change (e.g., entrenched bureaucracies, or constitutional amendments), or by making major investment decisions now that are costly to undo (e.g. large capital investments in energy sources or infrastructure). Letting climate change occur is in its own way a current choice

present force the future to bear some costs of policy action? Perhaps intertemporal penalties or intertemporal side payments could influence future governments? And are there institutions to enforce current policies on future governments, as courts enforce easements and servitudes on future property holders? Frankel suggests pre-commitment to tight targets, but as suggested above, it will still be important to learn over time and adapt; we would not want climate rules that rigidly bind future governments.

Second, there is a reverse problem of temporal mismatch: the future cannot bind the present to act now. Current emissions will affect future climate, but optimal action (e.g., development of new technology) takes time, such that costs must be borne before benefits are accrued. This yields an intertemporal externality, or a tragedy of the temporal commons, in which the present can impose costs on the future (similar to the problem of deficit spending). Current emissions impose damages on future populations. And current policies also affect the future costs of emissions abatement; if current policies do not stimulate investments in new technology, the declines in abatement cost now forecast to face future governments may not come to pass.

A third temporal aspect may also inhibit short-term policy action: perceptual updating. What is perceived as "normal" climate may change over time. As the climate changes over a century or more (and putting aside abrupt dramatic changes), each generation may experience only a small gradual change within its lifetime. Voters in any one period may not perceive the changing climate and the damages occurring over a longer period. (Increasing longevity may partly counteract this problem.) Even if significant damages can be forecast over a forthcoming century, within each generation the then-current climate may be viewed as "normal" and there may be less motivation to reverse that situation or to prevent future change. Whose valuations should be used to measure future damages? Giving full respect to future generations' own (adaptive) values could, ironically, imply lowering estimates of damages, compared to present values, and thus doing less to forestall future climate change than present values would suggest.

A fourth temporal challenge is that the governments of tomorrow will not necessarily be the same as those we know today. It seems likely

that binds future governments. There are irreversibilities, or better said, choices that are costly to reverse or modify, on all sides of the question.

that there will be changes in major power geopolitics. From 1783 to 1815, global geopolitics were radically changed and a new multipolar world order was created that persisted for a century. Something similar could happen from 1989 to 2020. Today, the Chinese and Indian economies are growing at almost 10 percent per year while the US grows at 3.5 percent and the EU at 2 percent. By 2020 or 2030, as China's economic growth and CO_2 emissions rise to exceed those of America, the United States may no longer be the Lone Superpower (a period that may come to be viewed as the short transition from the Cold War to a new era). Instead there may well emerge a more multipolar world of several great powers, including the United States, an organized EU, and a rising China and India. Russia could ally with Europe or remain a wild card, retaining large nuclear weapons stocks and exercising market power in natural gas and emission allowance markets. Japan and Brazil could also be major powers. The roles of the Middle East, Africa, South America, and others could also change. The key point here is that climate policy for the coming century will need to be a part of the geopolitical meta-architecture wrought by these powers in the future, not just the roles they play today.

Conclusions

Jeff Frankel's advocacy of a comprehensive cap-and-trade architecture is in my view the best design for international climate policy, even if I have questioned some of his proposals for allocation formulas and long-term targets. In light of the spatial and temporal challenges of designing policy architectures nested within the meta-architecture, there are strong grounds to pursue some policy experimentation – to try different approaches in plurilateral coalitions in order to provide a portfolio of adaptive policy options (Hahn 1998; Stewart and Wiener 2003; Pizer, this volume). Work by Carlo Carraro and others (see his entry in this volume) suggests that plural coalitions are likely to form in any case. The debate over policy instruments and architectures (e.g., cap-and-trade, taxes, and technology R & D) could be better informed by empirical experience with each. And such experimentation could also be a way to engage participation of major developing countries. For example, a cap-and-trade regime could be organized among the United States, Australia, China, India, Korea, Brazil, and Mexico, in parallel to the Kyoto and EU regimes but with improved elements such as fully comprehensive coverage of gases and sinks and no limits on trading (Stewart and Wiener 2003), to test their comparative performance over time. Jeff Frankel's proposals for policy design, participation, and targets could be tried by the Kyoto group. Other approaches could be tried as well. Through such experimentation in a global policy laboratory, we would learn and adapt, possibly merging the systems after a decade or so into a new global regime incorporating the best features of each.

References

Aldy, Joseph E. (forthcoming). "Divergence in State-Level Per Capita Carbon Dioxide Emissions," Resources for the Future Discussion Paper 06-07, forthcoming in *Land Economics* 83(3).

Buchanan, James M. (1987). "The Constitution of Economic Policy," *American Economic Review* 77: 243–250.

Coase, Ronald (1960). "The Problem of Social Cost," *Journal of Law and Economics* 3: 1–28.

Endres, Alfred, and Michael Finus (2002). "Quotas May Beat Taxes in a Global Emission Game," *International Tax and Public Finance* 9: 687–707.

Gruber, Lloyd (2000). *Ruling the World*, Princeton, NJ: Princeton University Press.

Hahn, Robert W. (1998). *The Economics and Politics of Climate Change*, Washington, DC: American Enterprise Institute Press.

Hammitt, James K. (1999)."Evaluation Endpoints and Climate Policy," *Climatic Change* 41: 447–468.

Keohane, Robert O. (1982). "The Demand for International Regimes," *International Organization* 36(2): 325–355.

Olmstead, Sheila, and Robert N. Stavins (2006). "An International Policy Architecture for the Post-Kyoto Era," *American Economic Review, Papers and Proceedings* 96(2): 35–38.

North, Douglass C. (1990). *Institutions, Institutional Change and Economic Performance*, Cambridge: Cambridge University Press.

Schmalensee, Richard (1998). "Greenhouse Policy Architectures and Institutions," in William D. Nordhaus (ed.), *Economics and Policy Issues in Climate Change*, Washington, DC: Resources for the Future Press, pp. 137–158.

Stewart, Richard B., and Jonathan B. Wiener (1990). "A Comprehensive Approach to Climate Change," *American Enterprise* 1(6): 75–80 (November–December).

(1992). "The Comprehensive Approach to Global Climate Policy: Issues of Design and Practicality," *Arizona Journal of International and Comparative Law* 9: 83–113.

(2003). *Reconstructing Climate Policy: Beyond Kyoto*, Washington, DC: American Enterprise Institute Press.

SueWing, Ian, A. Denny Ellerman, and Jaemin Song (2006). "Absolute vs. Intensity Limits for CO_2 Emissions Control: Performance Under Uncertainty," MIT Joint Program on the Science and Policy of Global Change.

Wiener, Jonathan B. (1995). "Protecting the Global Environment," in John D. Graham and Jonathan B. Wiener (eds.), *Risk vs. Risk: Tradeoffs in Protecting Health and the Environment*, Cambridge, MA: Harvard University Press, pp. 193–225.

(1999a). "Global Environmental Regulation: Instrument Choice in Legal Context," *Yale Law Journal* 108: 677–800.

(1999b). "On the Political Economy of Global Environmental Regulation," *Georgetown Law Journal* 87: 749–794. Reprinted in Carlo Carraro (ed.), *Governing the Global Environment*, Cheltenham: Edward Elgar, 2003, pp. 153–198.

(2001). "Something Borrowed for Something Blue: Legal Transplants and the Evolution of Global Environmental Law," *Ecology Law Quarterly* 27: 1295–1371.

(2002). "Designing Global Climate Regulation," in Stephen Schneider, Armin Rosencranz, and John-O Niles (eds.), *Climate Change Policy*, Washington, DC: Island Press, pp. 151–187.

Wigley, Thomas M. L., Richard G. Richels, and Jae Edmonds (1996). "Economic and Environmental Choices in the Stabilization of Atmospheric CO_2 Concentrations," *Nature* 379 (18 January): 240–243.

3 *Graduation and Deepening*

AXEL MICHAELOWA

Introduction

THE Earth is warming rapidly. Over the last twenty-five years, global surface temperatures have increased by almost 0.5°C and reached levels not seen at least for the last millennium. The impacts of global warming are increasingly visible and range from longer vegetation periods over retreat of mountain glaciers to a shrinking of the area covered by sea ice in the Arctic. The specter of extinction of polar bears due to disappearance of their habitat is a powerful symbol used by environmental organizations to mobilize humanity to fight climate change.

As global warming proceeds, the political discussion about its causes and countermeasures intensifies. Owing to the tireless efforts of thousands of scientists, the world's knowledge on climate change has been compiled in the Assessment Reports of the Intergovernmental Panel on Climate Change (IPCC). Since 1990, three Assessment Reports of thousands of pages each have been published and the fourth is due in 2007. Their message has become increasingly clear: global warming is due to humankind. Ever since the Industrial Revolution in the late eighteenth century, burning of fossil fuel and industrial processes have led to emissions of greenhouse gases (GHGs) such as carbon dioxide, methane, and nitrous oxide. Over the last two centuries, emissions of these gases have increased dramatically. The wealthy societies of the early twenty-first century are built on economic systems where greenhouse gas emissions are ubiquitous. It will be a huge challenge to achieve the transition to a low-carbon economy.

Once released to the atmosphere, the majority of greenhouse gases will stay there for centuries until natural processes destroy them. Some gases even have residence times of millennia. Therefore, international climate policy is a task for many generations. As it started only two decades ago, it is still in its infancy compared to international trade and monetary policies. Nevertheless, already a series of steps have been achieved that led from a very general framework to treaties defining detailed policy instruments. The United Nations Framework Convention on Climate Change (UNFCCC) agreed in 1992 defined the general outline of the regime. The Kyoto Protocol was achieved in 1997 and defines a structure of emissions commitments and policy instruments. A series of decisions taken in 2001 and confirmed in 2005 defined detailed rules for the operation of the Kyoto Protocol's instruments.

Compared to other international regimes such as the Law of the Sea that took almost three decades to negotiate, the progress made in international climate policy so far is relatively good. The climate policy regime is having annual meetings of its Conference of the Parties, has two subsidiary bodies and a permanent secretariat. So it has the institutional characteristics of a stable regime. According to Sandalow and Bowles (2001) the following seven parameters of a workable climate treaty are all fulfilled in the Kyoto Protocol:

1. In light of varying economic and political circumstances, nationally defined targets for total GHG emissions per time period are a better alternative than common fiscal or regulatory policies.
2. Where political goals are usually insufficient, legally binding commitments constitute effective incentives for action. The success of these has been demonstrated for instance by the Montreal Protocol.
3. Cost-effective abatement strategies should apply to all GHGs.
4. Multi-year compliance periods are appropriate as a buffer to inter-annual variations of GHG emissions caused by atypical weather events or industrial cycles.
5. International emission trading constitutes an effective means of meeting the objectives of the Convention in the most cost-effective way.
6. As a transitional phase to abatement at source, sequestration by afforestation has the dual advantage of reducing net GHG emissions cost-effectively and creating or maintaining environmental benefits.
7. According to current models of economic growth, GHG emissions from developing countries are projected to surpass current industrialized country levels. Mechanisms to reduce the associated GHG emission intensity should include investment into clean energy and energy efficiency, as well as options for developing countries to

adopt individual targets and participate in international emission trading.

In the discussions about the post-2012 climate policy regime, it is often assumed that a paradigm shift is necessary. While many observers criticize the Kyoto Protocol regime for its short-term orientation, it has always been geared toward a regular updating. However, negotiators have been sidelining a number of intractable issues instead of directly facing them, such as future global commitments and integration of some of the world's largest greenhouse gas emitters in its system of reduction commitments. Some path dependency is likely to evolve as the regime becomes more mature, but this would be the case in any policy regime.

The scenario "Graduation and Deepening" developed below organically orients the current regime toward strong emission reduction commitments and a rapid integration of developing countries, thus following the 7th criterion of Sandalow and Bowles. First, the assumptions underlying the scenario are described, followed by a description of its emission targets and policy instruments. Subsequently I will discuss how to overcome barriers to the implementation of "Graduation and Deepening."

Key components of Graduation and Deepening

Graduation and Deepening is not a short-term fix but a long-term concept for international climate policy. It can easily be adapted to changing circumstances such as unexpected scientific results with regard to future climate change or changes in economic systems. Graduation and Deepening builds on legally binding emission targets, the efficiency properties of market mechanisms and the cost reduction potential of addressing a wide range of greenhouse gases and mitigation options.

Definition of a greenhouse gas concentration target

Theoretically, a policy measure should be evaluated according to the present value of its benefits and costs. Optimally, the present value of net benefits would be maximized. In climate policy that has a time horizon of centuries, use of a benefit–cost approach is difficult due to the problem in setting a long-term discount rate, due to the need to valuate non-market impacts, and due to the assumption that distribution does not matter (for a good discussion of these points see Goulder [2003]). If economists cannot use a benefit–cost approach, they use the standard-price approach as a second best (Baumol and Oates 1971). The standard has to be set by the political process (which obviously generates an opportunity for lobbies to influence the standard level) and then instruments are introduced to equalize marginal costs of reaching the standard. For a global pollutant, equalization of marginal costs means that mitigation cost is minimized. This does not mean that the burden for everybody is the same, as an action to mitigate greenhouse gas emissions can be implemented by one actor but its costs covered by another one. The Kyoto Protocol has many features of a standard-price approach. The emission targets define the standard and the Kyoto mechanisms allow marginal cost equalization. However, the targets cover only a limited share of global emissions.

For a long-term cost-effective emission target path, an atmospheric concentration target has to be defined. No country has formally proposed such a target in the negotiations. In 1997 the European Union (EU) suggested long-term stabilization at 550 ppm (parts per million) CO_2.[1] The second step would be to define the target date and a path towards reaching the target. From there, the maximum level of global GHG emissions in a given period could be quantified setting the base for negotiations concerning the allocation of emission targets to all countries. A 450 ppm target is unlikely to be acceptable in the international negotiations while a 550 ppm target could be agreed as an *indicative* concentration target, as there is an increasing consensus that costs of reaching the 550 ppm target are much lower than of reaching 450 ppm (Edmonds and Sands 2003: 178). If the 550 ppm target is agreed, it should also be possible to achieve a consensus that global emissions should not peak later than 2030. Obviously, the emergence of new results regarding nonlinear effects of greenhouse gas concentration should lead to a revision of the concentration level and emission path over time. A process should be envisaged where each new IPCC assessment report triggers a reassessment of both concentration target and emission path underlying the global climate policy regime.

[1] In the document specifying the target, the EU also aims to limit the temperature increase to 2°C compared to the pre-industrial level. This is unlikely to be achieved with a 550 ppm CO_2 target as shown in Meinshausen (2006).

Expanding the circle of countries with obligations under the UNFCCC through thresholds

Article 3 of the UNFCCC defines the principle of common but differentiated responsibilities. It can be interpreted in the sense that countries should take up emission targets once they reach, for example, a level of wealth or of emissions comparable to the current Annex B countries. Such a threshold approach could lead to a "fair" contribution of countries to greenhouse gas mitigation. A generally agreed definition of "fairness" or "equal contribution" does not exist. Major elements for such a definition (see also Pichs *et al.* 2000) could be:

- historical responsibility;
- the need for (sustainable) development;
- capability in terms of finance (Jansen *et al.* 2001); and,
- capacity in terms of (cheap) mitigation potential (Claussen and McNeilly 1998).

Several options as to how thresholds may be defined can be derived from these elements, e.g., GDP per capita, cumulative past emission ("historical responsibility for climate change"), emissions per capita, etc. In order to reach a global emission path, the level of the threshold and the stringency of targets for countries that have passed the threshold are two distinct variables that can be set. The emission data source used for the following calculations is the World Resources Institute's (2006) Climate Analysis Indicators Tool; GDP per capita is taken from the International Energy Agency (IEA) (2005).[2] Micro-states of fewer than 100,000 inhabitants such as San Marino and dependent territories (e.g., Bermuda) are not analyzed. The latter should in any case be taken into account as part of their respective Annex B "mother" country.

If thresholds are to be based on financial indicators, one can – to a certain extent – take into account the capacity of a country to contribute to global GHG emission reduction or limitation. Financial indicators can be expressed in income per capita or in purchasing power parities of a reference period or averages of past periods per capita. This idea was first developed by Claussen and McNeilly (1998) but only applied in a fairly superficial manner. Gupta (1999), Gupta, van der Werff, and Gagnon-Lebrun (2001), and Gupta (2003) developed a far more detailed version with a "graduation profile" matrix comprising

[2] World Bank (2006) data unfortunately have major gaps and thus cannot be used.

per capita emissions and per capita income. There are nine possible entries in their matrix, each leading to a different target and policy package. They, however, do not quantify the thresholds. Based on the same idea but a simpler application, Graduation and Deepening is based on a graduation index complemented by overall emissions and institutional thresholds for our scenario.

Combined financial and emission thresholds: graduation index

A combination of per capita income and per capita emission thresholds captures both ability to pay and the "polluter pays" principle. Each criterion should get the same weight as both principles are equally important and are not directly correlated. If a "graduation index" is calculated where both figures are weighted with 50 percent, we get the results shown in Table 3.1. For the underlying per capita GDP (measured in purchasing-power parities) and emissions data, see Appendix. An index value of one corresponds to $10,000 GDP per capita and 10 tons CO_2 equivalent per capita, since this yields an approximately equal weight of each component on the world average.

Table 3.1 *Combined GDP and emissions per capita thresholds (year 2000)*

	Graduation index[1]	*Emissions ($MMTCO_2$)*	*Share of world emissions (%)*	*Emissions change 1990–2000 (%)*
Qatar	5.1	39.7		+142
United Arab Emirates	2.8	117.2		+74
Kuwait[2]	2.4	69.1		+165
Bahrain	2.0	16.6		+35
Brunei	1.2	7.2		+26
Singapore	1.9	55.8		+81
Average Annex B	*1.8*	*305.6*	*0.9*	+92
Israel	1.6	77.3		+64
Taiwan	1.5	230.4		+78
Saudi Arabia	1.5	340.6	1.0	+67
Trinidad and Tobago	1.4	24.9		+19
Korea	1.4	521.2		+78
Cyprus	1.3	8.0		+51
Oman	1.2	29.2		+81

	Graduation index[1]	*Emissions ($MMTCO_2$)*	*Share of world emissions (%)*	*Emissions change 1990–2000 (%)*
Malta	1.2	2.4		0
Lowest Annex II	***1.2***[3]	***1234***	***3.6***	***+72***
Bahamas	1.1	2.0		−9
Barbados	1.0	1.5		+7
Argentina	1.0	292.3		+16
South Africa	0.9	417.9	1.2	+17
Turkmenistan	0.9	64.1		−4
Botswana	0.9	15.1		+24
Libya	0.9	62.6		+26
Kazakhstan	0.8	161.0		−51
Malaysia	0.8	164.8		+79
Uruguay	0.8	25.5		+27
Venezuela	0.8	243.0		+20
Chile	0.7	76.4		+52
Gabon	0.7	10.1		+10
Iran	0.7	480.3	1.4	+67
Mexico	0.7	511.8	1.5	+19
Mauritius	0.7	4.0		+67
Mongolia	0.7	27.8		+7
Brazil	0.6	849.9	2.5	+75
Namibia	0.6	10.6		+23
Guatemala	0.6	21.5		+41
Costa Rica	0.6	12.5		+16
Turkey	0.6	362.0	1.1	+31
Macedonia	0.6	11.2		−9
Thailand	0.5	264.5		+51
Equatorial Guinea	0.5	2.5		+316
Panama	0.5	11.7		+43
Lowest Annex B	***0.5***	***4109***	***12.0***	***+21***

Notes:

[1] \$10,000 per capita and 10 tons CO_2 eq. per capita give the graduation index value 1.

[2] Very low 1990 level due to Iraqi occupation; compared to 1989 level, energy-related CO_2 emissions are constant!

[3] Switzerland.

To illustrate calculation of the graduation index we use the example of Qatar. It has a GDP of $33,833 per capita (GDP index component = 3.4) and annual per capita emissions of 67.9 tons CO_2 (emissions index component = 6.8). The graduation index of Qatar reaches 5.1 and is shown in the first line of Table 3.1.

Thresholds are defined by the per capita income and emissions characteristics of countries that currently are members of Annex B of the Kyoto Protocol. The highest threshold is set at the average per capita emissions and income of Annex B countries. Any country passing this threshold has the characteristics of a typical Annex B country and thus should be treated similarly to those countries (see discussion below). The second threshold is defined by Switzerland, the country included in Annex II (i.e., providing resources to the financial mechanisms of the UNFCCC) that has the lowest value of the graduation index due to a very low level of per capita emissions. The third and lowest threshold is set by Ukraine, the Annex B member with the lowest level of the graduation index, mainly due to a low per capita income.

Absolute emission thresholds for policy commitments

Many large emitters such as India and China do not graduate because their index levels are below the lowest threshold. In Graduation and Deepening, emitters above 50 million tons CO_2 equivalent annual emissions (see Table 3.2) are parties to a special Annex which allows them to choose between an *ex ante* intensity target with emission trading, or use of countrywide Clean Development Mechanism (CDM; see discussion in the policy strategies section). These countries could also give a voluntary commitment to implement all macroeconomically sensible policies with greenhouse gas benefits. A narrow approach would be to address all projects that are pure monetary no regrets. Krause (2000: 30f.) estimates the potential at 1.7 to 6.7 billion tons CO_2 for the period 2008–2012. A wider perspective would be to take positive local pollution externalities into account ("ancillary benefits," OECD 2000). In severely polluted areas, the externalities can be higher than the costs for greenhouse gas abatement and reach values of above $10/tons CO_2 equivalent.

These countries account for 29 percent of world emissions and thus combine more emissions than the entire set of graduating countries (16.5 percent). In contrast to general perceptions, their

Table 3.2 *Emitters above 50 MMTCO$_2$ in 2000*

	Emissions (MMTCO$_2$)	*Emissions change 1990–2000 (%)*	*TCO$_2$/ capita*	*GDP/capita ($ PPP)*
China	4962	+32	3.9	4059
India	1901	+41	1.9	2364
Indonesia	503	+25	2.4	2790
Pakistan	285	+38	2.1	1715
Nigeria	193	+18	1.5	763
Egypt	178	+66	2.8	3352
Colombia	160	+23	3.8	5535
Vietnam	132	+54	1.7	2057
Algeria	126	+22	4.1	5153
Philippines	131	+46	1.7	3827
Bangladesh	122	+34	0.9	1533
North Korea	111	−66	5.0	1262
Iraq	101	+15	4.4	1202
Sudan	100	+18	3.2	1767
Myanmar	83	+32	1.7	4664
Peru	70	+42	2.7	4457
Cambodia	69	+11	5.4	NA
Syria	67	+40	4.1	3005
Ethiopia	63	+29	1.0	813
Serbia	62	−39	5.8	3858
Tanzania	61	+17	1.8	520
Morocco	58	+29	2.0	3413
Kenya	54	+10	1.8	927
Azerbaijan	55	−16	6.9	2659
Congo, D.R	52	−2	1.1	643
Cuba	50	−13	4.5	6540

emissions increases during the 1990s have been smaller than those of the countries that graduate. Rapidly industrializing countries would quickly move up the per capita income ladder, meaning that they would graduate in a post-2017 commitment period. China would probably graduate more quickly than India. The provision of incentives such as countrywide CDM to embark on reductions would enable a "meaningful participation" of these countries right from 2013.

Organizational thresholds

Other than being derived from quantitative thresholds, graduation has to be linked to organizational parameters, i.e., on a country's membership in certain intergovernmental organizations. This is based on the observation that some organizations only admit countries that have reached a certain degree of economic development. In fact, Organisation of Economic Co-operation and Development (OECD) membership originally defined Annex II membership in the UNFCCC. Likewise, being a recipient of the World Bank's International Development Association (IDA) funds or food aid is a clear signal that a country is still a low-income, developing country. We therefore assume the following simple institutional graduation scheme: membership in the European Union, the OECD, or the IEA results in automatic inclusion in Annex B but least developed countries and recipients of IDA funds or food aid would be exempt from any target. Korea, Cyprus, and Malta would move up from the second level of graduation, Mexico from the third level.

Definition of emission targets according to graduation status

To achieve the emission path leading toward the concentration target and to convince developing countries to accept targets under the graduation formula, the targets of current Annex B countries have to lead to an intensification of the emission-reduction trends initiated in the first commitment period. Especially for countries in transition, targets would be considerably stricter taking into account the availability of hot air. As policymakers tend to use simple numbers,[3] target differentiation could be made in multiples of 3. Owing to complaints that the base year 1990 is arbitrary and inequitable, targets should be based on a simple reduction from "business-as-usual" (BAU). Under Graduation and Deepening, BAU would be proposed by review teams for Annex B countries with hot air in a similar process as adjustments to Assigned Amounts are made under the Kyoto Protocol. The final BAU figure would have to be agreed by the government of the respective country. For all countries without "hot air", BAU would be

[3] For example, the Kyoto target negotiations circled around a stepwise differentiation of one percentage point between the EU, USA, and Japan. The small project thresholds of the CDM all use the number 15, leading to an inconsistent but seemingly easy and just outcome.

defined by Kyoto Protocol first commitment period target levels. This would create the following target allocation (see Table 3.3), where the first column shows the target in percentage reduction compared to the first commitment period level. Columns two and three show the difference between the emissions the countries had in 2000 and their target level in 2013–2017. The column "emissions gap in 2000" refers to the difference between emissions in the year 2000 and the emission target for the second commitment period. BAU forecasts could have been used but were not chosen due to the political nature of many of those forecasts. Underlying emissions data are taken from WRI (2006).

Table 3.3 ***Target allocation for members of current Annex B in the commitment period 2013–2017 and target difference to current emissions***

	Target for emissions level (% change from 2008–2012)	*Emissions gap 2000 (% difference from target)*	*Emissions gap 2000 (MMTCO$_2$ difference from target)*
Ukraine[1]	−47	2.1	10
Russia[1]	−42	15.5	257
Australia	−12	24.8	95
EU-28		16.8	827
Canada		42.1	213
New Zealand	−6	15.2	11
USA		31.3	1653
Iceland	−3	0	0
Japan		19.8	217
Norway		14.9	7
Switzerland		5.9	3
Sum (compared to 1990)	−23.3	**17.9**	**2293**

Notes:
The column "emissions gap 2000" refers to the difference between emissions in the year 2000 and the emission target for the second commitment period. BAU forecasts could have been used but were not chosen due to the political nature of many of those forecasts. Underlying emissions data are taken from WRI (2006).
[1] Derived from hot air estimates in studies cited by Korppoo, Karas, and Grubb (2006).

Taking the 2000 levels, an annual demand of 2.3 gigatons CO_2 equivalent would be generated. Given that Annex B countries have difficulties in getting on an emissions trajectory leading them toward their targets of the 2008–2012 period, a substantial increase in that demand level is likely. However, there is still substantial potential for low-cost emission reductions in Russia, Ukraine, and Eastern Europe. Moreover, the accumulated volume of "hot air" in Russia alone is likely to surpass 2 billion tons CO_2 equivalent (Korppoo, Karas, and Grubb 2006) and about 2.5 billion tons in the EU accession countries. It would thus cover several years' demand.

The concentric circles of target stringency

We assume that all countries achieving the lowest graduation index threshold take up absolute targets, but with a base year of 2012 instead of 1990. There will be a decreasing degree of target stringency according to the level of the graduation index:

- Countries that join Annex B due to institutional graduation and countries whose graduation index is above Annex B average take the unweighted average reduction rate applied for current Annex B members with the exception of hot air countries, i.e., 6 percent;
- Countries whose graduation index is above the lowest Annex II country value take the lowest Annex B reduction rate, i.e., 3 percent;
- Countries whose graduation index is above the lowest Annex B country value stabilize emissions at 2012 levels;
- Countries on the large emitters list that lie below the graduation index threshold can use a countrywide CDM (see below).

Overall target allocation

Making a linear trend extrapolation of emissions growth from the 1990s to 2015, emission budgets and reduction needs for the graduating countries can be derived (see Table 3.4). This must be seen as a pessimistic variant, not taking into account saturation tendencies in countries at the top of the graduation index list.

The total annual demand would be around 1 billion tons CO_2 equivalent, i.e., about 3 percent of 2000 world emissions and about a third of the demand generated by the "core" Annex B countries.

Table 3.4 *Emission targets for Non-Annex B countries*[1]

	Estimated emissions 2015 ($MMTCO_2$)	*Emission target (compared to 2012)*	*Absolute emission reduction need*
Brazil	1806	1615	191
Korea	1131	949	182
Iran	963	866	97
Saudi Arabia	683	596	87
Mexico	658	590	68
Turkey	530	497	33
South Africa	524	503	21
Taiwan	500	446	54
Thailand	467	426	52
Argentina	362	348	14
Malaysia	360	321	39
Venezuela	316	301	15
United Arab Emirates	247	221	26
Qatar	124	107	17
Singapore	153	110	43
Israel	152	137	15
Chile	136	124	12
Libya	87	82	5
Oman	65	56	9
Turkmenistan	64	64	0
Uruguay	36	34	2
Guatemala	35	32	3
Trinidad and Tobago	32	30	2
Mongolia	31	30	1
Bahrain	26	24	2
Botswana	20.5	19.5	1
Panama	19.2	17.7	1.5
Costa Rica	15.5	14.9	0.6
Equatorial Guinea	14.4	11.9	2.5
Namibia	14.3	12.9	1.4
Cyprus	14.1	12.3	1.8
Brunei	10.1	9.4	0.7
Gabon	11.6	11.3	0.3
Mauritius	8.0	7.2	0.8
Bahamas	2.0	2.0	0

Table 3.4 *Emission targets for Non-Annex B countries[1] (continued)*

	Estimated emissions 2015 (MMTCO$_2$)	*Emission target (compared to 2012)*	*Absolute emission reduction need*
Malta	2.4	2.3	0.1
Barbados	1.7	1.6	0.1
Sum	**9622**	**8621**	**1001**
% of 2000 world emissions	**28.5**	**25.5**	**3.0**

Note:
[1] Kuwait is missing due to lack of a reliable growth rate.

Role of the flexible mechanisms in Graduation and Deepening

Attractiveness of IET compared to CDM

Non-Annex B countries can participate in International Emission Trading (IET) once they have joined Annex B and thus could avoid the higher transaction costs of the project-based mechanisms:[4]

- Project identification and baseline selection;
- Project approval;
- Monitoring and verification of project performances;
- Certification of GHG emission credits obtained as a result of CDM activities;
- Negotiating sharing of achieved credits.

It is quite obvious that for most developing countries and countries in transition climate change does not belong to the priority tasks and they do not intend to implement measures exclusively aimed at GHG emission reduction. For such countries, CDM is the instrument of choice as it helps to introduce new environmentally sound technologies in key economic sectors. That will promote both GHG emission reduction and sustainable development. In addition to these benefits, the participation in the CDM will further help to create national institutions to address climate change issues (Michaelowa 2003) and promote markets.

[4] Transaction costs for emission trading systems are about an order of magnitude lower than for project-based mechanisms (Michaelowa *et al.* 2003).

If a Non-Annex B country joins Annex B, the CDM projects it hosts automatically lose their CDM status. There are the following possibilities:

- No compensation for loss of Certified Emission Reductions (CERs) is given;
- Projects are converted into Joint Implementation (JI) and instead of CERs generate the same amount of Emission Reduction Units (ERUs);
- The investors receive Assigned Amount Units (AAUs) equal to the CERs;
- The investors are bought out.

The CDM market – real participation by developing countries?

Even if there is no expansion of Annex B, the CDM could already now lead to a considerable involvement of developing countries in mitigation activities. However, the Kyoto Mechanisms are competing against each other due to the decision taken at Marrakech that all types of emission rights are fully fungible. While CERs and ERUs formally can only be banked up to a limit of 2.5 percent of a country's emission budget and Removal Units (RMUs) from sinks projects are not bankable, countries will just use up RMUs, CERs and ERUs first, and bank AAUs. Moreover, there is no formal supplementarity threshold; in principle countries can buy as many emission rights abroad as they like.

The shares and revenues of the different mechanisms are strongly dependent on demand and supply. If one takes the targets for the first commitment period only and assumes the United States and Australia stay at the sidelines, the market is very lopsided. The withdrawal of the United States has reduced demand by over two-thirds. Russian hot air covers the residual demand alone; Ukrainian and Eastern European hot air add about the same amount. Thus the CDM and JI market depends on voluntary export restrictions of the hot air countries; an alternative would be voluntary import restrictions for hot air as already defined in the EU "linking directive" (EU Commission 2004). The linking directive specifies that only CERs and ERUs can be imported for use in the EU Emission Trading Scheme; credits from forestry projects are excluded. Member countries shall define caps for use of CERs/ERUs.

Current experiences with the CDM market show that price expectations of €15–20 per ton CO_2 can mobilize several billion tons of reductions relatively quickly. JI supply depends on the willingness of hot air countries to embark on positive cost measures whereas JI demand depends on the willingness to buy hot air. The latter so far has been limited but the situation may change once JI institutions in Russia and Ukraine are operational. The situation changes substantially if the targets of the second commitment period are known well in advance of the first period, involve more countries, and are relatively tough as in our scenario. Then banking of hot air as well as CDM (and to a lesser extent JI demand) will increase and several billion tons of reductions could be mobilized. In the context of a scheme to involve countries with high absolute but with low per capita emissions and low GDP, a CDM that would assign CERs to the introduction of policies such as a renewable energy portfolio standard or energy efficiency standard may be able to mobilize a high amount of reductions. The baseline for such a CDM would be the situation without the policy.

Other policy elements of Graduation and Deepening

There are a number of policy elements beyond the target setting that would improve coverage of emissions and efficiency of reductions under Graduation and Deepening.

Coverage of international transport

The Kyoto Protocol exempted international air and sea travel from the emission reductions commitment of Annex B targets. This exemption should be lifted. Emissions could be allocated to the countries where trips originate and end, or the International Maritime Organization (IMO) and the International Civil Aviation Organization (ICAO) would be allocated a distinct target (see Bode *et al.* 2002). Emissions from air travel should use a special conversion factor that includes the indirect effects due to the specific chemistry of aircraft emissions in the high troposphere. Using IPCC results, this factor amounts to 2 to 4 (IPCC 1999). Contrail effects are not included in this estimate but should enhance this factor further as recent findings have reconfirmed their significance (Travis, Carleton, and Lauritsen 2002). Our scenario assumes that IMO and ICAO become parties to the Kyoto Protocol and get an absolute stabilization target at 2000 levels.

Carbon sequestration by biospherical sinks

Biospherical sinks can be classified as terrestrial sinks or marine sinks. Many stakeholders see terrestrial sinks as a cheap option to comply with emission targets while contributing to "save the world's large forests." At the same time, the inclusion of terrestrial sinks is a chance for many developing countries to participate at the global market for emission certificates. Potentially, soil sinks can become highly important. Soil sequestration and general sustainable development practices in agriculture are a much easier "fit" than forest sinks and sustainable development. Marine sinks can involve artificial fertilization (see Michaelowa, Tangen, and Hasselknippe 2005). Graduation and Deepening allows the use of biological sinks, i.e., forests, soils, and seas without limit if they can be monitored. CDM sinks credits are temporary (expiry at the end of each commitment period but re-issuance if the sink continues to exist). CDM would include forest protection in the lines of the deforestation avoidance proposal submitted by Papua New Guinea at the Conference of the Parties in late 2005. Countries can use marine sinks in their exclusive economic zone for compliance.

Carbon capture and geological storage

Carbon dioxide could be separated from the effluent gas of fossil-fuel-fired plants, collected and stored in aquifers, coal seams and empty oil and gas reservoirs. In Graduation and Deepening, countries can use geological sinks on their territory but are liable for any leaks. CDM credits from storage are subject to the same rules as for biological sinks.

Coverage of substances generating tropospheric ozone

Additionally to the Kyoto basket of six greenhouse gases, tropospheric ozone, which has a share of radiative forcing of almost 15 percent, should be included in the international climate policy regime. Besides being a greenhouse gas, ozone acts as a major local pollutant with negative effects on health and agricultural production. Targets would have to be expressed in units of the well-monitored precursors NO_x, CO, and

VOC as the concentration of ozone is difficult to measure. These precursors cannot be directly included in the Kyoto basket as they do not generate radiative forcing and thus Global Warming Potentials (GWPs) are not applicable to them. A rough and easy way to take them into account would be to translate the overall Kyoto basket target to NOX, CO, and VOC, i.e., if a country has a reduction target of 5 percent for the Kyoto basket, NOX, CO, and VOC have to be reduced by the same percentage. Two NOX emission trading systems have been implemented in the United States and are functioning well. Problems of noncompliance with NOX caps in the RECLAIM program during the California electricity market crisis were mirrored by noncompliance of facilities subject to other forms of regulation (Burtraw *et al.* 2005).

Change of GWPs for conversion into CO_2 equivalent

In Article 5.3, the Kyoto Protocol fixed the use of 100-year GWPs as specified by the IPCC Second Assessment Report (SAR) for the first commitment period. Accordingly, the 2001 update of GWPs by the Third Assessment Report (TAR) has not been followed by the Kyoto Protocol. A procedure for updating GWPs to reflect the latest science should be developed. Also, the 100-year time frame may not be seen as adequate. Choosing a different time horizon might lead to significant differences in the GWP of a given greenhouse gas (see Table 3.5). In Graduation and Deepening we assume that GWPs are updated for the second commitment period to the values derived in the IPCC's Fourth Assessment Report (AR4) published in 2007, for example for HFCs (Gohar, Myhre, and Shine 2004). While GWPs have been criticized by economists, so far no convincing alternative metrics has been found (Fuglestvedt *et al.* 2003).

Policy strategies to make graduation and deepening happen

As in other policy fields that were initially believed to be unmanageable, an aggressive strategy on climate policy can become a mainstream position. Even oil-exporting countries may realize that phasing out coal production is in their interest and that keeping oil reserves in the long term for high-value-added chemical industry uses is better than shelling out oil in the short term. The following strategies could be used to promote Graduation and Deepening.

Table 3.5 *GWP changes over time*

Gas	*Average lifetime (years)*	*IPCC SAR Kyoto Protocol*	*IPCC TAR 20-year horizon*	*IPCC TAR 100-year horizon*	*IPCC TAR 500-year horizon*
CO_2	Variable, about 150	1	1	1	1
CH_4	12	21	62	23	7
N_2O	114	310	290	296	156
HFCs	0.3–260 (majority double-digit)	140–11,700	40–9400	122–12,000	4–10,000
PFCs	2600–50,000	6500–9200	3900–8000	5700–11,900	8900–18,000
SF_6	3200	23,900	15,100	22,200	32,400

Source: IPCC 2001: 388–389.

Weather catastrophes as political windows of opportunity

Extreme weather events such as Hurricane Katrina make the common man think about the impacts of unfettered climate change. They are thus an excellent opportunity to push policymakers to commit themselves to ambitious mitigation efforts. But people quickly forget, so the chance passes. An example of a chance grasped is the drought in the US Midwest in 1988 that led to the US support of developing an international climate policy regime. Likewise the German flood of 2002 which fell in the general election campaign was used by the Green Party to quickly put up posters announcing "We fight climate change." The Green result improved markedly compared to earlier polls and they were able to put an indicative 40 percent emission reduction target in the coalition treaty. Of course, initial political gains after a catastrophe can be eroded by stubborn lobbies of which France is a good example. The big storms of late 1999 led to the announcement of an ambitious climate policy strategy with a carbon tax as its cornerstone. However, within two years of debate, the tax had been successfully prevented by the emitter lobbies.

An analogy from the past where the window of opportunity was used is the sharp and costly SO_2 reduction in the early 1980s in Germany after reports of widespread forest dieback ("Waldsterben"). Despite later news that forest dieback was not as severe as thought, the program was continued.

The international negotiations would need a ratcheting structure of target offers that would allow countries to offer a target that could be strengthened at a subsequent negotiation session but not weakened. Then policymakers could ratchet up their offer after an extreme event.

Linking carrots and sticks

Graduation and Deepening includes carrots and sticks that shall cater to different critical groups.

The full inclusion of sinks allows countries with large geological reservoirs, vegetation and marine areas to cover a large part of their commitment. This should be attractive to the United States and Australia that have already embarked on large-scale sequestration research and development (R & D) programs. The "hot air" countries will also be compensated to some extent by sinks on their large territories for their more stringent target.

The full use of international flexibility is attractive for all parties under the regime. An effective stick could be the exclusion from the mechanisms if a country does not take up a target despite passing the graduation threshold. Graduation is also sweetened by the chance to exchange the relatively cumbersome project-based CDM for emissions trading and JI.

Grasping the fruit of past effort

Owing to the large-scale support of renewable energy, the situation could be reached when the cost gap to fossil fuels closes. For some technologies, there are clear indications that this is happening. For example, wind power in very good sites is only a hair's breadth away from being competitive with new coal-fired power plants. The increasing age of the fossil-power plant fleet makes it increasingly impossible to rely on fully depreciated Methuselah plants. Countries with a large renewable energy industry will become a lobby of strong climate policy efforts.

Conclusions

"Graduation and Deepening" builds on an increased political salience of climate change. A series of extreme weather events with major losses could induce the electorate in industrialized countries to embark on tough emission reductions after 2012. Political consensus could be achieved to orient the global emission path towards a stabilization at a concentration of 550 ppm CO_2. As industrialized countries show leadership, the country group that accepts quantified emission budgets expands. Per capita income and per capita emissions define differentiated thresholds for participation. Owing to an overall success of the Kyoto Mechanisms in the first commitment period, their basic structures will remain valid. The sequestration of carbon dioxide by biospheric sinks and technical storage in geological formations plays an increasing role.

References

Baumol, W., and W. Oates (1971). "The Use of Standards and Prices for the Protection of the Environment," *Swedish Journal of Economics* 73: 42–54.

Bode, S., J. Isensee, K. Krause, and A. Michaelowa (2002). "Climate Policy: Analysis of Ecological, Technical and Economic Implications for International Maritime Transport," *International Journal of Maritime Economics* 4: 164–184.

Burtraw, D., D. Evans, A. Krupnick, K. Palmer, and R. Toth (2005). "Economics of Pollution Trading for SO_2 and NO_X," *Annual Review of Environment and Resources* 30: 253–289.

Claussen, E. and McNeilly, L. (1998). *Equity and Climate Change*, Arlington, VA: Pew Center on Global Climate Change.

Edmonds, J., and Sands, R. (2003). "What Are the Costs of Limiting CO_2 Concentrations?," in J. Griffin (ed.), *Global Climate Change*, Cheltenham: Edward Elgar, pp. 140–186.

EU Commission (2004). *Directive 2004/101/EC of the European Parliament and of the Council of 27 October 2004 amending Directive 2003/87/EC establishing a scheme for greenhouse gas emission allowance trading within the Community, in respect of the Kyoto Protocol's project mechanisms*, Brussels.

Fuglestvedt, J., T. Berntsen, O. Godal, R. Sausen, K. Shine, and T. Skodvin (2003). "Metrics of Climate Change: Assessing Radiative Forcing and Emission Indices," *Climatic Change* 58: 267–331.

Gohar, L., G. Myhre, and K. Shine (2004). "Updated Radiative Forcing Estimates of Four Halocarbons," *Journal of Geophysical Research* 109: D01107.
Goulder, L. (2003). "Benefit-Cost Analysis and Climate Policy," in J. Griffin (ed.), *Global Climate Change*, Cheltenham: Edward Elgar, pp. 67–91.
Gupta, J. (1999). "North–South Aspects of the Climate Change Issue: Towards a Constructive Negotiating Package for Developing Countries," *Review of European Community and International Environmental Law* 8(2): 198–208.
(2003). "Engaging Developing Countries in Climate Change: (KISS and Make-Up!)," in D. Michel (ed.), *Climate Policy for the Twenty-First Century: Meeting the Long-Term Challenge of Global Warming*, Washington, DC: Centre for Transatlantic Relations, Johns Hopkins University, pp. 233–264.
Gupta, J., P. van der Werff, and F. Gagnon-Lebrun, eds. (2001). "Bridging Interests, Classification and Technology Gaps in the Climatic Change Regime," IVM Report E-01/06, Institute for Environmental Studies, Amsterdam.
IEA (2005). *CO_2 Emissions from Fossil Fuel Combustion 1971–2003*, Paris: IEA.
IPCC (1999). *Aviation and the Global Atmosphere*, Cambridge: Cambridge University Press.
(2001). *Climate Change 2001: The Scientific Basis*, Cambridge: Cambridge University Press.
Jansen, J., J. Battjes, F. Ormel, J. Sijm, C. Volkers, R. Ybema, A. Torvanger, L. Ringius, and A. Underdal (2001). "Sharing the Burden of Greenhouse Gas Mitigation: Final Report of the Joint CICERO-ECN Project on the Global Differentiation of Emission Mitigation Targets among Countries," CICERO Working Paper 2001:05, Oslo.
Korppoo, A., J. Karas, and M. Grubb (2006). *Russia and the Kyoto Protocol*, London: Chatham House.
Krause, F. (2000). "Solving the Kyoto Quandary: Flexibility with No Regrets," Working Paper International Project for Sustainable Energy Paths, El Cerrito, CA, November.
Meinshausen, M. (2006). "What Does a 2°C Target Mean for Greenhouse Gas Concentrations? A Brief Analysis Based on Multi-Gas Emission Pathways and Several Climate Sensitivity Uncertainty Estimates," in H. Schellnhuber, W. Cramer, N. Nakicenovic, T. Wigley, and G. Yohe (eds.), *Avoiding Dangerous Climate Change*, Cambridge: Cambridge University Press, pp. 265–280.
Michaelowa, A. (2003). "CDM Host Country Institution Building," *Mitigation and Adaptation Strategies for Global Change* 8: 201–220.

Michaelowa, A., M. Stronzik, F. Eckermann, and A. Hunt (2003). "Transaction Costs of the Kyoto Mechanisms," *Climate Policy* 3(3): 261–278.
Michaelowa, A., K. Tangen, and H. Hasselknippe (2005). "Issues and Options for the Post-2012 Climate Architecture – An Overview," *International Environmental Agreements*, 5(1): 5–24.
Pichs, R., R. Swart, N. Leary, and F. Ormond (2000). *Development, Sustainability and Equity*, Havana, Cuba: Proceedings of the Second IPCC Expert Meeting on DES.
Sandalow, David, and Ian Bowles (2001). "Fundamentals of Treaty-Making on Climate Change," *Science* 292: 1839–1840.
Travis, D., A. Carleton, and R. Lauritsen (2002). "Contrails Reduce Daily Temperature Range," *Nature* 418: 601.
World Resources Institute (2006). *Climate Analysis Indicators Tool (CAIT)*, Washington, DC: World Resources Institute, Available at www.cait.wri.org/.

Appendix

Table 3.A.1 ***Non-Annex B GDP (ppp) and per capita emissions data for 2000 and comparison to Annex B levels***

	GDP per capita	*Emissions per capita (TCO_2)*
Qatar	33,833	67.9
Singapore	23,600	13.9
***Average Annex B*[1]**	***21,916***	***14.1***
United Arab Emirates	20,688	36.1
Israel	20,587	12.3
Taiwan	19,645	10.4
Malta	17,500	6.1
Korea	16,353	11.1
***Lowest Annex II*[2]**	***16,311***	***7.3***
Cyprus	16,250	10.5
Bahamas[3]	16,210	6.7
Kuwait	15,682	31.6
Bahrain	15,143	24.8
Barbados[3]	14,716	5.8
Brunei	12,667	21.3
Saudi Arabia	12,570	16.4
Oman	12,542	12.1
Argentina	12,237	7.9
Mauritius[3]	10,450	3.4

Table 3.A.1 *Non-Annex B GDP (ppp) and per capita emissions data for 2000 and comparison to Annex B levels (continued)*

	GDP per capita	*Emissions per capita (TCO_2)*
South Africa	9434	9.5
Chile	9203	5.0
Mexico	9090	5.2
Uruguay	8879	7.7
Malaysia	8940	7.1
Costa Rica	8921	3.3
Trinidad and Tobago	8846	19.3
Botswana[3]	8674	9.5
Brazil	7366	5.0
Turkey	6811	5.4
Macedonia	6650	5.5
Thailand	6353	4.4
Tunisia	6229	3.2
Panama	6138	4.1
Dominican Republic	6119	3.7
Colombia	6113	3.8
Gabon	5923	8.1
Namibia	6052	5.6
Venezuela	5638	10.0
Equatorial Guinea[3]	5600	5.4
Iran	5573	7.5
Libya	5403	11.9
Peru	4737	2.7
El Salvador	4710	1.8
Paraguay	4585	4.9
Kazakhstan	4583	10.7
Lowest Annex B[4]	***4109***	***6.0***
Guatemala	3947	8.8
Turkmenistan	3696	13.8
Mongolia	1653	11.6

Notes:

GDP data from IEA (2005), per capita emissions from WRI (2006).

[1] Including Australia, United States, and Belarus.

[2] Greece (GDP), Switzerland (per capita emissions).

[3] 2002 GDP data from WRI (2006).

[4] Ukraine (GDP), Lithuania (per capita emissions).

Commentaries on Michaelowa

3.1 *Alternatives to Kyoto: the case for a carbon tax*

RICHARD N. COOPER

Axel Michaelowa's paper addresses how the world should proceed in a post-Kyoto Protocol period, which begins in 2013 but must be agreed before then. It is informative, ingenious, and constructive. But its proposal for extending emissions targets and enlarging the number of countries to which they apply is deeply flawed, partly by carrying forward the flaws inherent in the Kyoto Protocol. This comment will address the intellectual framework of the proposal, identify three fatal flaws (all inherent in the Kyoto Protocol), and suggest an alternative approach.

Michaelowa explicitly rejects a benefit-cost approach to public policy in dealing with global climate change in favor of an absolute (indicative) ceiling to atmospheric concentrations of greenhouse gases, mainly but not exclusively carbon dioxide. This approach implies an extreme degree of risk aversion with respect to climate change – any cost to avoiding it is worth the price – about which every economist should be skeptical. Moreover, ordinary citizens will in practice reject this approach – they will not be willing to bear any cost to reduce emissions enough to stabilize concentrations. Policy analysts should acknowledge this from the outset. The price citizens will be willing to pay will initially be modest; it may grow as hard evidence of the costs of climate change accumulate, but even then it will not become infinite, not least because those who will be expected to bear the brunt of the cost of reducing emissions may not be those who incur the greatest damage from climate change. This approach implicitly places climate change above all other social objectives, and it implies a degree of global communitarianism that does not exist today and is not likely to come into being within the next decade, when a post-Kyoto Protocol regime must be negotiated. Calling the framework "indicative" softens this strong formulation in tone but does not alter the substance until the possible limits are identified explicitly.

Emission permits need to be allocated if the trading regime envisaged is to function effectively. The proposal focuses on allocation of targets among countries, but not the allocation of national targets (= emission rights) within countries. On what principle should they be allocated? Many economists no doubt would favor national auctions. But auctions are in fact rarely used when valuable resources are to be distributed; allocation on the basis of historical emissions or some variant thereof, is usually preferred. Perhaps allocation can be undertaken honestly in Sweden or Germany or even (more doubtfully) in the United States. But they certainly cannot be undertaken honestly in many countries that under Michaelowa's proposal will be "graduated" into the class of countries assigned emission targets, and indeed they cannot be undertaken honestly in many countries in the Kyoto Protocol's Annex B. It is an invitation to favoritism, hence for corruption, the more so the more valuable the permits, and in the framework proposed they could be valuable indeed. Do we really want environmental programs to become the handmaiden of corruption in many countries, as this need for governmental distribution of valuable emission rights surely will?

The proposal envisages international trade in emission rights, something necessary in the Kyoto Protocol framework to minimize the economic costs of any given degree of reduction in emissions. But international trading entails potentially large transfers between law-abiding citizens in rich countries such as the United States, Canada, EU member states, and Japan to corrupt officials and their favored oligarchs in countries less meticulous about the rule of law – or directly to the governments of such countries. These transfers would not be conditioned on anything beyond willingness to sell emission rights that had been internationally agreed. The bottom line is that American and European citizens would be making unconditional transfers to Russia, Iran, and eventually (although not immediately in the next round) to Burma and North Korea. I would not want to have to defend such a proposal before the US Senate, whose assent would be required for ratification, or indeed before the German Bundestag. It is indefensible. I am aware that some advocates see large transfers from rich to poor countries as a positive advantage of a Kyoto Protocol-type trading regime, partly to draw poorer countries into the emission-control regime, partly because it involves redistribution from rich to poor. But if we have learned anything about unconditional or lightly conditioned transfers from rich to poorer countries during the past four decades, it is that they too rarely foster economic development, and they often enrich the powerful and the already rich in poorer countries.

China, India, and many other poor countries do not graduate into emission-target countries under the proposal, although their turn would eventually come with the continued growth in per capita income. But China is where the real action is with respect to climate change within the next decade or two. Despite vigorous programs to move to alternative fuels such as nuclear, hydro, natural gas, and wind power, if China continues to grow rapidly it will build more coal-fired electricity-generating plants in the next two decades than the United States and Europe put together. Once constructed, these power plants will last for half a century. If we are to take the problem of greenhouse gas emissions seriously, and give it the urgency implied by Michaelowa's framework, we all have an interest in these Chinese power plants. As a practical matter, I offer the judgment that atmospheric concentrations cannot be limited in the next few decades without sequestration of carbon dioxide from major emitters – carbon capture and storage (CCS) as it is increasingly called. The United States is devoting substantial research to this process, which is promising but still undeveloped, both as to the best technical approaches and with respect to keeping the cost of CCS within reasonable bounds. It is much cheaper to design a new plant with CCS in mind than to retrofit an existing plant for carbon capture. Thus here is an arena for practical international cooperation, with Americans and others providing the technical know-how, China providing the experimental ground, and the rich countries together paying the incremental costs – not for power generation, but for potential sequestration. No doubt some of the experiments will fail; we are still in uncertain terrain. But we can learn more quickly which methods are more effective and/or less costly with some full-scale experimentation, and in the meantime actually slow the growth in emissions that would otherwise take place. India and others could also offer similar opportunities, but quantitatively China dominates the field in the coming decades.

One can well ask, why try to extend the period of the Kyoto Protocol? The obvious answer is that it exists and has been accepted by most of the rich countries. But it does not include the United States or China, two countries whose cooperation is absolutely essential if greenhouse gas emissions are to be seriously limited. The proposed framework does not include China in the near future, and it is not much more likely to appeal to the United States than the Kyoto Protocol did. While President George W. Bush definitively killed the prospect of US adherence to the Kyoto Protocol (in an admittedly unnecessarily clumsy, indeed offensive, way), I conjecture that had Albert Gore – whose personal concern about climate change cannot be doubted – been elected US president in 2000, the United States still would not have adhered to the Kyoto Protocol. He could not have persuaded the required two-thirds of the US Senate that it is a good agreement, acceptable to the United States, and indeed President Clinton never submitted it to the Senate, although he could have done.

The other reason, I suspect, that some continue to push the Kyoto Protocol is that Europeans got a very good deal under the Kyoto Protocol, as is implicit in Michaelowa's new targets, which demand relatively more of Europe. Why, in late 1997, was 1990 chosen as a base year (and earlier years for two central European countries)? In part because it built highly inefficient East German industrial emissions into the European base, and British coal consumption, while on the way down, was known to have a considerable distance to go. Similarly for Russia and Ukraine, who had to be induced into the Kyoto Protocol with generous targets, even though by 1997 it was clear that the energy-intensive heavy industry of the Soviet era had collapsed and was likely to recover only partially even under optimistic scenarios. European growth was also slow throughout most of the 1990s, making it easier for Europe to meet a target based on 1990. In strict negotiating terms, the United States was simply out-negotiated (and carried a surprised and embarrassed host country, Japan, with it). Table 3.1.1 reports the 1990 emissions, the targeted reduction for Annex B countries in the Kyoto Protocol, an estimate of emissions in 2010 made by the US Department of Energy in 2004 (excluding any new measures to meet the Kyoto Protocol targets), and the percentage reductions required by the Kyoto Protocol from projected 2010 emissions. It can be seen there that the largest reductions are required by Canada, Australia, the United States, and Japan, in that order, all 25 percent or more, while required reductions in western Europe are only 12 percent, while eastern Europe, Russia, and Ukraine have targets well above their projected emissions. We knew more about growth during the past decade in early 2004 than we did in late 1997, but the main patterns were already known during the Kyoto negotiations. It is not surprising Australia and the United States withdrew; Japan would have "lost face" not to ratify an

Table 3.1.1 ***Carbon dioxide emissions (MMTCO$_2$)***

	1990	*2010*[p]	*KP required Cut (percent)*	*KP required Cut (MMT)*	*Cut as % of 2010*
USA	4989	6559	7	1919	29
Canada	473	686	6	241	35
Japan	987	1239	6	311	25
Australia/NZ	294	455	~(8)	137	30
Western Europe[a]	3412	3567	~8	428	12
Eastern Europe	1104	797	8	(219)	(27)
Russia	2405	1792	0	(613)	(34)
Other FSU[b]	1393	808	0	(585)	(72)
Total (excluding USA and Australia)	9774	8889		(437)	(5)

Notes:
[p] Projection in EIA, 2004.
[a] EU, Iceland, Norway, and Switzerland.
[b] Only Estonia, Latvia, Lithuania, and Ukraine covered by Annex B.

agreement that bore the name Kyoto, but Japanese officials felt let down by their US negotiating partners, whom they had counted on to protect their position. And Canada had a vigorous debate over ratification, and the then Liberal government imposed a cap on the price of emission rights in order to get the treaty through Parliament, even though such a cap was outside the Kyoto Protocol framework and might lead to non-attainment of Canada's Kyoto target. The subsequent conservative government has effectively abandoned Canada's Kyoto target.

One way or another, the energy-consuming public is going to have to pay higher prices – under the proposal, significantly higher prices – to cut demand for fossil fuels and to induce emission-reducing technical changes in the energy system. Barring some technical breakthroughs in energy production or consumption that are not now foreseen, higher prices are unavoidable. Advocates of significant action in the near future to reduce emissions have been reluctant to acknowledge this ineluctable fact. (If instead they expect fast-acting technical progress sufficient to keep the cost low, they should be explicit about that assumption, and think of hedging strategies if it turns out to be incorrect.) This strategy of concealing or seriously downplaying an important consequence of proposed actions will not work in open societies where skepticism of government claims has grown significantly.

A strategy more likely to be successful is to acknowledge that carboniferous energy needs to become more expensive, and to accomplish the required increase in prices with an internationally agreed tax, revenues to accrue to each tax-levying country, to avoid the issue of large unconditional transfers among countries. Many countries would welcome the additional revenue; countries where this is not the case could use the revenues to lower other taxes. This proposal – an alternative to Michaelowa's – is discussed in more detail in the following section. It is not assured of success. But in my judgment it has a better chance of actually reducing greenhouse gas emissions than does the proposal in Michaelowa's paper.

A global carbon tax

There are negative and positive arguments for introducing a tax on emissions of greenhouse gases (GHGs). The negative argument is that the leading alternative, quantitative goals with a trading regime in emission rights, is almost certainly politically unsustainable on a global basis. Key developing countries must be seriously involved in any effective effort to reduce GHG emissions. On US Department of Energy projections, for instance, China's CO_2 emissions will reach those of Europe before 2010 and those of the United States by 2015 (US EIA 2006). Emissions from India, Brazil, and others are also significant and growing rapidly. Yet it is difficult to imagine a set of effective national quantitative targets that China and the United States could both agree on, to take only the leading emitters among rich and poor countries. Kyoto excludes developing countries. Kyoto's advocates acknowledge that, but aver it is only the first step. What does the next step look like? Michaelowa's proposal is a constructive effort to specify the next step but contains the weaknesses noted above.

Furthermore, "cap-and-trade" will involve the allocation of valuable rights. The prospect of such allocation might be attractive to domestic businessmen, who are always looking for government handouts (witness any tax bill), but it will necessarily be a highly political process, unless the rights are auctioned, which will be resisted strongly by the business community. While the domestic process is merely unattractive, and in a sense deeply corrupting, the international allocation with trading will be politically impossible. What Senator, once he or she understands the full implications of a trading regime, can vote for a procedure which could result in the unconditional transfer of billions of dollars, even tens of billions, to the government of communist China, or to Castro's Cuba, or even to Putin's Russia? Not only is it politically impossible, at least in the United States, but I would argue that large unconditional transfers to governments are in general highly undesirable, shifting attention in receiving countries away from the need for fiscal discipline and thoughtful benefit-cost analysis of the balance to be struck between taxation and government expenditure.

The key alternative, if action to reduce GHG emissions is to be taken, is to focus on level of effort rather than on quantitative targets: concretely, on the introduction, within an internationally agreed framework, of a domestic tax on GHG emissions, revenues to accrue to the government of each country where the emissions occur. The focus initially would be on fossil fuels, cement, and other industrial processes that result in emissions of carbon dioxide. Methane is more difficult under any regime and can be added later after experience is garnered with CO_2.

The proposal involves international agreement on a regime for a common tax to be levied on the major sources of emissions of carbon dioxide, and on the selection of the common tax rate, both initially and subsequently. The tax would be *incremental* to existing taxes (and subsidies), including those on fossil fuels, on the grounds that whatever taxes exist were introduced for reasons unrelated to global climate change, that global climate change is a newly recognized problem for purposes of collective action, and that all parties should add new incentives for the reduction of emissions. (Allowance could be made for taxes that have been introduced in a few European countries following agreement on the Kyoto Protocol whose explicit rationale was to reduce CO_2 emissions.)

A uniform incremental CO_2 tax would introduce an incentive, worldwide, to reduce carbon emissions. The response to the tax would of course differ from country to country. Where emissions can be reduced at a cost lower than the tax, such reductions can in time be expected to take place. Where the cost of reducing emissions exceeds the tax, the tax will be paid. In either case the cost of fossil fuels will be raised everywhere, in proportion to their carbon content. A uniform tax thus is economically efficient, in that reductions will be greatest where the cost of such reductions is least, worldwide. The universal presence of the tax will also avoid geographic relocation of industries to avoid the tax – a potential problem under the Kyoto Protocol and its extensions – except where such relocation is in fact economically efficient.

Introduction of such a tax raises a number of issues, which will be taken up in turn: the level of the tax, and procedures for changing it; compliance; enforcement; macroeconomic effects; possible differential treatment; use of revenues; and how to treat sequestration – activities that deliberately withdraw atmospheric CO_2.

One objection sometimes raised to a tax is that we will not know initially what the quantitative impact will be. Entirely true. But the Kyoto Protocol targets also bear little direct relationship to the underlying problem, namely, the growing concentration of GHGs in the atmosphere. It is, as its advocates insist, only a first step. The tax would similarly be a first step, with a much clearer path to what the second and subsequent steps would look like.

The initially agreed tax should be at a level sufficient to attract serious attention to tax-avoiding emission reduction, say $50 a ton of carbon. (This would amount to nearly $14 per ton of CO_2, the unit of measurement used in the Kyoto Protocol, and would amount to roughly a 100 percent tax on coal, with lower tax rates per useful BTU (British Thermal Unit) for oil and still lower for natural gas.)

The world would gain experience over time with the impact of this tax on emissions, while it is also learning more about the climate system and refining its estimates and its preferences concerning the prospects for climate change. Provision would be made for a review of the rate of tax after, say, the first ten years, and quinquennially thereafter, taking into account both greater knowledge about the impact of the tax and about the evolution of climate in response to continuing GHG emissions.

Compliance would be easy to assess. Every country has a known mechanism for promulgating new tax rates and regulations. We would know whether a country had responded to the international agreement by changing its tax regulations in accordance with it. Administratively, the tax would best be levied at the choke points for fossil fuels: main gas and oil pipelines, or refineries, and main coal shipments by rail or barge, plus allowance for pit-head power production. But this practical detail could be left to each country.

Promulgating new taxes and actually collecting them are two different things, for any tax. Enforcement of tax collection raises complicated questions, as indeed would enforcement of emission ceilings. Almost all countries (Cuba, North Korea, Taiwan, and Hong Kong, along with a number of mini-states, are the exceptions) are now members of the International Monetary Fund (IMF), and as such their economic policies, including fiscal policies, are subject to detailed annual surveillance by the IMF staff. Under a carbon tax agreement, the IMF could be asked to pay special attention during these reviews to sources of revenue, and in particular to carbon tax revenues. Each country's revenue books would be open to inspection, and its tax officials available for questioning. Countries' tax systems would also be monitored to assure that the carbon tax was not nullified by changes in other taxes which indirectly favored CO_2-emitting activities, a concern that has been expressed by Wiener and others. Of course any country that desired to cheat could do so, but that is a problem with any regime to limit emissions, and many officials would have to be brought into the conspiracy. Furthermore, physical readings of the largest sources of emissions, such as power plants, could be taken (e.g., by satellite and by on-site inspection) as part of the compliance regime.

What about the erosion of impact of the carbon taxes through other tax relief or subsidies to the emitters? Again, the IMF could be asked to scrutinize any major tax change for consistency with the carbon tax regime. The process would be a consultative one, initially bilateral between each country and the IMF. Presumptive cases of violation could be referred to special panels, WTO-style, for further investigation and scrutiny. Publicity would be given to significant violations. Exports from countries with egregious and quantitatively significant violations could, by panel finding, be made subject to countervailing duties by importing countries, even under existing legislation, once the tax on CO_2 emissions was judged internationally to be a cost of business, subsidization of which would be treated as a conventional export subsidy.

Any significant change in taxation can have disruptive macro- and microeconomic effects. Provision should be made in all countries for phasing in the tax, starting low and gradually rising to the full agreed and pre-announced rate. Macroeconomic effects could be minimized by making the tax fiscally neutral (which would involve making a guess in each country what its initial impact on emissions would be), either by increasing expenditures or by reducing other taxes. Many governments would need the additional revenue, and for this reason ministers of finance everywhere would welcome such a tax. Where the revenue is not needed, or where an increase in the total tax burden is politically insupportable, the new revenues could be used to reduce other taxes.

The revenues are likely to be substantial, but not overwhelming. The Clinton Administration calculated in 1998 that if the Kyoto Protocol were to be extended to China, India, Mexico, and South Korea (each of which was given a notional target equal to its business-as-usual trajectory), the trading price that would achieve the Kyoto targets would be $23 a ton of carbon, equivalent to a tax of that rate, about half the rate suggested above. With estimated worldwide emissions in 2010 under effective Kyoto targets of 7 billion tons of carbon, this tax would yield worldwide revenues of $160 billion, about 0.4 percent of gross world product in that year.

Developing countries, as noted above, must be fully embraced by the carbon tax regime if there is to be any hope of limiting atmospheric GHG concentrations. However, developing countries could be granted a longer period of time to introduce the tax, so long as the period was not so long as to induce uneconomic relocation of economic activity to countries that had not yet introduced the tax. Five years might be an appropriate delay, to be followed by the phase-in period.

Even though the carbon tax would increase the price of fossil fuels, growth need not be seriously affected, since the revenues could be used for expenditures or tax-reductions that contribute to growth. Decisions about use of the carbon tax revenues would be left entirely to each country, so long as they were not used to undermine the purpose of the tax, which is to reduce CO_2 emissions.

Reduction of emissions may not always be the most efficient way to limit growing atmospheric GHG concentrations. Sequestration of CO_2 from the atmosphere should be included in the menu of permissible actions. Subsidies (at the agreed CO_2 tax rate) could be given for sequestration, or tax rebates where the sequester is also the emitter. Again, this process would be up to each country to implement, subject to international surveillance.

References

Clinton Administration (1998). *The Kyoto Protocol and the President's Policies to Address Climate Change: Administration Economic Analysis*, White House, Washington, DC, July 1998.

Cooper, Richard N. (2004). "A Carbon Tax in China?" Washington, DC: Climate Policy Center.

(2005). "The Kyoto Protocol: A Flawed Concept," in John Maxwell and Rafael Reuveny (eds.), *Trade and Environment: Theory and Policy in the Context of EU Enlargement*, Chelthenham: Edward Elgar, pp. 17–36.

Nordhaus, William D. (2006). "After Kyoto: Alternative Mechanisms to Control Global Warming," *American Economic Review, Papers and Proceedings* 96(2): 31–34.

US Energy Information Administration (2006). *International Energy Outlook 2006*, Report DOE/EIA-0484(2006), Washington, DC: Department of Energy.

Wiener, Jonathan B. (2001). "Policy Design for International Greenhouse Gas Control," in Mike Toman (ed.), *Climate Change Economics and Policy: An RFF Reader*, Washington, DC: Resources for the Future Press, pp. 205–215.

3.2 *Beyond Graduation and Deepening: toward cosmopolitan scholarship*

JOYEETA GUPTA

Introduction

The climate change problem has sparked extreme positions in most policymakers and researchers, each presenting his or her standpoint as the most rational. Is climate change a serious problem, as some seem to think it is; or is it not a serious problem as climate skeptics argue? Is it possible to define a long-term objective through a scientific-driven endeavor or do only fools rush in where angels fear to tread? Is climate change an opportunity to promote the rule-of-law project on the international agenda, or given the anarchy at international level, is the only sensible option one of focusing primarily on narrowly defined national interests? Is the neoliberal agenda and its focus on market mechanisms and small government adequate to address the problem of climate change, or do we need to go beyond that? While intellectuals battle with each other about whether climate change threatens civilization on the playground of academic debate, the reality is that the most vulnerable individuals, societies, and ecosystems are likely to be threatened by the potential impacts of climate change on ice caps, glaciers, sea-level and rainfall patterns. While we debate on whether the impacts are caused by climate variation or climate change, more species and more individuals are marginalized. To whom are we as scientists accountable? To our disciplines, governments, ideologies, or to those who depend on our advice? These are some of the basic issues that the climate change problem challenges us with.

At the Architectures for Agreement Workshop held in May 2006 at the Kennedy School of Government, where the papers for this volume were debated, I was once more reminded that there is a scientific reality

This paper has been written in the context of the VIDI project on intergovernmental and private environmental regimes and compatibility with good governance, rule of law and sustainable development, supported by the Netherlands' Organisation for Scientific Research.

that exists within the United States, and that there are competing scientific realities that exist in Europe and in the developing countries. Science is so strongly anchored in ideologies, disciplines, perceptions, and in the spatial location of each individual scientist, that it is inevitable that scientific discussion is likely to be heated and parochial, protecting the terrains that each of us knows best within the national and ideological contexts in which we feel safe. It was a privilege to be invited to be part of this group of diverse thinkers, and to note that debate is promoted even though consensus may not be forthcoming.

Against this background, I will first point to what I think are the strengths of the ideas proposed by Axel Michaelowa, and then move on to explore some other issues that deserve more prominence in the debate on the architectures for agreement for a post-Kyoto world.

The strengths of Graduation and Deepening

Introduction

Michaelowa presents a follow-up regime to the Kyoto Protocol which aims at elaborating further on the targets-and-timetables aspect of the climate change negotiating process. He focuses on the period 2013–2017 and elaborates in great detail how the targets-and-timetables agenda could be promoted further in a way that it tries to increase incrementally the number of countries subject to targets, while making a more substantial contribution to reducing greenhouse gas emissions. He elaborates further on a model I had developed but also simplifies it somewhat (Gupta 2003).

The long-term objective

The point that I am most sympathetic to in his paper is his conviction that there needs to be a long-term objective with respect to the climate change problem. In the literature one can discern three schools of thought. Some argue that science has nothing useful to say about when climate change becomes dangerous to humanity either because it does not threaten civilization or because it cannot be objectively derived (see Barrett, Victor, this volume). Others argue that science can provide evidence on the basis of which such ideas can be further developed (e.g., Pachauri 2005; Schneider and Lane 2006) and there are still others who submit that where the science is uncertain, the stakes high, decisions urgent, there are scientific methods to generate information about when climate change can become dangerous to society (see the rising literature on post-normal science; see also Gupta and van Asselt 2006 with respect to climate change). Michaelowa pegs his approach on the target developed by the European Commission in 1996 to limit global warming to no more than a two-degree rise in temperature from pre-industrial levels. We are already in the realm of uncertainty since such high concentration levels have not previously been experienced in human history. Three reasons justify the attempt to determine when climate change may become dangerous to society. First, the potential impacts of climate change are likely to be serious in different parts of the world and affect the poor and vulnerable the most, even if they have made the least contribution to the problem. Second, one needs to have a sense of the dimensions of the problem in order to assess whether we are in a position to deal seriously with the challenge of climate change and whether we are taking action fast enough. Finally, science is expected to be able to think beyond nationalities and loyalties; to understand the global dimensions of the problem, especially where politicians and policymakers are mandated only to represent their national interests at negotiations.

Since climate change is an unstructured problem (Hisschemöller *et al.* 2001); it can be addressed only through a learning strategy that seeks to confront, evaluate, and integrate diverse views. Such a learning strategy is expected to yield ideas to deal with intractable problems using innovative exercises such as backcasting and transition management (Rotmans *et al.* 2000). At a more practical level, scholars argue that we need "anchors" amid uncertainty (Van der Sluijs 1997) and that we need to develop a best practice for constructing serviceable truths to guide us in a process of developing specific norms to derive some degree of certainty out of uncertainty (Guston 2001). Such serviceable truths must be scientifically acceptable and support decision making. Legal and other scholars focus on the notion of the precautionary principle as a strong guiding tool to manage uncertain problems (Trouwborst 2006).

Principles

From my perspective, I am also delighted that the author has focused on two key principles as a cornerstone of his approach, which were also incidentally cornerstones of my own approach – namely the polluter pays principle and the ability to pay principle. Neither of these two principles are mentioned as such in the United Nations Framework Convention on Climate Change (UNFCCC), but they are implicit in the "leadership" concept in the preamble and the "common but differentiated responsibilities" principle in Article 3 of the Convention. The importance of these two principles is that they assign responsibilities to those countries that have contributed to the problem but also take into account their ability to take action. This is significant not only with respect to establishing the responsibility of current powers, but also because of the necessity of putting future polluters on notice. The notion of treating like countries alike is very important for the legitimacy of the regime.

This principle could imply that the polluter (country or individual) is obliged to pay for the pollution beyond a certain per capita level. Such payment would ensure a continuous incentive to the polluter to reduce his pollution and at the same time generate resources for adaptation and/or technology transfer. It could be applied to all states that pollute above a specific amount per capita; and it would put rapidly industrializing developing countries on notice. Those opposing the idea may submit that the principle does not make scientific sense of the temporal distance between cause and effect (i.e., current impacts are caused by past emissions and present emissions will cause future impacts) and, hence, the causality is unclear. They would submit that an optimal tax rate is difficult to determine and that it is not politically feasible in the Western world. However, these problems may not be insurmountable. Causality issues can be dealt with statistically (Allen 2003) or through legal precedent (see the following section). A second-best solution for determining an optimal tax rate can be devised and political feasibility is a function of time and social pressures.

Short-term target

Michaelowa emphasizes the need for a short-term quantitative target for the period 2013–2017 and articulates the target as 23 percent reductions from the 1990 level. The purpose of the target is twofold. It incrementally pushes the regime further and by demonstrating the promised leadership of the developed countries, it gives incentive to the developing countries to take action too. I think this too is an important feature of the proposal, because ultimately the only way to an effective reduction of emissions is by setting targets that send a significant signal to industry and society and to the different countries of the world that this is the natural progression of policymaking in this field.

Comprehensive, and yet some minor weak points

Finally, I would like to support the comprehensiveness of the approach developed in the proposal. The proposal attempts at including emissions from international transport, carbon sequestration by biospherical sinks, carbon capture and geological sinks. It also looks at policy moments and ideas that offer opportunities for changing political positions, such as the impacts of weather catastrophes on politics, the use of carrots and sticks, and the need to profit from past research. Notwithstanding the strong points of the proposal, there are some weaknesses. From my perspective, the proposal is not predictable enough. Although it indicates a potential possible future path of development, it does not highlight how the system is likely to develop in the future. The index approach treats countries with high emissions and low income per capita in the same way as countries with low emissions and higher income per capita and thus does not capture the key differences between the two types of countries. The proposal with respect to including air transport is weak, and most likely institutionally impossible. International agencies cannot be meaningfully ascribed targets that have to then be fulfilled by nation-states. The sequestration and capture issue perhaps raises more questions than it answers. The treatment of institutional thresholds is extremely superficial and adds nothing to the proposal. Notwithstanding these weaknesses, the proposal as a whole puts forward a good and solid idea, one that is worth exploring further in international contexts. Many of the weaker ideas are elements that need to be further debated and discussed and the chapter serves to table these as agenda items for the future. Michaelowa's proposal could be attractive to most developed and developing countries because it attempts to further develop the Kyoto Protocol regime. It could eventually be attractive to the United States because he takes a comprehensive approach to sources and sinks; he has arguably low targets for the United States and because he attempts to involve all countries in the negotiating process.

Beyond Graduation and Deepening

Introduction

This section briefly dwells on a number of other ideas about how the climate change regime could develop further. It might be appropriate here to state that in my view we need both a centralized multilateral system to deal with the problem of climate change as well as a range of other initiatives of like-minded actors and states which support and/or compete with each other in a century-long effort to address the problem. We need the former to ensure that we are on track toward our goals and to review and determine all the efforts needed to deal with the problem in a cooperative manner. We need the latter to ensure that all opportunities for dealing with the problem are encouraged, recognized, and where possible emulated. I will begin by highlighting the Keep-It-Simple, Stupid approach to targets and timetables (Gupta 2003) and will then go on to elaborate on some other ideas. These include mainstreaming climate change in other UN agencies as a first step toward mainstreaming climate change in national policy; finding ways and means of welcoming US unilateral endeavor as part of the multilateral effort to address climate change; using litigation as a tool to increase the pressure on governments; encouraging local action worldwide to supplement inter-state action and to generate a broad-based support for climate action. Then I move on to discuss the need for the rule of law at global level.

Keep-It-Simple, Stupid

The Keep-It-Simple, Stupid proposal divides the world into twelve categories of countries based on their emissions per capita and their income per capita. This division allows for simplicity and predictability in that countries know in advance that if their income or emissions per capita increases beyond a certain level they will "graduate" into a new classification, and after a stable membership of about three years, they will be "included" in that category. Each classification of country is subject to a set of responsibilities, which includes targets and timetables but is not limited to that. It submits that the current classification is too simplistic besides not necessarily treating like countries alike. It includes a number of new types of targets such as percentage of official expenditure devoted to renewable energy and it allows for exceptions to be made. Furthermore, the system also identifies the highest gross emitters and puts them on a fast-track approach to reducing the (rate of) growth of their emissions (Gupta 2003). Since the system is predictable and attempts to ensure that all countries come on board, there will be an incentive for the developing countries to participate in it. If such a system starts to work, this also allows room for the United States to rejoin the process, especially as there are a number of measures being adopted by the Federal government (see "Unilateral US measures" section below) as well as by lower governments (see "Local action" section below).

Mainstreaming and linking up with other regimes

There is a tendency to see climate change as an abstract esoteric issue; but in fact it is a vital everyday issue which will affect water supplies, food security, health, housing (especially in vulnerable areas), energy supply, transport, employment, culture, and lifestyles. Hence, it is essential to, where possible, mainstream climate change limitation and adaptation policy into other policy areas at the international level, and where not possible to at least emphasize the links with these other areas. Much research has been done on this issue, but this is an approach that has not been discussed much at the Workshop or in this book. Climate change could also be linked to trade (e.g., Aldy, Orszag, and Stiglitz 2001; Biermann and Brohm 2005), investment (e.g., Cosbey *et al.* 2004; Peterson 2004) and development regimes (Pershing, this volume). Though it is not difficult to research the issue of linkages, synergies, and conflicts with other policy domains (e.g., Gupta and Hisschemöller 1997; Hisschemöller and Gupta 1999; van Asselt, Gupta, and Biermann 2005; Victor, this volume); there are some practical challenges (Oberthür 2002).

Unilateral US measures

Although the United States is not party to the Kyoto Protocol, it has initiated a number of agreements with other countries. These include the Methane to Markets Partnership with Australia, China, Colombia, India, Italy, Japan, Mexico, Ukraine, and the United Kingdom, which has an associated funding of 50 million dollars. The purpose of this agreement is to collect waste methane from coal mines and landfills and to market this gas as energy. There is also the US initiative on the International Partnership for a Hydrogen Economy with fifteen countries and the EU to promote the introduction of a Hydrogen Economy (Sindico and Gupta 2004). The Carbon Sequestration Leadership Forum with fifteen other countries and the EU promotes sequestration of carbon while through its Tropical Forest Conservation Act[1] the United States is entering into agreements with several developing countries to discourage illegal logging.[2] The Generation IV International Forum includes ten countries[3] that want to move toward the fourth phase in nuclear power; from prototypes, through current operating plants and advanced reactions to generation IV plants. The purpose is to ensure optimum use of natural resources, address nuclear safety, waste and proliferation issues, and the concerns of the public.[4] Another partnership of seventeen countries, several industries, and NGOs aims to promote renewable energy and efficiency.[5] The United States is entering into a number of bilateral and regional agreements on climate change with Australia, Brazil, Canada, China, Belize, Costa Rica, El Salvador, Guatemala, Honduras, Nicaragua, Panama, the EU, India, Japan, Mexico, New Zealand, Republic of Korea, the Russian Federation, and South Africa. And finally, the United States initiated in July 2005 the Asia Pacific Partnership for Clean Development and Climate that include Australia, China, India, South Korea, and Japan.

Will all these measures taken by the White House demonstrate to the world that the United States is serious about dealing with the problem of climate change? The United States itself argues that: "We remain fully engaged in multilateral negotiations under the UNFCCC, and have created or worked to revitalize a range of international climate initiatives within the last two years."[6] One could potentially argue

[1] The Tropical Forest Conservation Act, Foreign Assistance Act, Part I, Section 118 – Tropical Forests; available also at www.usaid.gov/our_work/environment/compliance/ faa_section_118.htm.

[2] Climate Change Fact Sheet: The Bush Administration's Action on Global Climate Change, Fact Sheet released by the White House, Office of the Press Secretary, Washington DC, November 19, 2004, found at www.state.gov/g/oes/rls/fs/46741.htm.

[3] Argentina, Brazil, Canada, Euratom, France, Japan, Republic of Korea, Republic of South Africa, Switzerland, United Kingdom, and United States.

[4] For more information, visit the website at www.gif.inel.gov/.

[5] For more information, visit the website at www.reeep.org/.

[6] See White House Climate Change Fact Sheet at www.whitehouse.gov/news/releases/2005/05/20050518–4.html.

that the range and width of the agreements initiated by the United States demonstrate that it is serious about promoting international dialogue, if not action, in the area of climate change. However, there is also the risk that many of these agreements do not mean much. The international community could encourage such initiatives and proactively test these to see if they are consistent with the climate change regime and include these as part of the global multilateral process to deal with climate change.

Litigation

Another interesting recent development is the rise of the collaboration between legal scholars and non-state actors to develop legal cases against polluters. The increasing academic interest in the field (Penalver 1998; Marburg 2001; Weisslitz 2002; Allen 2003; Grossman 2003; Verheyen 2003; Burns 2004; Gillespie 2004; Thackeray 2004; Hancock 2005; Jacobs 2005; Lipanovich 2005; Mank 2005) as well as the rise in climate and climate-related litigation, not only within the United States and Australia, but also in other parts of the world including Argentina, Germany, and Nigeria, indicate that many feel that governments and industry could be forced to take action by courts. There is also a case before the Inter-American Commission on Human Rights initiated by the Inuit people. While the initial judgments are not always positive for the plaintiffs, and it may be a while before we see the direction in which courts will lean, the fact that the courts are becoming actors in the process of policymaking will give a new dimension to the issue. These cases explore different causes of actions.

Local action

Another impressive arena of action is what is happening at local and provincial level. Worldwide several hundred municipalities are taking action on climate change and have joined some international forum – such as the International Coalition of Local Environmental Initiatives. Cities in Canada, South Africa, Europe, Latin America, and Asia have joined the process to create grass-roots engagement with the climate change process and to see if they can learn from other cities and adapt existing policies to take into account climate change concerns. In Australia and the United States, several cities and provinces are proactively engaged in such policymaking. For example, California is regulating carbon dioxide emissions from automobiles and has established emission targets for itself and aims to regulate the emissions from power plants.[7] Some of the North-Eastern and mid-Atlantic states are in the process of finalizing a Regional Greenhouse Gas Initiative to create a cap-and-trade system for power plants in these states.[8] Increasingly too the scientific literature is focusing in this area arguing that it may be more scientifically justified, efficient, effective, and legitimate to take action at local level (e.g., Clark 1985; Wilbanks and Kates 1999; Shackley and Deanwood 2002).

The national communications of DCs

While there is a general tendency to assume that the developing countries are not subject to any specific obligations under the climate change agreements, there is also a general tendency to assume that developing countries are not taking any action. Both of these assumptions are incorrect. The first 122 national communications of the developing countries to the climate change secretariat indicate the variety of measures that are being taken, have been taken, and are planned on being taken by these countries with respect to climate change. These include Brazil's national Alcohol Program which has saved 550 million barrels of oil over a period of twenty-five years while avoiding 400 million tons of CO_2 emissions, China's Greenlight Program which has saved 17.2 million megawatt-hours of electricity in 1996–1998, and the renewable energy program in India which avoided 330 million tons of CO_2 emissions, while reforestation of indigenous forests in Lesotho increased CO_2 uptake of 185 tons per hectare.[9] China submits that it has taken measures to decouple economic growth from energy consumption since 1980 and India argues that "by consciously factoring

[7] See www.climatechange.ca.gov for details.

[8] See www.rggi.org for more details.

[9] Sixth Compilation and Synthesis of Initial National Communication from Parties Not Included in Annex I to the Convention, Addendum on Sustainable Development and the Integration of climate change concerns into medium- and long-term planning, FCCC/SBI/2005/18/Add.1, 25 October 2005.

in India's commitment to the UNFCCC, [it has] realigned economic development to a more climate friendly and sustainable path."[10] Ways and means to build on the rich information in these national communications should be explored further.

Rule of law

My last point focuses on a philosophical issue. As we increasingly ascribe local problems to failure in local governance and recommend the adoption of good governance including the rule of law to national governments (Carothers 1998; Santiso 2001), we need to also find ways of promoting the rule of law and good governance at the international level. There is much opposition to promoting the rule of law and good governance framework at the international level on the grounds that power will dominate international relations (Koskenniemi 1990; Watts 1993). And yet there is a need to limit the role of power politics at global level if we wish to promote sustainable development, and the only way to do this is through promotion of the rule of law not just in procedural terms (Esty 2006) but also in substantive terms. For as Fitzpatrick (2003) argues, politics will need law to sustain its legitimacy. One of my favorite recent quotes is that of Madeline Albright, who submits that President George W. Bush talks about promoting the rule of law in countries but is "allergic to treaties designed to strengthen the rule of law in such areas as money-laundering, biological weapons, crimes against humanity and the environment" (Whittell 2002: 8). As Kofi Annan puts it: "Those who seek to bestow legitimacy must themselves embody it; and those who invoke international law must themselves submit to it."[11]

Conclusion

I have argued in favor of key elements of Michaelowa's proposal and have added some additional dimensions of my own. I do not think there is any one exclusive approach to addressing the climate change problem. We will have to pursue all the approaches that exist, whether it is in

[10] Government of India 2004: xiii.

[11] Secretary-General's address to the General Assembly, New York, September 21, 2004, www.un.org/apps/sg/sgstats.asp?nid=1088.

terms of persuading like-minded countries that they need to push the Kyoto regime further through targets and timetables, convincing countries like the United States that their multiple initiatives at international level need to lead to concrete and measurable results, convincing local actors in recalcitrant states to take more action to deal with climate change, or whether in terms of developing policy or challenging others in local and international courts and recognizing that the developing countries are taking action and that within their national communications there may be room to identify cost-effective new steps in the future. In the ultimate analysis, we need to create a mass movement, and the role of research in this process is to identify the policy instruments that can be adopted by a range of different actors in order to deal with the problem. We need to adopt a cosmopolitan as opposed to a "nationalistic" research agenda.

References

Allen, M. (2003). "Liability for Climate Change: Will It Ever Be Possible to Sue Anyone for Damaging the Climate?" *Nature* 421: 891–892.

Aldy, J. E, P. R. Orszag, and J. E. Stiglitz (2001). "Climate Change: An Agenda for Global Collective Action," paper presented at Pew Center on Global Climate Change Workshop on the Timing of Climate Change Policies, Washington, DC, October 2001, AEI-Brookings Joint Center on Regulatory Studies.

Asselt, H. van, Gupta, J., and F. Biermann (2005). "Advancing the Climate Agenda: Exploiting Material and Institutional Linkages to Develop a Menu of Policy Options," *Review of European Community and International Environmental Law* 14(3): 255–264.

Biermann, F., and R. Brohm (2005). "Implementing the Kyoto Protocol without the United States: The Strategic Role of Energy Tax Adjustments at the Borde," *Climate Policy* 4(3): 289–302.

Burns, W. G. C. (2004). "The Exigencies that Drive Potential Causes of Action for Climate Change Damages at the International Level," *American Society of International Law Proceedings* 98: 223–227.

Carothers, T. (1998). "The Rule of Law Revival," *Foreign Affairs* 77(2): 95–106.

Clark, W. C. (1985). "Scales of Climate Impacts," *Climatic Change* 7(1): 5–27.

Cosbey, A., H. Mann, L. E. Peterson, and K. von Moltke (2004). *Investment and Sustainable Development: A Guide to the Use and Potential of International Investment Agreements*, Manitoba: International Institute for Sustainable Development.

Esty, D. C. (2006). "Good Governance at the Supranational Scale: Globalizing Administrative Law," *Yale Law Journal* 115(7): 1490–1562.

Fitzpatrick, P. (2003). " 'Gods Would Be Needed. . .': American Empire and the Rule of (International) Law," *Leiden Journal of International Law* 16(3): 429–466.

Gillespie, A. (2004). "Small Island States in the Face of Climate Change: The End of the Line in International Environmental Responsibility," *UCLA Journal of Environmental Law and Policy* 22: 107–129.

Government of India (2004). "India's Initial National Communication to the United Nations Framework Convention on Climate Change," New Delhi: Ministry of Environment and Forests.

Grossman, D. A. (2003). "Warming up to a Not-so-Radical Idea: Tort Based Climate Change Litigation," *Columbia Journal of Environmental Law* 28: 1–61.

Gupta, J. (2003). "Engaging Developing Countries in Climate Change: (KISS and Make-Up!)," in D. Michel (ed.), *Climate Policy for the Twenty-First Century: Meeting the Long-Term Challenge of Global Warming*, Washington, DC: Centre for Transatlantic Relations, Johns Hopkins University, pp. 233–264.

Gupta, J. and Hisschemöller, M. (1997). "Issue-linkages: A Global Strategy Towards Sustainable Development," *International Environmental Affairs* 9(4): 289–308.

Gupta, J. and H. van Asselt (2006). "Helping Operationalise Article 2: A Transdisciplinary Methodological Tool for Evaluating when Climate Change is Dangerous," *Global Environmental Change* 16(1): 83–94.

Guston, D. (2001). "Towards a 'Best Practice' of Constructing 'Serviceable Truths,' " in M. Hisschemöller, J. Ravetz, R. Hoppe, and W. Dunn (eds.), *Knowledge, Power and Participation*, New Brunswick, NJ: Transaction Publishers, Policy Studies Annual, pp. 97–119.

Hancock, E. E. (2005). "Red Dawn, Blue Thunder, Purple Rain: Corporate Risk of Liability for Global Climate Change and the SEC Disclosure Dilemma," *Georgetown International Environmental Law Review* 17(2): 223–251.

Hisschemöller, M., and J. Gupta (1999). "Problem-Solving Through International Environmental Agreements: The Issue of Regime Effectiveness," *International Political Science Review* 20(2): 153–176.

Hisschemöller, M., R. Hoppe, W. N. Dunn, and J. R. Ravetz (2001). "Knowledge, Power and Participation in Environmental Policy Analsysis," in M. Hisschemöller, J. Ravetz, R. Hoppe, and W. Dunn (eds.), *Knowledge, Power and Participation*, New Brunswick, NJ: Transaction Publishers, Policy Studies Annual, pp. 1–28.

Jacobs, R.W. (2005). "Treading Deep Waters: Substantive Law Issues in Tuvalu's Threat to Sue the United States in the International Court of Justice," *Pacific Rim Law and Policy Journal* 14: 103–128.

Koskenniemi, M. (1990). "The Politics of International Law," *European Journal of International Law* 11(1): 4–32.

Lipanovich, A. (2005). "Smoke before Oil: Modelling a Suit Against the Auto and Oil Industry on the Tobacco Tort Litigation is Feasible," *Golden Gate University Law Review* 35: 429–489.

Mank, B. C. (2005). "Standing and Global Warming: Is Injury to All, Injury to None?" *Environmental Law* 35: 1–83.

Marburg, K. L. (2001). "Combating the Impacts of Global Warming: A Novel Legal Strategy," *Colorado Journal of International Environmental Law and Policy* (2001 Yearbook): 171–180.

Oberthür, S. (2002). "Clustering of Multilateral Environmental Agreements: Potentials and Limitations," *International Environmental Agreements: Politics, Law and Economics* 2(4): 317–340.

Pachauri, R. (2006). "Avoiding Dangerous Climate Change," in H. J. Schellnhuber (ed.), *Avoiding Dangerous Climate Change*, Cambridge: Cambridge University Press, pp. 3–5.

Penalver, E. M. (1998). "Acts of God or Toxic Torts? Applying Tort Principles to the Problem of Climate Change," *Natural Resources Journal* 38: 563–569.

Peterson, L. E. (2004). *Bilateral Investment Treaties and Development Policy-Making*, Manitoba: International Institute for Sustainable Development.

Rotmans, J., R. Kemp, M. van Asselt, F. Gells, G. Verbong, and K. Molendijk (2000). *Transities en Transitiemanagement: De Casus van een Emissiearme Energievoorziening*, Maastricht: ICIS.

Santiso, C. (2001). "Good Governance and Aid Effectiveness: The World Bank and Conditionality," *Georgetown Public Policy Review* 7(1): 1–22.

Schneider, S. H., and J. Lane (2006). "An Overview of 'Dangerous' Climate Change," in H. J. Schellnhuber (ed.), *Avoiding Dangerous Climate Change*, Cambridge: Cambridge University Press, pp. 3–5.

Shackley, S., and R. Deanwood (2002). "Stakeholder Perceptions of Climate Change Impacts at the Regional Scale: Implications for the Effectiveness of Regional and Local Responses," *Journal of Environmental Planning and Management* 45(3): 381–402.

Sindico, F., and J. Gupta (2004). "Moving the Climate Change Regime Further Through a Hydrogen Protocol," *Review of European Community and International Environmental Law* 13(2): 175–186.

Thackeray, R. W. (2004). "Struggling for Air: The Kyoto Protocol, Citizen's Suits Under the Clean Air Act, and the United States Options for

Addressing Global Climate Change," *Indiana International and Comparative Law Review* 14: 855.

Trouwborst, A. (2006). *Precautionary Rights and Duties of States*, Dordrecht: Martinus Nijhoff Publishers.

Van der Sluijs, J. P. (1997). "Anchoring Amid Uncertainty: On the Management of Uncertainties in Risk Assessment of Anthropogenic Climate Change," PhD thesis, University of Utrecht.

Verheyen, R. (2003). Climate Change Damage in International Law, Dissertation zur Erlangung des Dr. iur; Universität Hamburg, Fachbereich Rechtswissenschaft.

Watts, A. (1993). "The International Rule of Law," *German Yearbook of International Law* 36: 15–45.

Weisslitz, M. (2002). "Rethinking the Equitable Principle of Common but Differentiated Responsibility: Differential versus Absolute Norms of Compliance and Contribution in the Global Climate Change Context," *Colorado Journal of International Environmental Law and Policy* 13: 473–509.

Wilbanks, Thomas J., and Robert W. Kates (1999). "Global Change in Local Places: How Scale Matters," *Climatic Change* 43(3): 601–628.

Whittell, G. (2002). "Albright Attacks US Foreign Policy as Schizophrenic," *The Times*, May 21, p. 8.

Yamin, F., J. B. Smith and I. Burton (2006). "Perspectives on 'Dangerous Anthropogenic Interference'; or How to Operationalize Article 2 of the UN Framework Convention on Climate Change," in H. J. Schellnhuber (ed.), *Avoiding Dangerous Climate Change*, Cambridge: Cambridge University Press, pp. 81–91.

PART II

Harmonized domestic actions

4 *Fragmented carbon markets and reluctant nations: implications for the design of effective architectures*

DAVID G. VICTOR

POLICY analysts and diplomats worldwide are now focused on the design of the regime that could replace or extend the Kyoto Protocol in 2012 when the Protocol's main commitments expire. While the Kyoto Protocol has several achievements to its credit – in particular, it has played a role in sustaining political attention on the need for policies to control emissions of the gases that cause global warming – in many respects the Protocol is in deep trouble. In Australia, Canada, and the United States – three countries where the Kyoto commitments would have been most demanding – governments have largely abandoned the treaty. In Japan and Europe governments are implementing some limits on emissions, but much effort is now focused on a shell game of accounting that will probably yield formal legal compliance with the treaty's strictures through the trading of credits that do not reflect actual reductions in emissions. The Russian government nearly abandoned the treaty for fear that it would actually require changes in behavior. The integrated international market for emissions envisioned under the Protocol has not yet materialized; instead, at least six different carbon markets have emerged – each with their own rules and prices. The Clean Development Mechanism (CDM), a scheme intended to engage developing countries by subsidizing projects that would allow them to build less-carbon-intensive energy systems, is tied in red tape that will be familiar to any student of the US Environmental Protection Agency's limited emission trading

Thanks to Sheila Olmstead and Carlo Carraro, along with Joe Aldy and Rob Stavins, for comments that led to a massive reworking and sharpening of an earlier draft. I am grateful to Sarah Joy for research assistance, Michael Wara for sharing interim results from his study of CDM, and to Tom Heller for discussions on many joint projects on climate change policy. This paper also draws heavily on Victor, House, and Joy 2005.

programs – notably the offsets program – of the 1970s (Hahn and Hester 1989). Most of the pipeline of CDM projects is dominated by schemes to achieve largely false credit for reductions in industrial gases; energy projects account for just 17 percent of the effort (Wara 2006). And while nobody thought that the Protocol would be the final word in efforts to control emissions, it is hard to see that it is even the first word in a viable framework for the future.

This paper suggests an alternative framework. I proceed by examining the four key elements of the Kyoto architecture: universal participation, binding targets and timetables for emissions of greenhouse gases (GHGs), integrated international emission trading, and compensation to encourage participation by developing countries. For Kyoto's enthusiasts these four elements are the bedrock of the Kyoto system and, by their account, must be replicated in any future climate treaty.

My argument is that Kyoto's troubles are rooted in these four elements of conventional wisdom. The patient followed the doctor's orders scrupulously, and by that regimen he has incapacitated himself. I make this argument in the first part of this essay.

The second part of the essay offers an alternative vision that is based on more extensive use of nonbinding agreements among smaller groups of important countries, allowance for fragmented emission trading systems, and new mechanisms for engaging developing countries.

Throughout, my argument rests on the broad argument that serious action to address climate change must be anchored in capable institutions. And the most capable institutions exist at the level of nation-states (and some regions, notably the European Union). International institutions – especially global institutions rooted in the United Nations – are relatively weak. They play important roles as codifiers, coordinators, attractors of attention, and suppliers of a few collective goods. Too much attention has focused on the global institutions and not enough on the more diverse national and regional bodies that actually get things done. This mismatch between diplomatic focus and real action explains much of Kyoto's trouble, and it explains why the architecture that I advance makes much heavier use of the institutions that are intrinsically more capable than global binding treaties. It also explains why my approach is more fragmented, as that reflects the reality of how authority is allocated in the international system. Carbon markets are likely to be fragmented rather than integrated – a world that is second best in theory but first best when the theories are updated to reflect how property rights can be assigned, monitored and enforced.

Conventional wisdoms

Global warming is perhaps the most difficult problem to be confronted within the international system. It is caused by a large number of countries (Figure 4.1) with highly diverging interests – some (mainly the industrialized world) want action to control emissions and are willing to pay for it. Others (mainly in the developing world) are willing to accept action if others pay for it. Because countries enjoy a broad freedom of action in their international affairs, collective action is especially difficult to achieve because every participant must see that it benefits individually from the joint effort. As with most common pool resources, the incentives to defect are strong – especially as enforcement is difficult and governments have strong incentives not to honor their commitments. Furthermore, while the benefits of action accrue far in the future in the form of an uncertain amount of a lighter footprint on the world's climate system, the costs begin to accrue today. Achieving deep cuts in the emissions of these gases implies a radical reorganization of the world's energy systems and perhaps key elements of the world economy – a journey that one does not begin on a whim. With all these obstacles, it is not surprising that effective action has been a long time coming.

The Kyoto framework, ironically, is designed in a way that seems to maximize the difficulties in collective action. These difficulties are evident in all four of Kyoto's main architectural features – universal participation, binding targets and timetables for emissions of greenhouse gases, integrated international emission trading, and compensation to encourage participation by developing countries. I address each in turn – and for each I criticize the conventional wisdom and suggest some alternatives.

Universal participation

The standard argument for universal participation is that it creates legitimacy and avoids "leakage." The former is a woolly concept that has not been subjected to careful empirical tests. The legitimacy that comes from giving all nations a voice is probably overrated in importance. Indeed, most of the world's effective international institutions began with large doses of discrimination and inequality – they include the GATT, the IMF, the UN Security Council, and the G8.

Economists have focused, instead, on the problem of "leakage" – the phenomenon that tight regulations on some countries will cause industrial activities to migrate ("leak") elsewhere in the world where regulations are lax. For any aggressive climate control regime, leakage is clearly a problem that demands attention. But if we assume that the first few decades of efforts to control emissions will be marked by modesty, learning, and innovation – hardly the hallmarks of aggression – then the problem of leakage becomes much less severe. Indeed, research on the "leakage" hypothesis done two decades ago when there were large concerns about industrial flight due to differences in environmental regulation show that such fears are largely overblown (e.g., Low 1992). Indeed, economics is probably poised to overstate the leakage problem because it analyzes these issues using equilibrium models that already do not reflect the huge variation across borders in the factors that affect industrial location (not least, the huge variation in retail energy prices), and analysts who focus on carbon policy tend to imagine that most industrial decisions are driven by such policies when in reality they are not. Already there is a substantial ($20–30 per ton CO_2) difference in carbon costs between the USA and EU with no evidence that this is affecting industrial decisions. In short, the benefit of paying close attention to "leakage" does not seem to be large.

The costs of efforts to address leakage, however, are substantial. Leakage-inspired agreements that include all major contributors necessarily include countries that have quite different interests. As evident in Figure 4.1, for example, the top emitters include Organisation of Economic Co-operation and Development (OECD) nations (notably the EU but also, to a lesser degree, the USA) that generally favor some emission controls. They also include countries (e.g., China) that reject any limits. And they include countries (e.g., Russia) that might actually perceive benefit from climate change and thus could have a special interest in undermining effective emission controls. Such dispersion in interests is hardly a new problem in international affairs, and much of the process of negotiating effective agreements is one of adjusting terms and geometry to find a Pareto-improving bargain. If concerns about leakage dominate those negotiations – and thus membership is

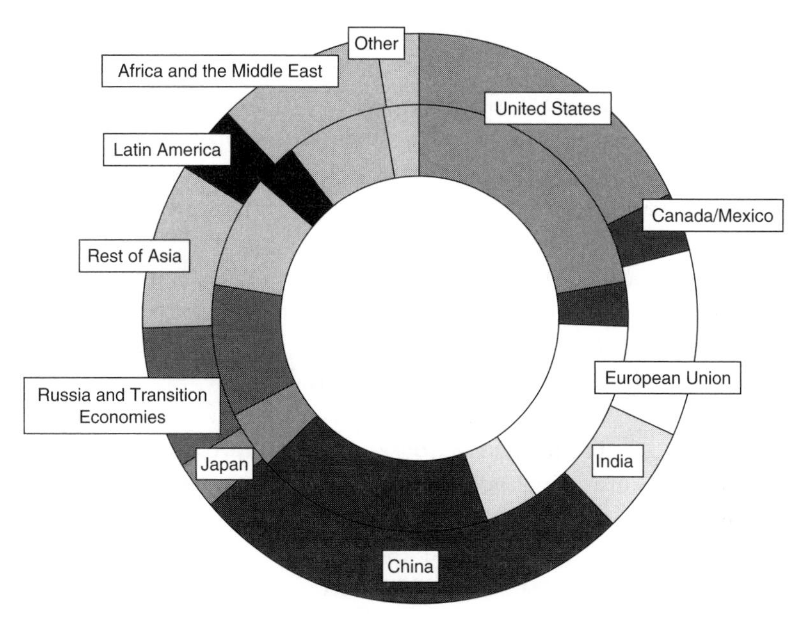

Figure 4.1. Top emitters 2004 and 2030
Global greenhouse gas emissions: 2004 (26,079 Mt CO_2) and 2030 (40,420 Mt CO_2)

Source: International Energy Agency 2006.

high – it becomes increasingly difficult to assemble the subtle package deals that are needed to ensure Pareto-improvement.[1]

Binding targets and timetables

A second area of conventional wisdom lies in the types of commitments that are selected and the way that they are codified. The common

[1] For the same reasons, efforts to build an international regime to control climate change on a shared "objective" are likely to fail because countries in fact do not have shared assessments of the danger and opportunity. Article 2 of the United Nations Framework Convention on Climate Change (UNFCCC) lays out exactly that objective (it calls for avoiding "dangerous anthropogenic interference in the climate system"), and considerable diplomatic and scientific effort has focused on putting Article 2 into practice. Those efforts are built on an unrealistic vision of politics. It is perhaps understandable that diplomats, whose job description includes the manufacture of a seemingly logical order to what is in fact a highly political process of sausage making, would create something like Article 2; why sober analysts would devote so much attention to the impossible task of actually assessing the true meaning of Article 2 is mystifying.

assumption is that legally binding instruments are best because governments take those most seriously. Because such instruments are formally binding they are usually negotiated by diplomats – especially diplomats with legal training. And for pollution problems, the conventional wisdom holds that the best kinds of commitments are "output" measures – that is, targets and timetables for emissions – rather than "input" measures such as the effort or policies that each country will pursue. Output measures, it is thought, assure that governments will not slack off; if they are held accountable for a particular level of pollution then the environment will be protected best. Again, the evidence for these propositions is scant.

The assumption that binding instruments are best is part of the folklore of global environmental law, for which treaties are assumed to hold a prized place. In reality, the decision about legal status is just one of several design choices that affect outcomes. I will concentrate on the interplay between legal status and another critical design choice – the style of diplomacy. As shown in Table 4.1, diplomatic style can involve professional diplomats (usually lawyers) who specialize in negotiating texts with an eye to compliance. Alternatively, negotiations can occur at the level of senior political leaders who make aspirational commitments, often without an eye to the particulars of compliance.

Table 4.1 ***Level of participation and legal status of commitments***

	Style of negotiation	
	Senior politicians	*Diplomats, lawyers, and bureaucrats*
Legal status of commitments		
Nonbinding	• North Sea pollution • Tokyo Round "codes"	• Technical standards groups (e.g., Codex)
Binding	• Arms control treaties	• Whaling • Montreal Protocol • Kyoto

Most of the canon of international environmental law is based on the lower-right cell of the matrix in Table 4.1. But all four of the cells offer relevant experiences. In general, instruments in the lower-right cell work best when it is clear how to comply with commitments and when locking countries into place is of special importance. By contrast, instruments in the upper-left cell – nonbinding agreements crafted by accountable political leaders – work best when the purpose of an international agreement is to signal a direction for effort but it is unclear exactly what the effort will cost and whether it is politically or technically achievable. Nonbinding political agreements (upper-left cell) are more flexible and less prone to raise concerns about noncompliance, and thus they allow governments to adopt ambitious targets and far-ranging commitments. In contrast, binding legalistic agreements (lower-right cell) are usually crafted to assure compliance – especially when the agreement includes countries whose internal legal procedures assure that international commitments are enforceable such as through direct application in domestic law. A binding commitment might be useful for codifying an effort that is already in hand (or which requires actions that are easy for governments to deliver). But uncertain, strenuous efforts at cooperation are easier to organize when the commitments are not formally binding.

In addition to trade-offs on these two dimensions – accountability and legal status – there is also a third dimension of choice, which I will call the measure of accountability. Commitments that are set in terms of outputs usually work best when there is a clear chain of causation that links actions by governments (who are usually the actors that subscribe to international commitments) and the ultimate output of pollution. Classic arms control agreements usually had terms set in outputs (e.g., number of missiles) because governments were confident that they could control their own behavior. Similarly, the agreements on the ozone layer were expressed in outputs (consumption of ozone-depleting substances) because the limited number of industrial firms and strong mechanisms for regulation made it possible to control production and trade and thus assure compliance. By contrast, the global warming problem is marked by extreme difficulty in connecting government actions to particular outputs (emissions) because the factors that determine are largely outside the near-term control of governments, such as the state of the economy and technology and investment decisions by firms. Governments can solve this problem by capping emissions – such as through an emission trading program – but only by accepting large uncertainties in the cost of compliance. As those uncertainties rise so does the credibility of the commitments since a program that is excessively costly will raise the risk that that country will simply abandon its commitment – as the United States did with its Kyoto commitments in 2001 and as Canada is doing now in 2006. Such policy choices about regulatory instruments – whether outputs or, alternatively, a specified level or type of effort – are familiar to students of the "prices vs. quantities" literature in economics (e.g., Weitzman 1974; Roberts and Spence 1976; Pizer 1998).

These trade-offs are evident in the case histories of lesser-known experiences with controlling environmental pollution where architects actually deployed decision-making procedures, legal agreements, and regulatory instruments that differ from the conventional wisdom. The experience with international cooperation in the North Sea, the Baltic Sea, and with acid rain in Europe are all examples where nonbinding instruments backed by senior politicians proved to be more effective than binding alternatives (Roginko 1998; Skjærseth 1998; Wettestad 1998).[2] In those cases, there had been efforts to use binding instruments to address the problems at hand, but those efforts often fell short. In the North Sea and Baltic Sea regimes, the addition of ministerial-level conferences that included ambitious (but nonbinding commitments) helped, in part, to break the logjam. In the European acid rain regime, more-ambitious nonbinding commitments to control NO_x (a leading cause of acid rain) were adopted by a smaller number of countries alongside a binding convention to address the same pollutant. In all three of these cases the nonbinding efforts alone did not lead to more effective cooperation. Rather, at least three elements were necessary for effectiveness. First, the commitments required high-level attention – usually at ministerial level – to improve accountability and implementation. Second, the nonbinding commitments worked because they were embedded within institutions that could mobilize detailed performance reviews, which are especially important when commitments concern areas of activity where it is difficult to gauge the best implementation strategies at the outset. In the North Sea and Baltic Sea regimes, notably, the nonbinding commitments along with extensive review helped to focus attention and effort on the difficult-to-manage problem of land-based pollution runoff. Third, the commitments and review should be part of an ongoing relationship so that the shadow of future interactions (and linkages across issues) impose discipline on current behavior.

[2] See also generally Victor, Raustiala, and Skolnikoff (1998).

Integrated emission trading

Having embraced the earlier elements of conventional wisdom – global participation and binding targets for emission outputs – a relatively small step of logic is needed to arrive at the third pillar of conventional wisdom: integrated global emission trading. So long as governments worldwide are involved in the effort to control emissions, it is best to use a flexible market-based system to ensure that the costs are as low as possible. The logic for this position is sound, but its application in the international system are problematic for at least three reasons. One set of reasons relate to all that is said above – namely, the other elements of conventional wisdom that lead to emission trading are not robust under scrutiny.

A second problem with emission trading relates to allocation of emission credits, which is particularly difficult when regulating gases such as carbon dioxide at the global level. Where membership rules allow for unit veto and some countries are risk-averse (i.e., the developing countries, Russia, and others who don't see a strong interest in controlling emissions), allocations are prone to padding. Government officials from these reluctant countries arrive at a negotiation with a brief to incur no cost, and they know that the treaty will be rejected back at home if it is seen to violate this maxim. So they imagine their worst-case scenario for emissions and they demand an allocation equal to (or higher than) that level. That is what Russia did in Kyoto; that's what the developing countries will do when they are forced to accept emission targets. Since nobody knows the countries' true levels of future emissions, each new entrant to this negotiation makes it harder to gain a meaningful agreement on the total size of the pie and its allocation. Elsewhere I have called this problem "negative sum bargaining" (Victor 2001). Ironically, as the system is expanded – so that gains from trade are largest and so that leakage is minimized – the very process of expansion undermines the ability of the scheme to impose a meaningful limit on emissions.

This problem is hard enough to address, but the international institutions within which binding allocations of emission credits would be negotiated are structurally weak and are marked by long delays between agreement and implementation. As in Kyoto, a long time passes from the point of negotiation (in that case, 1997, with negotiations constrained by the decisions of 1997 dragging on to 2001) and the actual implementation (2008 to 2012). During this delay, errors in forecasting grow; many of those errors magnify the financial flows that will arise when the emission trading system begins to equilibrate. Many analysts have called for even longer time horizons because an optimal response to the challenge of managing carbon requires a long-term approach; that call is logically sound yet practically troublesome because it would magnify the likely financial imbalances that would arise as the trading system equilibrates. These imbalances matter because if they create political trouble in just one (or a few) countries then the pressure to defect will grow and the deterrent against defection is quite weak. As that one country exits the system then the financial imbalances and burdens will shift and that will put pressure on other countries. I find it striking to compare this system with the early days of the GATT. In an emission trading system, errors in allocation of emission credits or the exit of one country creates instabilities that magnify the pressure for others to exit; in the early days of the GATT, the reciprocal and therefore self-enforcing nature of tariff concessions created pressures that magnified the benefit of remaining within the regime.

None of this means that international trading mechanisms are completely infeasible. It does mean, however, that they rest on a much more fragile basis than is conventionally thought – because the process of allocation will be extremely sensitive to underlying interests, and because the operation of the market will be sensitive to the capabilities of the institutions that oversee and enforce the trading system. For the most part, studies of international trading have not looked at these issues.[3] Rather, they have treated states as black boxes – emission credits would be allocated to governments (itself a Herculean task) and then governments would establish the internal procedures needed to make their markets work. A rich literature has arisen around efforts by governments to put those internal procedures into effect, but so far there is strikingly little connection between that "domestic" literature and the thinking about international emissions markets. Putting those two strands together requires looking carefully at how real governments implement emission trading systems. When that is done, we find that the internal characteristics of states are extremely important – they are perhaps the dominant reason why international trading systems are likely to be fragmented even as there are strong economic incentives for integration.

[3] Among the few exceptions is Hahn and Stavins 1999.

The internal characteristics of states will influence trading in at least two ways. First, the national economy might not be organized around market principles, and thus there might be rampant inefficiencies that already affect the economy. Thus, even if governments (or individual enterprises) are forced to trade emission credits, they might not focus their efforts on the least-cost way to control emissions. In China, for example, large central coal-fired generators get their capital from central allocations at nearly zero cost. By contrast, gas-fired generators pay something closer to the real cost of capital – in part because gas is not an incumbent industry and in part because both the equipment and some of the gas is priced on international markets. In these settings it will be a lot less costly and more effective to focus on the underlying market failures – such as pursuing capital reforms and sundry other tasks that are core to the Chinese project for economic reform – rather than allocating emission credits.[4]

Second, national regulatory institutions might not have the administrative capabilities to implement an emission trading system. For a trading system to work it is necessary to allocate credits to the institutions that actually govern decisions about technology and behavior, which in nearly every economy is a large number of mainly energy enterprises. Making a trading system work requires, then, the capacity to monitor the behavior of these enterprises and to enforce compliance. Those are not easy tasks. They are akin to what Western governments have had to do when overseeing banking regulation – an area where even highly capable governments have failed, such as the United States did with the savings and loan crisis. An international trading system implies that these functions would be performed with similar competence across all the jurisdictions that are part of the trading system. The closest analogy is perhaps Europe's efforts at creating the European

[4] There is a tendency among Western analysts to assume that such problems do not affect the more market-oriented economies of the OECD. While that assumption is probably generally true, there are some specific instances that reveal how much additional reform is still needed. For example, lignite-fired power plants in Germany receive allocations under the EU Emission Trading Scheme (ETS) that are equal to their expected emissions. And a growing number of European analysts expect that allocations will be reset every five years, which means that so long as the lignite lobby stays strong it can probably count on special allocations. In that world, political incentives favor building lignite power plants (despite their high CO_2 emissions), and the practical effect of the emissions markets is severely hobbled.

Monetary Union (EMU); indeed, the analogy with a currency is apt since, in effect, the creation of integrated and demanding emission trading systems is not much different from creating a common currency. It is sobering to see how difficult this has been for Europe, despite having a long history of cooperation, strong common institutions and trading relationships, and even a common central bank. It took many years to build the necessary institutions. Even then, when enforcement proved inconvenient for politically powerful members – such as Germany and France, both of which have violated the EMU's deficit rules – the strictures were never applied fully. Those difficulties should be a warning about how rapidly we could expect a common international carbon currency to arise (Victor and House 2004).

These two characteristics – an efficient market-oriented organization of the broader economy and the capacity to administer an emission trading system (e.g., allocation of permits, monitoring and enforcement) – appear to be highly correlated. That's no accident since a market-oriented economy demands that government develop the capacity to intervene through regulation and market surveillance rather than through direct control. For the most part, markets emerge only where government has developed the capacity to act at arm's length. These characteristics are also highly correlated with willingness to pay for controlling emissions. In general, it is rich and highly industrialized market-oriented economies with capable governments that are most concerned about climate change.

These correlations tell us something about how emission trading is likely to emerge. There is a "zone" of countries that have the will and capability to create meaningful emission markets. These countries all have intense trading and investment relationships with each other. Their institutions tend to recognize each other – even to the point of allowing extraterritorial application of law. For these countries it is a *relatively* small step to imagine that they would extend their trading relationships to include a nascent currency of tradable emission credits. What defines the zone, however, is not their common interest in controlling emissions – indeed, their interests vary considerably as is evident when comparing the domestic policies in the EU, United States and Australia – but rather their institutional capabilities and the extent to which other members in the zone are confident of those capabilities.

I borrow the term "zone" from Anne-Marie Slaughter's work on the "zone of law" – a concept she developed to explain why some courts

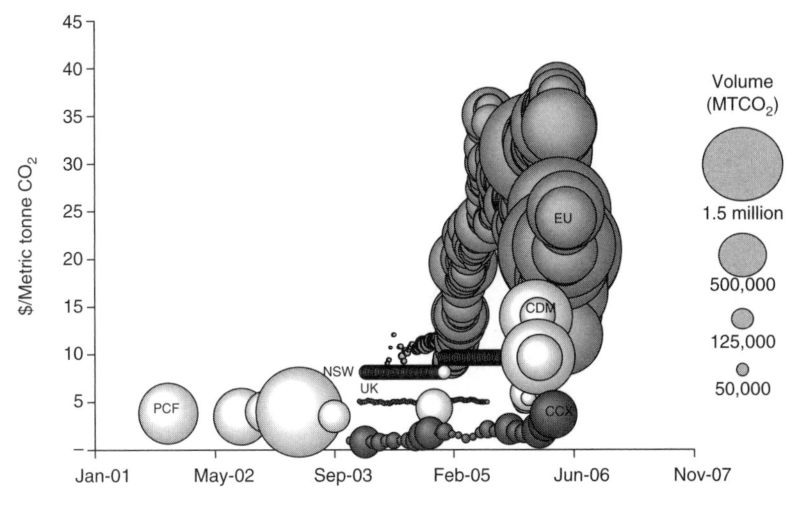

Figure 4.2. Traded carbon prices and volumes

recognize the jurisdiction and decision of other courts while others do not (Slaughter 1992). Her argument was that a "zone of law" existed where courts engaged in such mutual recognition for countries where courts were confident that the judiciary followed certain core tenets, such as independence of judges. On the strength of that zone of law, various other elements of mutual recognition usually also arose – for example, limited forms of regulatory cooperation. By contrast, such systems of mutual recognition were much more difficult to craft with countries that were outside the zone.

Thus, what is likely to occur is not an integrated international emission trading system but, rather, a series of fragmented markets. Indeed, real carbon markets are emerging in this manner (Figure 4.2). These markets are most active in countries where political interest in regulation is the greatest – today, the European Union. Thinner markets with weaker rules – such as the Chicago Climate Exchange – have arisen in markets where political interests and regulatory effort are weaker (Victor, House, and Joy 2005).

Europe's ETS has been most successful and provides insights into the challenging logistics of market implementation. The EU ETS entered into force on January 1, 2005 as a response to the Kyoto Protocol's

GHG emissions reductions targets. The scheme caps GHG emissions from about 11,500 power and industrial installations across twenty-five countries and six major industrial sectors. These installations account for roughly 45 percent of the EU's CO_2 emissions (Nicholls 2005). Emissions are allocated within nations via National Allocation Plans (NAPs). Brussels exerts some discipline on the NAPs through the application of common standards – for example, rules that prevent national governments from allocating permits in ways that favor national industries at the expense of foreign companies. It is nonetheless revealing that the permits, for the most part, are not auctioned – indeed, there are caps on the fraction that governments would be allowed to auction. During the first phase of the ETS (2005–2007), 5 percent of a nation's permits may be auctioned, while 10 percent may be auctioned in the second phase (2008–2012). These are free new assets and, not surprisingly, the NAPs have been highly politicized and rife with problems.

A 2005 WWF study found that the incentive structures of the NAPs of Europe's six largest economies favored those new plants that burn coal over those that use natural gas (ILEX Energy Consulting 2005). This perversion exists where allocations are distributed based on installments' economic need rather than on environmental targets. For example, grandfathering provisions in Germany's NAP give coal-fired plants a competitive advantage over gas-fired plants.

Much has been written about the ETS's strong provisions for monitoring and enforcement. Indeed, those provisions are noteworthy.[5] My view, however, is that the critical mechanism in the ETS is the "linking directive" – the set of rules that govern where and how credits can be

[5] Thus far the ETS has had little monitoring and enforcement experience, as the scheme is still in its early stages. The basic approach, however, is quite clear and likely to be effective. Emissions of installations that fall under the scheme are monitored under EU guidelines and primarily through records of fuel purchases. Self-reported emissions must be audited by an independent third party. The EU ETS enforcement mechanism consists of a series of fines for noncompliance (40€/TCO_2 during phase one, and 100€/TCO_2 from 2008 onward). Installations must make up their missed emission reductions during the next year – thus the fines are *not* a safety valve (in fact, they work in the opposite fashion by *amplifying* price spikes rather than dampening them). If permit prices begin to approach the fine level, it seems likely that there will be a political intervention to adjust the system as it is hard to imagine that European industry would tolerate such a burden when their competitors face, in effect, a zero price on carbon.

imported from other jurisdictions – notably the developing countries (CDM) and the countries in transition (JI [Joint Implementation]). The most efficient outcome would eliminate the linking directive and allow European firms to acquire Kyoto-valid permits anywhere in the world. Under that scenario, the price would drop from the high levels (shown in Figure 4.2) to something between $3 (roughly transaction costs) and perhaps $5 or $7. Instead, the linking directive erects a wall around Europe, with Brussels and the member states controlling how wide the doors in the wall are opened to the outside world. The very concept of trading has been inverted from its original goals. In the areas where the gains from trade are least – that is, within the EU – trading is relatively unfettered. Across the jurisdictions where the gains from trade would be the greatest – that is, from within the "zone" to the outside – the rules that limit trading are most onerous. Many observers view this as a political compromise that was necessary to placate Greens who demanded visible pain (in the form of higher prices for carboniferous energy) and thus sacrifice.

I suggest that this also reflects the nature of the institutions that are necessary for implementing a trading system. Indeed, we should expect fragmented trading systems to arise everywhere else as governments start to tackle the climate problem. That future landscape will be dominated by multiple prices, a proliferation of barriers against trading, and eventually the slow emergence of integrated systems only within the "zone" – that is, where governments have confidence in each other's institutions and level of effort. If we think about trading as the creation of a common currency then the root cause of this outcome is quite clear – governments want to avoid a "Gresham's Law for greenhouse gases," and they are willing to honor (in the form of mutual exchange) emission credits only from jurisdictions that have the institutional capacity and record of commitment that is necessary to assure that they aren't simply printing money.

This process of evolution might take a long time – perhaps a generation, maybe longer – even within the zone. That's because each country is likely to implement its trading system with an eye to its own procedures, capabilities, and political economy. Thus the EU system gives prominence to the national member governments and their NAPs, and because of the long history of Brussels' regulation of large industrial sources, it focuses especially on just the 45 percent of emissions that comes from those sources. The US system might be more broad-based. If the US system includes a safety valve, then troubles will arise for trading across the Atlantic if the US valve is set at a level that is much lower than the EU's price cap. (That outcome is highly likely.) Japan's trading system, if it exists at all, will arise within an industrial organization that is much more highly regulated – it is unclear if prices will really matter in that system, and that will be a source of tension and opportunity if a Japanese system becomes stitched together with the EU or US system.

Compensation for developing countries

The fourth part of the conventional wisdom is that developing countries will participate only if they are paid. The logic for this position is sound – developing countries generally place a low priority on slowing global warming when compared with more urgent priorities, and they are wary about committing resources when the industrialized countries have not done much on their own. The result of this logic is the Kyoto Protocol's CDM.

Politically, a large CDM has become necessary because the CDM has been offered as the main instrument for engaging the developing countries. If that mechanism offered nothing for those nations, then the task of convincing these governments would be much tougher in the future. Even so, most of the CDM projects have concentrated in a handful of countries that have organized themselves well to process CDM applications efficiently and are also generally attractive places to invest in the developing world – these include India, China, South Africa, and Brazil. Most of the rest of the countries have fared much worse, and the political overtones that bedevil the CDM are evident in the recent efforts to re-jigger CDM investments so that more of the money flows to the rest. Another main political driver for a large CDM, however, is Europe's desperation. On current trajectories the EU will not meet its Kyoto targets, which is a politically unacceptable outcome. Nor is it acceptable to open the doors through the ETS wall to all forms of JI projects from the transition countries, notably Russia – as most of those projects are seen as "hot air." So the CDM is the only way to make the books balance. As investors have seen that arithmetic unfold, they have scrambled to develop CDM projects that have the most sure-fire chance of approval and can be imported into the ETS most quickly. As soon as the ETS shortfall is fully saturated, the CDM market is likely to stagnate.

In addition to the political story, it appears that the important substantive story about CDM is that actual project pipeline is very different from what was imagined when the CDM was invented in the months leading to the final negotiations in Kyoto. It has not become an instrument for large-scale investment in clean energy systems in developing countries – such projects, in fact, account for just a few percent of the total credits in CDM pipeline. Some 65 percent of the credits come from two industrial gases (HFC-23, a byproduct in the manufacture of refrigerant HCFC-22, and N_2O, a byproduct in the manufacture of adipic acid that is used for making nylon). Perhaps half of these reductions are actually the result of accounting tricks, and if the purpose of CDM is to target a subsidy for emission cuts in developing countries then it has become a highly inefficient mechanism. The projects that are reducing emissions of HFC-23 cost in total about $70m, and the value of the subsidy being provided through CDM is about $1billion.[6] Yet the various rigidities that exist inside the CDM and its main market (the ETS) provide very strong incentives not to rock the boat. The EU needs the credits to assure compliance with its Kyoto commitments. The main developing countries that host the projects welcome the money and investment. The traders gain from trading volumes. And for the environmentalists it is politically tricky to attack the only mechanism that exists to engage the developing countries. With these powerful forces at work, the CDM apparatus could continue for some time. Some efforts are under way to reform the CDM, but such offset systems fundamentally are prone to fail because they require the technically and politically impossible task of making a baseline assessment – that is, the level of CDM credit must be assessed as the reduction below some baseline. Thus projects that are intrinsically marginal gain favor (e.g., industrial gases that are destroyed by special equipment bolted to the end of a pipe), while projects that actually put countries on different development pathways are impossible to credit because it is impossible to determine *ex ante* the carbon emission profile of a whole country.

Toward a new architecture

So far I have suggested that the existing climate change architecture suffers, at least partly, because it has embraced conventional wisdoms

[6] The analysis of the CDM pipeline is based heavily on Wara 2006.

that do not apply to the case of climate change. In part, a better architecture for addressing the climate issue would begin by adopting alternatives to these conventional wisdoms. In this section I focus on how that architecture might be constituted.

From the discussion of conventional wisdoms it is clear that a new architecture for addressing the climate problem must have at least three features. First, it must allow for variable geometry of participation. In particular, it must allow discrimination so that the most serious negotiating efforts can concentrate on the countries whose participation matters most. That criterion probably means that none of the existing global institutions, including notably the UN, would suffice. Such institutions are usually built on principles of universal participation; discrimination is particularly difficult in such settings.

Second, the architecture must allow for a variety of efforts that are tailored to each key member's capabilities and interests – rather than a single integrated system within which all members must adopt similar instruments. In particular, the system must include a mechanism that entices the participation of developing countries.

Third, the institution must offer a framework within which ambitious political commitments can be made while offering sufficient accountability so that commitments become connected to action. This standard will be the toughest to satisfy, for it raises the questions of enforcement and accountability that are most difficult for international institutions to address.

There are no easy solutions for meeting these criteria. However, it is encouraging to see that, in history, there are several examples of this type of cooperation. The best examples come from economic cooperation – such as trade, monetary, and development policies – not from issues that are classically thought to be "environmental." Part of the trouble in designing an effective response to the problem of climate change may be that the tools for collective action that have been applied are drawn too much from the environmental experience – in particular, the cases of environmental cooperation marked by relatively low levels of uncertainty in the ability of governments to implement their commitments and relative ease in assembling packages of Pareto-improving commitments (Victor and Coben 2005). Such cases tend to favor the kinds of instruments (e.g., emission targets and timetables) and negotiating processes (e.g., binding regulatory conventions) that have been applied in Kyoto. Conceptualizing the climate change issue as one of economic cooperation might help to mobilize attention to better precedents.

Among the precedents from economic cooperation, perhaps the most effective example is the World Trade Organization (WTO). That regime for cooperation began by focusing, through the original GATT, on a limited number of countries whose interests (and capabilities) were sufficiently aligned to allow gains from cooperation. Over time, experience and success have allowed deeper and wider cooperation (and also led to negotiations that extend over much longer time periods because they are more complex). Widening and deepening occurred at the same time, rather than in sequential order. The GATT round that ended in the early 1990s with the creation of the WTO has included much more than simply the tariff bindings that were the core of the first GATT agreement. Similarly, the EU emerged from a more focused cooperation (on infrastructures and key commodities such as coal and steel) among a limited number of countries. With experience and the confidence of success the EU has expanded and deepened. The recent expansion to include twelve new countries, and the agenda for talks with Turkey, may test the limits of EU expansion. Both of these cases – the WTO and the EU – are examples of successful cooperation emerging within a regime of variable geometry and through practical efforts that, with time and effort, became stitched together into a more integrated system. In both these cases, cooperation was forged not mainly through "top-down" central institutions but through packages of proposals that each individual country offered. Collective agreements were forged through negotiations around the adequacy of those packages. Mechanisms for peer review, coupled (eventually) with central institutions such as the WTO's dispute resolution mechanism and the European Court of Justice, helped to ensure that the collective effort had integrity. Elsewhere, I have called this cooperation from the "bottom up" – that is, cooperation steered by common institutions but rooted, fundamentally, on practical actions implemented by key countries and jurisdictions (Victor 2004). The two examples cited here are not the only ones. Tom Schelling (this volume) has suggested that NATO budget allocations and the Marshall Plan are good precedents. Chayes and Chayes (1991) have noted that the process of negotiating Article IV exceptions within the IMF also offers a model marked by individual countries proposing packages of measures and then subjecting themselves to periodic scrutiny to probe whether the packages are sufficient and being implemented to plan.

What would this mean for the design of an effective regime to address climate change? The key is to craft an institution that allows for variable geometry and effort. That probably cannot emerge within the UNFCCC process because that institution, rooted in the UN, is too large and inclusive. The UNFCCC could play a role in this process as the main global forum for addressing climate issues, but as efforts become serious and negotiations complex, it is hard to see that a universal forum can be the only mechanism.

The most interesting idea for a new institution is former Canadian Prime Minister Paul Martin's concept for a forum of leaders from the twenty key countries (L20). Martin (2005) has offered a general vision; a series of meetings have applied the concept to major issues in world affairs, including climate change and energy (www.l20.org). Whether by creation of a new institution such as the L20 or reform of an existing forum such as the G8, such a standing body would offer a way to craft deals among the smaller number of countries that matters most. The exact membership of this institution might, in practice, depend on a variety of factors not simply emissions (e.g., see the ranking in Table 4.2). For example, if the G8 occupied this role then the most likely membership would be the "G8 + 5" – that is the five additional countries (Brazil, China, India, Mexico, and South Africa) that have become semi-permanent fixtures of G8 meetings although not yet formal members. That arrangement would tend to elevate Brazil's importance relative to its emissions of carbon dioxide from fossil fuels while excluding other big emitters such as South Korea or Iran. While Brazil's industrial emissions are relatively small due to the large role for hydropower in that country, Brazil is important nonetheless because of its role in tropical forestry.[7] (By the same logic, however, Indonesia and Malaysia are both very important – yet excluded from the G8 + 5 group.) There is probably no magical formula that will instruct the proper membership of the club of greenhouse emitters, but what matters most is that the club be small enough to be functional.

Small size matters because the style of negotiating collective agreements will be very demanding. As with GATT/WTO "rounds," each country could propose its own package of policies and measures that

[7] It should be noted that Brazil's emissions estimates might be revised upward as more is learned about the large emissions of greenhouse gases from tropical dams. For a review of the issues and citations to the main literature, see Cullenward and Victor (2006).

Table 4.2 *Top emitters*

Rank	*Country*	*2002 CO_2 emissions (1,000s of metric tons)*	*% total 2002 world emissions*	*Cumulative % total 2002 world emissions*
1	United States	5,838,118	24	24
2	EU25	3,674,876	15	39
3	China	3,513,103	15	54
4	Russian Federation	1,432,913	6	60
5	India	1,220,926	5	65
6	Japan	1,203,535	5	70
7	Canada	517,157	2	72
8	South Korea	446,190	2	74
9	Mexico	383,671	2	76
10	Iran	360,223	1	77
11	Australia	356,342	1	79
12	South Africa	345,382	1	80
13	Saudi Arabia	340,555	1	81
14	Brazil	313,757	1	83
15	Ukraine	306,807	1	84
16	Indonesia	306,491	1	85
17	Thailand	231,927	1	86
18	Turkey	207,996	1	87
19	Malaysia	151,630	1	88
20	Kazakhstan	147,921	1	88
21	Egypt	143,697	1	89
22	North Korea	143,216	1	89
23	Argentina	133,322	1	90
24	Uzbekistan	122,330	1	91
25	Pakistan	108,677	0	91

would constitute its contribution to the collective effort. In the case of climate change, most countries that are keen to control emissions probably would propose some combination of emission caps or taxes. Some, however, might decide to propose alternatives that are better suited to their regulatory and political environment. The EU, for example, is attempting to meet its Kyoto commitments through a combination of emission caps and trading for the industrial sector and other regulatory instruments for the rest of the economy such as transportation and buildings.

By itself, this style of nominating policy packages is unlikely to be effective because governments might simply nominate what they were planning to implement anyway. Or, they might nominate fanciful ideas that have little chance of real implementation. Whether simply restating the status quo or propagating a fantasy, collective action will not emerge. Thus the process of negotiating packages must include a mechanism for review and scrutiny. There are many precedents that could guide the creation of the needed capacity for the case of climate change. WTO accession talks, for example, involve review of policy packages by the WTO secretariat and also, especially, interested WTO members. That approach, in effect, shifts the requirements for institutional review to the countries that are most interested in and capable of supplying review functions – for WTO accessions that is usually the United States and the European Union. A second approach to creating the needed institutional capacity is to rely on a rotation so that no single country is saddled with the function on a permanent basis. Such a rotation exists in the G8 and in the North Sea and Baltic Sea ministerial conferences – the host country supplies the functions of review. That model has worked quite well for the North Sea ministerial conferences, especially, because all host countries have been motivated and competent. Host country review has been less impressive in the G8 (where it is erratic); it would have failed in the Baltic Sea ministerial conferences if not for an active program to build up the capacity of the east European countries when they hosted the meetings. A third model, however, is probably most appropriate for the climate area: a semi-permanent qualified review staff. That model is followed in the IMF. In the EU, as the agenda has become more complicated the functions of policy review have, de facto, shifted from the rotating EU presidency to the permanent bureaucracy. The standard reasons for a permanent secretariat usually involve the benefits of a reliable and competent staff and the gains that arise from continuity. Those reasons are valid, but there is another that may be even more important: confidence. Countries do not easily submit themselves to allow review of their policies, and the system will fail if those reviews are conducted in a way that does not balance the collective need for serious review with the individual sensitivities of sovereignty. The experiences of the WTO and the North Sea conferences, among many others, show that it is possible to balance those conflicting needs in a way that (with time and confidence) creates a truly effective review mechanism.

The review mechanism would perform at least three major functions. First, it would service the negotiations – making it possible to compare the level of effort implied in disparate policy packages that governments propose. Such comparisons would require analytical support such as energy modeling – much as is done by reviewing countries during WTO accession talks. Second, the review mechanism could play a role in checking to see whether countries are actually implementing what they have promised (and, if not, the reasons for the shortfall and the adequacy of new measures that countries might implement to compensate for areas where they fall short). In international institutions there are weak versions of this review function – among the very weakest are the country reviews being conducted under the UNFCCC at present, but similarly weak reviews include the OECD science policy and environmental policy reviews. This review function might begin in this weak mode because it will be difficult for countries to gain confidence until they see the institution at work. The experience with the OECD science policy and environmental policy reviews suggests that even in the weak mode the external scrutiny can exert a significant influence. A stronger version of this function is seen in the North Sea and Baltic Sea conferences where, in advance of every ministerial meeting, a detailed review of each country's efforts helped to set the scene for discussions about the actions that would be needed next. The IMF reviews offer an example of this function being performed in an even more stringent fashion.

The third function of the review process involves checking the collective consequences of the individual efforts. If all the core countries implement their policy packages, what might be the impact on total emissions and eventually concentrations? The WTO offers a partial model for this function, as the WTO engages in country policy reviews with an eye to the overall effects on the world's trading system. The North Sea and Baltic Sea conferences offer perhaps the best model as such country reviews have been conducted in light of common goals – for example, cutting nitrogen pollution in the North Sea by 50 percent, or eliminating pollution "hot spots" in the Baltic Sea. In some earlier work, colleagues at the International Institute for Applied Systems Analysis (IIASA) and I compiled highly detailed studies of such review experiences; what is most striking is that there are many such examples – especially when looking beyond the simple canon of international environmental law to the slightly more obscure (yet probably more relevant) precedents such as the North Sea and Baltic Sea cooperative efforts (see generally Victor, Raustiala, and Skolnikoff 1998). Such collective reviews will require some goals to use as benchmarks – such as long-term concentrations, rates of change, or emissions. Such goals will be much easier to adopt in a nonbinding framework where countries are less sensitive about formal compliance than in a binding treaty. Indeed, in the North Sea and Baltic Sea cases (and many others), efforts to set meaningful goals through binding treaties failed; once nonbinding instruments were available it was much easier to agree on goals.

My proposal for an architecture is based on this "club" of countries that would negotiate policies through the process of proposal, review, and scrutiny just outlined. In addition, my proposal has two other elements. One element is the need for binding and more-universal agreements as a complement to the club approach. Binding agreements would be used for particular topics where countries have confidence in their ability to implement actions and where binding law has a special value in "locking in" a set of commitments. More universal agreements would be used to extend the basic approaches that emerge from the core club. They would also serve the function, already performed well by the UNFCCC, of providing standards and information that have universal application – for example, procedures and data for emission inventories.

Another major element of my proposal is a new way to engage with developing countries. At root, the problem with developing countries is that their interests vary with those in the industrialized world and yet their participation is essential to the long-term success of any effort to control greenhouse gases. The club approach will make it easier to engage them because smaller groups with a more intense focus on serious policy packages will facilitate the tailoring that could get key developing countries involved. However, clever tailoring is not enough; a mechanism is also needed to address the problem of diverging interests. I have suggested that the approach of compensation is not working; the CDM, fundamentally, is unable to direct compensation to the kinds of activities in developing countries that would have large leverage on emissions in the future.

An alternative approach would focus not on compensating these countries for implementing policies that they don't favor but, instead, finding policies that align with their interests (e.g., Heller and Shukla,

2003). Examples include clean natural gas infrastructures in China, which would help the Chinese address local air pollution problems while also cutting by half the emissions of CO_2 when compared with coal. Other examples include provision of proliferation-resistant nuclear power technologies and fuels to India, which would help shift the Indian economy away from coal and diversify the sources of electric power. Elsewhere, colleagues and I have calculated the huge leverage on carbon emissions that would arise from such an approach (e.g., Victor 2006; Jackson *et al.* 2006).

Such an approach is not the classic "free lunch" or "low-hanging fruit" approach to climate policy, for both these policy examples would require effort. Expanding the use of gas in China will require an accommodation with fuel suppliers – notably Russia. Allowing India to obtain commercial nuclear power technologies will require legislative action to lift export controls in the United States as well as broader international agreement on new fuel cycles. What matters, though, is that the style of these negotiations and the interests involved are radically different from the standard discussions about climate change. The developing countries have a much stronger (and positive) fundamental interest in adopting such policies, and thus the problem is not compensation to get them to do something they otherwise abhor. The key participants in making these arrangements are very different from those who have dominated climate change negotiations so far, as these participants include the investors in new power plants and energy infrastructures, fuel ministries, and the industrial-planning arms of these governments. Climate change negotiations, by contrast, are dominated by foreign ministries and environmental ministries. Each of the deals that would be needed to make these climate-friendly infrastructure investments feasible will have its own characteristics and could be extremely complex to negotiate, and thus this approach would contrast sharply with the CDM. Rather than hundreds of small and marginal projects, this style of engaging developing countries would focus on just a handful of large pivotal actions involving just a few critical countries.

Conclusions

The Kyoto Protocol has faced tough times not just because the problem of global warming is exceptionally difficult to address but also because the architecture chosen was inappropriate for the task at hand. Binding treaties negotiated through processes dominated by diplomats do not offer good prospects for serious cooperation. Such instruments work best when governments can be confident of their ability to comply with their commitments and when it is relatively easy to negotiate those commitments. Yet the climate problem is marked by high uncertainty about the ability to implement promises, especially when promises are codified as emission outputs rather than policy efforts. Universal participation, another maxim of the Kyoto Protocol, is important when leakage is a serious concern. Yet, in practice, these early decades in the global effort to address climate change are marked by the diversity that arises from experimentation and generally low carbon prices – conditions where leakage is not a paramount problem.

I have suggested an alternative approach that is rooted in a small club and is based on a much more intensive negotiating and review process. This club review scheme, such as the "L20" concept proposed by former Canadian Prime Minister Paul Martin, allows for much finer tailoring of policy commitments around the efforts of a few key countries. While the idea is relatively new in the debate about collective action to address climate change, such club approaches have been used effectively in many other areas of international economic policy, such as in the WTO and the IMF, and a few areas of environmental policy, such as in the successful collective efforts to clean up the North Sea and the Baltic Sea.

References

Ausubel, J. H. (1991). "Does Climate Still Matter?" *Nature* 350: 649–652.

Barrett, Scott. (1994). "Self-Enforcing International Environmental Agreements," *Oxford Economic Papers* 46: 878–894.

Broecker, W. (1987). "Unpleasant Surprises in the Greenhouse?" *Nature* 328: 123.

Chayes, A. and A. Chayes (1991). "Compliance without Enforcement: State Behaviour under Regulatory Treaties," *Negotiation Journal* 7(3): 311–330.

Cullenward, Danny, and David G. Victor (2006). "The Dam Debate and Its Discontents," *Climatic Change* 75(1–2): 81–86.

European Gas Markets (2006). "Focus on EU Emissions Trading Scheme," *European Gas Markets*, January 31, 2006.

Hahn, Robert W., and Robert N. Stavins (1999). "What Has Kyoto Wrought? The Real Architecture of International Tradable Permit Markets," Resources for the Future Discussion Paper 99–30, Washington, DC.

Hahn, Robert W., and Gordon L. Hester (1989). "Where Did All the Markets Go? An Analysis of EPA's Emissions Trading Program," *Yale Journal on Regulation* 6: 109–153.

Heller, Tom, and P. R. Shukla (2003). "Development and Climate: Engaging Developing Countries," in *Beyond Kyoto: Advancing the International Effort Against Climate Change*, Arlington VA: Pew Center on Global Climate Change, pp. 111–140.

International Energy Agency (2006). *World Energy Outlook 2006*, Paris: IEA.

ILEX Energy Consulting (2005). "The Environmental Effectiveness of the EU ETS: Analysis of Caps" Final report submitted to WWF, October, UK: ILEX Energy Consulting, available at www.panda.org/downloads/170envteffectivenessav60.pdf.

Jackson, Michael P., Sarah Joy, Thomas C. Heller, and David G. Victor (2006). Greenhouse Gas Implications in Large-scale Infrastructure Investments in Developing Countries: Examples from China and India, Program on Energy and Sustainable Development Working Paper 54, Stanford University.

Keohane, R. O. (1984). *After Hegemony: Cooperation and Discord in the World Political Economy*, Princeton, NJ: Princeton University Press.

Low, Patrick (1992). "International Trade and the Environment," *World Bank Discussion Papers*, No. 159 (January).

Martin, Paul (2005). "A Global Answer to Global Problems," *Foreign Affairs* 84(3): 2.

National Research Council, Committee on Abrupt Climate Change (2003). *Abrupt Climate Change: Inevitable Surprises*, Washington, DC: National Academies Press.

Nicholls, Mark (2005). "A Permanent Revolution," *Environmental Finance* (November): S18–S20.

Oppenheimer, M. (1998). "Global Warming and the Stability of the West Antarctic Ice Sheet," *Nature* 393: 325–332.

Pizer, William A. (1998). "Prices vs. Quantities Revisited: The Case of Climate Change," Resources for the Future Discussion Paper 98–02 (revised), Washington, DC.

Roberts, Marc J., and Michael Spence. (1976). "Effluent Charges and Licenses under Uncertainty," *Journal of Public Economics* 5: 193–208.

Roginko, Alexei. (1998). "Domestic Implementation of Baltic Sea Pollution Commitments in Russia and the Baltic States," in Victor, Raustiala, and Skolnikoff (eds.), *The Implementation and Effectiveness of International Environmental Commitments*, Cambridge, MA: MIT Press, pp. 575–638.

Schelling, Thomas C. (1992). "Some Economics of Global Warming," American *Economic Review* 82(1): 1–14.

Skjærseth, Jon Birger (1998). "The Making and Implementation of North Sea Commitments: The Politics of Environmental Participation," in Victor, Raustiala, and Skolnikoff (eds.), *The Implementation and Effectiveness of International Environmental Commitments*, Cambridge, MA: MIT Press, pp. 327–380.

Slaughter, Anne-Marie (1992). "Liberal States: A Zone of Law," paper presented at the 1992 Annual Meeting of the American Political Science Association, Chicago, IL.

Victor, David G. (2001). *Collapse of the Kyoto Protocol and the Struggle to Slow Global Warming*, Princeton, NJ: Princeton University Press.

(2004). *Climate Change: Debating America's Policy Options*, New York: Council on Foreign Relations.

(2006). "The India Nuclear Deal: Implications for Global Climate Change," Testimony before the US Senate Committee on Energy and Natural Resources, July 18.

Victor, David G., and Lesley Coben (2005). "A Herd Mentality in the Design of International Environmental Agreements?" *Global Environmental Politics* 5: 24–57.

Victor, David G., and Joshua C. House (2004). "A New Currency: Climate Change and Carbon Credits," *Harvard International Review* (Summer): 56–59.

Victor, David G., Joshua C. House, and Sarah Joy (2005). "A Madisonian Approach to Climate Policy," *Science* 309: 1820–1821.

Victor, David G., Kal Raustiala, Eugene B. Skolnikoff, eds. (1998). *The Implementation and Effectiveness of International Environmental Commitments*, Cambridge, MA: MIT Press.

Wara, Michael (2006). "Measuring the Clean Development Mechanism's Performance and Potential," Working Paper, available at www.pesd.stanford.edu.

Weitzman, Martin L. (1974). "Prices vs. Quantities," *Review of Economic Studies*, 41: 477–491.

Wettestad, Jørgen (1998). "Participation in NO_X Policy Making and Implementation in the Netherlands, UK, and Norway: Different Approaches but Similar Results?" in Victor, Raustiala, and Skolnikoff (eds.), *The Implementation and Effectiveness of International Environmental Commitments*, Cambridge, MA: MIT Press, pp. 381–430.

Wildavsky, Aaron (1988). *Searching for Safety*. New Brunswick, NJ: Transaction Books.

Young, Oran R. (1989). *International Cooperation: Building Regimes for Natural Resources and the Environment*. Ithaca, NY: Cornell University Press.

Commentaries on Victor

4.1 *Incentives and institutions: a bottom-up approach to climate policy*

CARLO CARRARO

Incentives and institutions. These are the two cornerstones of David Victor's analysis. Incentives and institutions have often been neglected in the study of climate policy, which mostly focused on concepts like optimality or cost-effectiveness. Most economic analyses of climate policies have indeed focused on a global target and on the optimal or cost-effective way to achieve it, but they largely neglected the incentives for negotiating countries to agree on the target, disregarded the policy instruments that can provide these incentives, and did not pay adequate attention to the institutions that are necessary to implement the policy instruments and enforce the target.

The paper by David Victor fills this gap. It carefully analyses both incentives and institutions, first in relation to the Kyoto Protocol, then to identify some features of an alternative policy architecture. As for the Kyoto Protocol, Victor emphasizes the weakness of international institutions that should implement the Protocol rules and achieve the Protocol targets. He correctly criticizes the Protocol's objective of achieving universal participation, its focus on binding targets, and the lack of effective measures to get developing countries involved in the cooperative effort to control climate change.

However, the main contribution of David Victor's paper is not his analysis of the Kyoto Protocol. His main lesson is that the design of any future agreement on climate policy must start from analyzing each country's incentives to participate in the agreement, and then move to identify policy instruments and institutions that provide adequate incentives to reluctant countries.[1]

He draws this lesson from a careful analysis of recent events in international climate negotiations and domestic policymaking. His

[1] This point is not new (refer, for example, to Chapter 10 in the IPCC's [2001] Third Assessment Report), but is not yet adequately considered in many policy analyses and in actual policymaking.

main conclusions are not new. Similar results have been achieved in the recent theoretical literature on international environmental agreements (Carraro and Siniscalco 1993; Barrett 1994; see Carraro and Marchiori 2003 for a survey). However, rather than using a general and sometimes unrealistic game-theoretic framework, Victor develops his arguments by looking at the reality of existing policy institutions and markets. From this viewpoint his analysis is novel, well rooted, and more convincing than previous ones. In addition, many features of the new policy architecture that he proposes are also interesting and original.

Nevertheless, to understand the importance and the robustness of David Victor's analysis, let me first summarize the main motivations and results of the game-theoretic literature. This would also help to highlight the similarities and differences between theoretical works and David Victor's contribution.

Among the transnational policy issues, environmental protection is a limiting case. In areas such as global warming, ozone layer depletion and biodiversity, spillovers, as well as the absence of clear property rights, create strong incentives to free ride, which undermine cooperation. Hence, the difficulty of reaching agreements that are both effective and widely accepted.

The above problems are not new to economists and have been analyzed in the area of externalities and public goods. What is new is the context in which these problems take place. Currently, climate change control is managed as a global common property good, but there is no institution which possesses powers to regulate it by means of supranational legislation, economic instruments, or by imposing a system of global property rights. Hence, the necessity to design negotiation mechanisms leading to self-enforcing agreements, i.e., agreements to control climate change which are voluntarily signed by a "large" group of countries (large enough to keep climate change under control).

Economists are quite skeptical about the possibility of achieving self-enforcing agreements on climate change. Early contributions (Hardin and Baden 1977) would have characterized the climate control game among countries as a prisoner's dilemma, inevitably leading to the so-called "tragedy" of the common property goods. But in the real world, a large number of international environmental agreements on the commons have been signed, often involving subgroups of negotiating countries and sometimes involving transfers and links to other policies (trade, technological cooperation, etc.). This is why the "prisoner's dilemma" approach is unsatisfactory and new conceptual models have been developed, which proved more helpful in understanding the logic of international cooperation in the presence of positive spillovers. These new models were developed in the last decade within a non-cooperative game-theoretic framework and provide interesting indications on the likely outcomes of climate negotiations (Carraro and Marchiori 2003; Finus and Rundshagen 2003 for surveys of the literature).

The main results of this game-theoretic literature can be summarized as follows:

- The presence of asymmetries across countries and the incentive to free ride make the existence of global self-enforcing agreements, i.e., agreements which are profitable to all countries and stable, quite unlikely (Carraro and Siniscalco 1993; Barrett 1994).
- When self-enforcing international environmental agreements exist, they are signed by a limited number of countries (Hoel 1992, 1994; Barrett 1994).
- When the number of signatories is large, the difference between the cooperative behavior adopted by the group of signatories and the non-cooperative one is very small (Barrett 1997).
- The grand coalition, in which all countries sign the same environmental agreement, is unlikely to be an equilibrium (Finus and Rundshagen 2003).
- The equilibrium coalition structure is not formed by a single coalition (a single group of signatories). In general, more than one coalition forms at the equilibrium (Bloch 1997; Yi 1997).
- Coalitions of different sizes may emerge at the equilibrium, even when countries are symmetric (Ray and Vohra 1997; Carraro and Marchiori 2003; Yi 1997). [2]

The lesson that can be drawn from these results can be phrased as follows. A global agreement is unlikely to be signed by all the relevant countries. Several parallel agreements are going to emerge over time.

[2] The specific results on the size of the coalitions depend on the model structure and in particular on the slope of countries' reaction functions, i.e., on the presence of leakage. If there is no leakage and countries are symmetric, then the Nash equilibrium of the multi-coalition game is characterized by many small coalitions, each one satisfying the properties of internal and external stability (Carraro and Marchiori 2003).

Domestic measures and/or policies implemented by small groups of countries are going to be adopted to control climate change.

These predictions are consistent with David Victor's ones. For example, he writes: "The integrated international market for emissions . . . has not yet materialized; instead at least six different carbon markets have emerged," and "Too much attention has focused on global institutions and not enough on the more diverse national and regional bodies." For the future, he sees "An extensive use of non-binding agreements among smaller groups of countries and allowance for fragmented emission trading systems." And also: "The future landscape will be dominated by multiple (permit) prices."

Similarly to Victor, several authors working within a game-theoretic framework have used their results to question the design of the Kyoto Protocol (Carraro 1998; Bloch 2003; Finus and Rundshagen 2003; Yi 2003; Bretteville, Hovi, and Menz 2004; Buchner and Carraro, forthcoming). They argue that the Kyoto Protocol is unlikely to be signed by all relevant players and that the emergence of alternative, parallel climate blocs is likely. Some indications that multiple regional or sub-global climate blocs could be the appropriate way to address the difficulties emerging in climate negotiations can also be found in the political science literature (see, for example, Egenhofer and Legge 2001; Egenhofer, Hager, and Legge 2001; Stewart and Wiener 2003; Reinstein 2004; Carraro and Egenhofer, forthcoming).

The basic idea in all these contributions is that a bottom-up, country-driven approach to defining national commitments should be adopted. Instead of top-down, global negotiations on national emission targets, each country or group of countries would determine its contribution to a cooperative effort to curb greenhouse gases (GHGs) and choose the partners with whom it intends to cooperate. In a process analogous to trade negotiations, each country would put its offer of commitments on the negotiating table and invite proposals from other countries for similar commitments.

A fragmented climate regime characterized by the formation of climate blocs (regional coalitions for example) would then emerge in much the same way as is now emerging in trade negotiations. This should not be surprising. As Victor correctly says, in substance, even though not in form, the Kyoto Protocol already reflects agreements among several different coalitions. It incorporates special provisions for several different groups of countries. The Non-Annex B countries have no commitments and can benefit from emission reduction investments through the Clean Development Mechanism (CDM). The most vulnerable Non-Annex B countries can also receive financial assistance for adaptation from the levy imposed on the CDM (and possibly on the other mechanisms). The European Union has the ability under Article 4 to redistribute the emission reduction burden. Australia had obtained/negotiated a special provision on land use emissions in Article 3.7.[3]

In addition, the lesson that can be derived from trade negotiations consistently tells us that progress on trade liberalization can be achieved mostly through regional agreements, at least in the coming years.[4] In international trade, the "resurgence" of regionalism has thus become a crucial subject, underscored by the formation of competing customs unions and the debate about free trade areas. Substantial attention has been focused on the efficiency and implications of these regional or sub-global cooperations (Krugman 1991; Baldwin 1993; Casella 1995; Bond and Syropoulos 1996; Yi 1996a, 1996b, and 1998; Bloch and Ferrer 1999).

In particular, several authors have pointed out that Regional Trade Agreements (RTAs) may seem to be contradictory, but they can often actually support the WTO's multilateral trading system (Sampson and Woolcock 2003). Regional agreements have allowed groups of countries to negotiate rules and commitments that go beyond what was previously possible multilaterally. In turn, some of these rules have paved the way for agreements within the WTO. Services, intellectual property, environmental standards, investment and competition policies are all issues that were raised in regional negotiations and later developed

[3] As has been stressed by Egenhofer and Legge (2001), "it is increasingly becoming clear, [that] the Kyoto Protocol is less a global agreement than a set of differing regional approaches."

[4] The strong increase in the number of trade bloc agreements registered with the World Trade Organisation is discussed in Tjornhom (2000) and Boonekamp (2003). Some 250 regional trade agreements (RTAs) have been notified to the GATT/WTO up to December 2002, of which 130 were notified after January 1995. About 200 RTAs are currently in force. An additional 70 to 100 are estimated to be operational although not yet notified. RTAs, which includes bilateral free trade agreements between countries that are not in the same region, have become so widespread that all but one WTO member are now parties to one or more of them. Indeed, as of August 2006, all 146 WTO Members, with the exception of Mongolia, participate in or are actively negotiating regional trade agreements.

into agreements or topics of discussion in the WTO.[5] For these reasons, on 6 February 1996, the WTO General Council created the Regional Trade Agreements Committee. Its purpose is to examine regional groups and to assess whether they are consistent with WTO rules. The committee is also examining how regional arrangements might affect the multilateral trading system, and what the relationship between regional and multilateral arrangements might be.

A similar process may be envisaged for the case of climate change control. Some domestic or regional initiatives to reduce GHG emissions today may pave the way to a global agreement tomorrow.

The parallelism with trade negotiations is used also by David Victor to claim that a new policy architecture, no longer based on binding agreements and on a global target, would be more effective in reducing global GHG emissions. The basic ingredients of this policy architecture would be as follows:

- The coordination of a variety of efforts. Countries would agree on things to do rather than on emission reduction targets.
- A variable geometry of participation. Some countries would agree on more efforts than others.
- A sufficient accountability system to ensure that commitments become connected to action.

How can this be achieved? David Victor does not neglect the negotiation process and the related institutions. He emphasizes how more and better cooperation on GHG emission control could be achieved by limiting the number of negotiating countries to the most important ones (e.g., the twenty top polluters). He also highlights the importance of issue linkage and transfers (through economic cooperation). And the necessity of a mechanism for review and scrutiny.

[5] The groupings that are important for the WTO are those that abolish or reduce barriers to trade within the group. The WTO agreements recognize that regional arrangements and closer economic integration can benefit countries. It also recognizes that under some circumstances regional trading arrangements could hurt the trade interests of other countries. Normally, setting up a customs union or free trade area would violate the WTO's principle of equal treatment for all trading partners ("most-favoured-nation"). But GATT's Article 24 allows regional trading arrangements to be set up as a special exception, provided certain strict criteria are met. In particular, the arrangements should help trade flow more freely among the countries in the group without barriers being raised on trade with the outside world. In other words, regional integration should complement the multilateral trading system and not threaten it.

Again all this is not completely new. There is, for example, a large literature on both issue linkage and transfers (among many others, two recent contributions are Buchner *et al.* 2005, and Carraro, Eyckmans, and Finus 2006) that highlights under what conditions these mechanisms can enhance participation incentives and lead more countries to agree on GHG emission abatement.

What is new in David Victor's paper is the careful analysis of what these mechanisms could be in practice. For example, how can scrutiny be designed or where economic cooperation can effectively reduce GHG emissions (e.g., by building natural gas infrastructures in China). What David Victor does is to give substance and realism to ideas and results that were developed via mathematical models and therefore were unable to deal with the details of the actual policy process. We have seen that the theoretical economic literature supports David Victor's analysis. But the other way around is also true. David Victor's analysis gives meaning and appeal to the theoretical results.

It is also important to stress the relevance that Victor gives to the role of institutions. A bottom-up approach is to be favored not only because the underlying participation incentives inevitably leads to a fragmented climate policy regime, but also because the institutions which are capable of implementing an effective climate policy do not yet exist at the international level but sometimes exist at the domestic and regional level. Therefore, a club approach, in which cooperation takes place on specific dimensions where (a few) participating countries have institutions that guarantee compliance and effectiveness, becomes the appropriate one.

There are three elements of David Victor's proposal that deserve additional scrutiny. It is clear that in terms of incentives and institutions a bottom-up approach is the only one with chances to succeed in curbing GHG emissions. A set of coordinated efforts with a variable participation geometry is likely to be the future of climate policy. However, by relying only on self-interests, this approach may not succeed in achieving the large emission reductions that most scientists believe to be necessary to control climate change. Is there a way of assessing the amount of emission reduction that a bottom-up approach is likely to achieve in the next fifty years?

A second element concerns equity and burden sharing. If countries agree on different sets of efforts, how can the costs of these efforts be assessed and compared? For example, is there a way to claim that the effort to develop new energy technologies (e.g., in the United States) is larger or costs more than the effort to replace coal power plants (e.g., in Germany or China) or the effort to accept higher temperatures at home and at work in the summer (as recently suggested by the Japanese Minister of the Environment)? Is there a metric of efforts?

Finally, David Victor's policy architecture does not mention adaptation. Adaptation, whatever we do to reduce GHG emissions and whatever the anthropic influence on climate change, will be needed. The related large investments are likely to crowd out other investments, including those to reduce emissions. Is there a link between mitigation and adaptation in David Victor's proposal? Can the set of coordinated efforts proposed by Victor also include the effort to adapt to climate change? Given the policymakers' discount rate, the costs of adaptation (to be paid far in the future) is likely to be smaller than the costs of mitigation (to be paid in the coming years), unless some catastrophic impacts of climate change are expected. Does this mean that a coordination of efforts to adapt our economic systems and lifestyles to climate change will crowd out most efforts to reduce GHG emissions?

These three questions are relevant for all policy architectures and not only for the one proposed by Victor. The answer to these questions is a necessary complement to all policy proposals on GHG mitigation. And the three questions are strictly interlinked. For example, if countries' incentives and institutions lead to a set of regional or sub-global agreements that are unable to provide a sufficient amount of emission abatement, then more adaptation efforts would be necessary. Are there incentives and institutions to provide the right amount of adaptation investments? As another example, it is well known that some countries are more vulnerable than others to climate change. How can the costs of adapting to climate change be distributed in a fair and acceptable way among world countries? Is an adaptation fund the appropriate answer to this question? How should contributions to the adaptation fund be designed?

Let me conclude by summarizing my own version of David Victor's policy architecture (which is also my own favored policy architecture). Let me start from the number of negotiating countries: as recently proposed by the Canadian Prime Minister, twenty is probably the right number, but a slightly smaller number would also be appropriate. Then, the issues on which these countries will negotiate have to be defined. Technological cooperation, climate-related trade rules, carbon taxation, carbon sinks, contribution to a global adaptation fund, forestry preservation, biofuels, development aid, energy infrastructures are some examples. For each issue a number of countries, not necessarily the same countries, not necessarily the same number, decide to cooperate. This decision implies both that some measures are adopted in each country (or small group of countries) to achieve a common issue-specific objective, and that a system of monitoring and enforcement is established amongst the signatory countries. A regular verification process of how different measures are implemented in each country or group of countries (and of their impacts and costs) can also be agreed upon. Regular meetings can be organized to update the set of countries cooperating on each issue. Whenever the verification process identifies an insufficient effectiveness of the adopted measures or an unequal sharing of the burden of controlling climate change, new measures or a new distribution of existing measures have to be negotiated. Given that all measures are implemented domestically or within a bilateral or regional cooperative setting (e.g., the European Union or Mercosur), the establishment of new global or supranational institutions would not be necessary.

This framework implies that climate change is no longer an environmental problem to be dealt with exclusively through specific environmental policy measures. It is a global economic problem to be dealt with by global economic policy measures. Whatever the issue that world leaders are going to analyze and discuss, climate change should be a dimension of their own analysis and discussion.

References

Baldwin, R. (1993). "A Domino Theory of Regionalism," NBER Working Paper No. 4465.

Barrett, S. (1994). "Self-Enforcing International Environmental Agreements," *Oxford Economic Papers* 46: 878–894.

(1997). "Towards a Theory of International Co-operation," in C. Carraro and D. Siniscalco (eds.), *New Directions in the Economic Theory of the Environment*, Cambridge: Cambridge University Press, pp. 239–280.

Bloch, F. (1997). "Non-cooperative Models of Coalition Formation in Games with Spillovers," in C. Carraro and D. Siniscalco (eds.), *New Directions in the Economic Theory of the Environment*, Cambridge: Cambridge University Press, pp. 311–352.

(2003). "Non-cooperative Models of Coalition Formation in Games with Spillovers," in C. Carraro (ed.), *The Endogenous Formation of Eeonomic Coalitions*, Cheltenham: Edward Elgar, pp. 35–79.

Bloch, F., and H. Ferrer (1999). "Trade Fragmentation and Coordination in Bilateral Oligopolies," CORE Discussion Paper No. 9908.

Bond, E., and C. Syropoulos (1996). "The Size of Trading Blocs, Market Power and World Welfare Effects," *Journal of International Economics* 40: 417–437.

Boonekamp, C. (2003). "The Changing Landscape of RTAs, WTO Secretariat, Trade Policies Review Division, prepared for the Seminar on 'Regional Trade Agreements and the WTO'," WTO Secretariat, Geneva, 14 November 2003.

Bretteville, C., J. Hovi, and F. C. Menz (2004). "Regional versus Global Cooperation for Climate Control," mimeo, Cicero, Oslo, Norway.

Buchner, B., and C. Carraro (forthcoming). "Parallel Climate Blocs: Incentives to Cooperation in International Climate Negotiations," in R. Guesnerie and H. Tulkens (eds.), *The Design of Climate Policy*, Cambridge, MA: MIT Press.

Buchner, B., C. Carraro, I. Cersosimo, and C. Marchiori (2005). "Back to Kyoto? US Participation and the Linkage between R&D and Climate Cooperation," in A. Haurie and L. Viguier (eds.), *Coupling Climate and Economic Dynamics*, Dordrecht: Kluwer Academic Publishers, pp. 173–204.

Carraro, C. (1998). "Beyond Kyoto: A Game-Theoretic Perspective," paper presented at the OECD Experts Workshop on *Climate Change and Economic Modelling: Background Analysis for the Kyoto Protocol*, Paris, September 17–18.

(2005). "Institution Design for Managing Global Commons," in G. Demange, D. Ray and M. Wooders (eds.), *Group Formation in Economics. Networks, Clubs and Coalitions*, Cambridge: Cambridge University Press, p. 354–380.

Carraro, C., and C. Egenhofer, eds. (forthcoming). *Climate and Trade Policy: Bottom-up Approaches towards Global Agreement*, Cheltenham: Edward Elgar.

Carraro, C., J. Eyckmans, and M. Finus (2006). "Optimal Transfers and Participation Decisions in International Environmental Agreements," *Review of International Organizations* 1(4): 379–396.

Carraro, C., and C. Marchiori (2003). "Stable Coalitions," in C. Carraro (ed.), *The Endogenous Formation of Economic Coalitions*, Cheltenham: Edward Elgar, pp. 156–198.

Carraro, C., and D. Siniscalco (1993). "Strategies for the International Protection of the Environment," *Journal of Public Economics* 52: 309–328.

(1998). "International Environmental Agreements: Incentives and Political Economy," *European Economic Review* 42: 561–572.

Casella, A. (1995). "Large Countries, Small Countries and the Enlargement of Trade Blocs," National Bureau of Economic Research Working Paper No. 5365.

Demange, G., D. Ray, and M. Wooders, eds. (2005). *Group Formation in Economics: Networks, Clubs and Coalitions*, Cambridge: Cambridge University Press.

Egenhofer, C., and T. Legge (2001). "After Marrakech: The Regionalisation of the Kyoto Protocol," CEPS Commentary.

Egenhofer, C., W. Hager, and T. Legge (2001). "Defining Europe's Near Abroad in Climate Change: A Russian–EU Alliance – Sub-global Bargaining to Further International Environmental Agreements," CEPS Discussion Paper.

Finus, M., and B. Rundshagen (2003). "Endogenous Coalition Formation in Global Pollution Control: A Partition Function Approach," in C. Carraro (ed.), *The Endogenous Formation of Economic Coalitions*, Cheltenham: Edward Elgar, pp. 199–244.

Harden, Garrett, and John Baden (1977). *Managing the Commons*, New York: Freeman and Co.

Heal, G. (1994). "The Formation of Environmental Coalitions," in C., Carraro (ed.), *Trade, Innovation, Environment*, Dordrecht: Kluwer Academic Publisher, pp. 301–322.

Hoel, M. (1992). "International Environmental Conventions: The Case of Uniform Reductions of Emissions," *Environmental and Resource Economics* 2: 141–159.

(1994). "Efficient Climate Policy in the Presence of Free-Riders," *Journal of Environmental Economics and Management* 27: 259–274.

IPCC (2001). *Climate Change 2001: Mitigation*, Cambridge: Cambridge University Press.

Krugman, P. (1991). "Is Bilaterism Bad?" in E. Helpman and A. Razin (eds.), *International Trade and Trade Policy*, Cambridge, MA: MIT Press, pp. 9–23.

Ray, D., and R. Vohra (1997). "Equilibrium Binding Agreements," *Journal of Economic Theory* 73: 30–78.

(1999). "A Theory of Endogenous Coalition Structures," *Games and Economic Behavior* 26: 286–336.

Reinstein, R. A. (2004). "A Possible Way Forward on Climate Change," mimeo, Reinstein & Associates Inc.

Sampson, G., and S. Woolcock (2003). *Regionalism, Multilateralism, and Economic Integration: The Recent Experience*, Tokyo: United Nations University Press.

Stewart, R., and J. Wiener (2003). *Reconstructing Climate Policy*, Washington, DC: American Enterprise Institute Press.

Taylor, M. (1987). *The Possibility of Cooperation*, Cambridge: Cambridge University Press.

Tjornhom, J. (2000). "Dynamic Trade Bloc Formation: Building Blocs or Stumbling Blocs?" University of Minnesota, Department of Economics Working Paper, November version.

Yi, S.-S. (1996a). "Endogenous Formation of Customs Unions under Imperfect Competition: Open Regionalism is Good," *Journal of International Economics* 41: 153–177.

(1996b). "Open Regionalism and World Welfare," *Eastern Economic Journal* 22(4): 467–475.

(1997). "Stable Coalition Structures with Externalities," *Games and Economic Behaviour* 20: 201–223.

(1998). "Free Trade Areas and Welfare: An Equilibrium Analysis," Dartmouth College, Department of Economics.

(2003). "Endogenous Formation of Economic Coalitions: A Survey of the Partition Function Approach," in C. Carraro (ed.), *The Endogenous Formation of Economic Coalitions*, Cheltenham: Edward Elgar, pp. 80–127.

4.2 *The whole and the sum of its parts*

SHEILA M. OLMSTEAD

David Victor's contribution to this volume is an excellent treatment of the barriers to broad, collective action on climate change and some of the more subtle advantages of a smaller-scale, more fragmented approach. My commentary begins with a summary of Victor's criticisms of the framework offered by the Kyoto Protocol, followed by my reflections on these issues. I then outline his parameters for a new climate policy architecture and offer some critiques of his proposed approach. I conclude with a summary of the key components of a post-Kyoto climate policy architecture from my own perspective, and some contrasts of this perspective with Victor's.

Victor's assessment of problems with the Kyoto framework

The Kyoto Protocol came into force in February 2005. Its impacts on greenhouse gas (GHG) emissions will be trivial, but scientific (Watson 2001) and economic (Kolstad and Toman 2005) analyses suggest the need for a credible international approach. Kyoto's targets apply only to the short term (2008–2012) and only to industrialized nations, thus it will impose relatively high costs for the modest short-term benefits it offers (Aldy, Barrett, and Stavins 2003). Most economists see the agreement as deeply flawed (Cooper 1998; McKibbin and Wilcoxen 2002; Olmstead and Stavins 2006), and almost all agree that the Kyoto Protocol is not sufficient to the overall challenge of climate change.

Victor joins the chorus of analysts criticizing the Kyoto framework, as he has in previous work (Victor 2001). His critique in this volume focuses on four points: (1) universal participation; (2) the focus on targets and timetables; (3) emissions trading; and (4) compensation for developing countries.

Regarding the first point, he disapproves of the conventional wisdom that broad participation creates legitimacy. He also suggests that while, in theory, universal participation can avert "leakage," the migration of carbon-intensive industry from countries in which GHG emissions are regulated to those in which they are not, in practice this is an insignificant concern, in comparison to the costs of achieving universality. On the second point, Victor reminds us that so-called "binding" international agreements (like those with targets and timetables for GHG emissions reduction) are usually crafted to assure compliance, often codifying efforts already under way and in the interest of participating countries, and that such agreements are unlikely to truly "bind" for difficult, costly actions. On emission trading, Victor is concerned that any trading system will be plagued by (1) the padding of initial emissions allocations, reducing the ability to achieve meaningful progress; and (2) the inevitable long delays between agreement and implementation, leading to forecasting errors that will magnify controversial financial flows that will occur once trading begins. Finally, Victor's assessment of the Clean Development Mechanism (CDM), Kyoto's framework for compensation to developing countries, is strongly negative. I address each of these points in the paragraphs that follow.

The critique of universal participation

Victor's skepticism regarding the importance of universal participation for the legitimacy of an international climate policy architecture is well founded. The process initiated by the Rio Summit in 1992 and continuing through the Kyoto Protocol's negotiation and ongoing implementation has been nothing if not inclusive – as of April 2006, 163 countries had ratified the Protocol. Nonetheless, as Victor points out, the Protocol is deeply flawed. The world's largest contributor to the current stock of atmospheric carbon, the United States, is not a party, and some of the most important participants from the perspective of future emissions (countries like China, India, and Brazil) are not required to reduce emissions under the current agreement. In the context of regulating GHG emissions, some countries do matter more than others. To support his point, Victor draws on other examples of relatively successful international institutions that historically have been much more exclusive than current climate efforts (such as the GATT, IMF, and G8). This is an important point, and one that has not been made often enough in the ongoing academic and political discussions of climate policy alternatives.

Although universal participation may not be desirable, broad participation by major industrialized and developing countries is essential to address this problem effectively and efficiently (Olmstead and Stavins 2006). While on an ethical basis, it can be argued that industrialized countries should take the first steps, there are two major reasons why key developing countries must participate: opportunities for low-cost emission reductions; and the possibility that nonparticipating developing countries will shift to more carbon-intensive growth paths, increasing the costs of their joining an agreement later. This second phenomenon, leakage, is an issue upon which Victor and I disagree.

On the issue of leakage, Victor suggests that empirical analysis does not support the idea that stringent emission regulations in some countries will alter international trade patterns, resulting in the export of emission-intensive industries to countries without stringent regulations (the so-called "pollution havens hypothesis"). It is true that comprehensive literature surveys through the mid-1990s found no evidence for impacts of environmental regulations on trade patterns, although studies certainly established that such regulations were costly (Jaffe *et al.* 1995; Levinson 1996). Nonetheless, the underlying theory and intuition are strong, and considerable further exploration of this issue has established some empirical evidence for the phenomenon.

Tests of the theory are complicated by unobservable heterogeneity and endogeneity (Brunnermeier and Levinson 2004). Unobserved industry and country traits can be correlated with the likelihood of regulation and the export of pollution-intensive goods. Environmental regulation often appears in high-income countries with substantial international trade – does trade influence pollution regulation, or vice versa? Recent studies accounting for these statistical problems have established pollution haven effects where trade is between high and low-standard countries, and industries are mobile (Ederington, Levinson, and Minier 2005).[1] The relevant empirical question in this context is whether GHG-intensive industries under meaningful international climate policies fit this description. To the degree that they do, we should worry about what will happen to the carbon intensity of outlying economies if nonuniversal climate policy coalitions do arise.

[1] Small effects have been found in trade among US states, as well (Levinson 2003).

In a related point, Victor notes in support of his argument regarding leakage that the current differences in carbon costs between the United States and the European Union (EU) have not affected industrial location decisions. It is too early to judge the effects of EU emission trading in this regard, as trading has occurred only since January 2005. The extent to which differences in EU–US carbon costs will affect trade flows and carbon intensity depends on expected prices over the next ten or twenty years for firms considering long-lived capital investments. In addition, this is simply not a good example of the types of trade flows with which those interested in climate policy ought to be concerned – those between industrialized and developing countries. Even more important than contemporary shifts in emissions are the effects of leakage on future carbon intensity in developing countries, given the long life of capital investments in manufacturing processes, for example. It would be a mistake to conclude that, because leakage is not occurring due to the current, very young, scattered regimes in developed countries, it will be an unimportant phenomenon under future climate policy scenarios.

The critique of targets and timetables

Victor's criticism of Kyoto's designation of targets and timetables is one of the strongest points in his paper. He reminds us of the very serious collective action problem faced by any attempt at addressing climate change on a global scale. If taking action to address climate change is unduly costly for individual countries, they will not commit to taking action due to the possibility of free riding; if countries do commit, anticipating low costs, they will abandon their commitments if costs escalate. Victor offers a number of useful examples to support his point that there really is no such thing as a binding international agreement that involves substantial sacrifice on the part of individual countries with widely varying interests, in the name of a global public good. This theme arises in other chapters of this volume, as well.

The critiques of emissions trading and compensation for developing countries

There are strong links between Victor's criticisms of emission trading and of regimes for compensation of developing countries. He opposes the concept of international emission trading on many grounds. I agree that participating countries in any international climate policy regime have strong incentives to pad their initial emission allocations, eating away at meaningful commitments. In another point, Victor notes that the internal characteristics of states (particularly those with non-market economies) will affect trading, pushing the eventual allocation of emission reductions away from the cost-minimizing allocation. This is undoubtedly true – market-based pollution control policies rarely result in the theoretical ideal of cost-minimization in real-world application (Hahn and Stavins 1992). But they virtually always result in very substantial cost reductions, in comparison to non-market approaches, and the same would be expected of international GHG emissions trading, were it to occur on a large scale.[2]

The cost of achieving significant GHG emissions reductions is a very substantial barrier to high-cost country participation, unless the countries with high and low abatement costs are engaged simultaneously in a climate policy regime. Without some participation by low-cost countries, there is no agreeable way for high-cost countries to pay for reductions where they are needed (and can be afforded) most. This point is illustrated most effectively in the 1997 Byrd–Hagel resolution, in which the US Senate declared in a vote of 95–0 its unwillingness to consent to US participation in an international climate agreement that did not involve, essentially, targets and timetables for developing as well as industrialized countries. Market-based approaches, including emission trading and carbon taxes, may have all the weaknesses that Victor highlights, but they are still the least-cost substantive alternatives for engaging both high- and low-cost countries simultaneously in GHG emission reduction.

Many successful emissions trading programs have been developed in the United States. Tradable permits were used to phase out leaded gasoline in the 1980s, at a savings of more than $250 million per year over an equivalent prescriptive approach (Stavins 2003). The active market in permits for sulfur dioxide (SO_2) emissions from US power

[2] Even trading solely among industrialized countries was predicted to have lowered US compliance costs with a (now hypothetical) Kyoto commitment by 50 percent; participation by the major developing countries would have increased that figure to 75 percent (Edmonds *et al.* 1997).

plants saves $1 billion annually (Ellerman *et al.* 2000). The upstream lead-rights system in the first example is a better model for climate change than the downstream SO_2-trading regime. For some countries, systems of domestic carbon taxes may be more appealing than emissions trading. A hybrid approach, in which tradable permits are paired with a government promise to sell additional permits at a stated price, acting as a "safety valve" or cap on eventual permit prices, may also be promising (McKibbin and Wilcoxen 2002; Pizer 2002). Rather than dismissing market-based approaches, it would be useful to discuss the relative appropriateness of these different types of market-based instruments.

Victor's critique of compensation for developing countries amounts to a critique of the CDM. Indeed, few economists would consider the CDM a "success" relative to the possibility of large-scale trading among today's largest emitters and tomorrow's. The small contribution of the CDM to potential emission reductions may be due, in part, to its weaknesses relative to real emissions trading – the issues of accountability and liability, and many others. Compensation to developing countries in the form of generous emissions trading allocations has, in its favor, at least the benefit of lowering other trading participants' costs. Victor suggests financial flows of another flavor in his proposal for a new architecture, which I will address below. If such flows lack the cost-reducing benefits of compensation through emissions allocations, they will likely be politically infeasible in the United States, and in many other countries as well.

Victor's proposal: a new climate architecture

In place of the Kyoto framework, Victor offers a new international climate policy architecture with three main components. First, a successful new architecture will have what Victor calls a "variable geometry of participation" – only selected countries will participate, to varying degrees, and the most serious negotiations will "concentrate on countries whose participation matters most." Second, the architecture will comprise a variety of efforts, rather than one integrated system. Within this framework, one or more mechanisms will "entice the participation" of key developing countries. Finally, a successful climate policy architecture will involve both ambitious commitments and accountability on the part of participating countries.

Accommodating varied national interests

In this contribution and other recent papers on the topic, Victor argues that the current, fragmented efforts at climate change mitigation have arisen for two main reasons. First, there is wide variation in national interests regarding climate change and its mitigation. Second, the most capable institutions for addressing this challenge are nation-states; thus efforts arise at the national level and vary with interests.

Victor maintains that the "bottom-up" nature of this process has significant merits, and that it can build some of the institutions and commitments that are essential predecessors to an eventual larger-scale approach (Victor, House, and Joy 2005). He offers the evolution of the GATT/WTO and the European Union (EU) as examples of the direction in which current national and regional efforts on climate policy may lead. Tom Schelling, in his contribution to this volume, offers some additional historical examples. The key component of this evolution, in Victor's estimation, will be groups of nation-states acting out of mutual interest, with efforts over time becoming "stitched together into a more integrated system."

Engaging developing countries

Mindful of the problem of divergent national interests, Victor suggests an alternative to the CDM for the engagement of developing countries in a climate change mitigation effort: financial contributions for the implementation of climate-friendly policies these countries favor for other (self-interested) reasons. He offers examples such as investment in clean natural gas infrastructure in China, or assistance with safe nuclear power in India.

This approach, while enticing for developing countries, may be the opposite for industrialized countries which, unlike in the case of generous emission allowances within an international trading regime, obtain no direct return for these flows in the form of own compliance-cost reductions. The types of investments that Victor suggests could very well be funded by the rents generated for developing countries through generous permit allocations in a trading regime, thus the end result of the two approaches for carbon intensity in developing countries could be similar. But the incentives for participation by industrialized countries (the envisioned investors) are fundamentally different.

Victor joins others in his worry that the financial flows required by international carbon trading with unequal allocations will imply politically unacceptable wealth transfers from industrialized to developing countries (Cooper 1998; Pizer 2002; McKibbin and Wilcoxen 2002; Newell and Pizer 2003; Nordhaus 2005). Many of these critiques have endorsed international carbon taxes for this and other reasons. Nonetheless, the wealth-transfer feature of the permit approach allows participating countries to take advantage of the cost savings from permit trading, while offering the incentives for the low-cost countries to join in the first place.

Victor suggests that developing countries will have a stronger interest in his compensation approach than in one that offers "compensation to get them to do something they otherwise abhor." I do not argue with this, but it would seem to be a straw man, at least in the comparison to proposed international emissions trading regimes. Countries would choose any policy mechanism they like to achieve commitments under an international emission trading regime, although the choice of domestic non-market approaches by some participants would increase the aggregate costs of international climate change mitigation (Hahn and Stavins 1999).

The current European emissions trading regime can be seen as an example of the feasibility of simultaneous engagement of countries with high and low abatement costs using permit allocations as incentives for low-cost countries. Admittedly, Europe is homogeneous in comparison to world nations as a whole, and income and abatement cost differences across European borders are much smaller than these differences globally. Nonetheless, the EU, strongly opposed to international emissions trading during Kyoto Protocol negotiations (Frankel 2005), has implemented trading, and some income redistribution has occurred through permit allocations. The EU Bubble clearly allocates more emissions to the lower-income periphery countries, and fewer to high-income countries. Jeffrey Frankel's chapter in this volume, as well as previous work, demonstrates this progressivity of emission commitments under the EU Bubble, which results in an income elasticity of required reductions of about 0.10 (Frankel 1999).

Ambitious commitments and accountability

Within the bounds of the particular institution or institutions that arise under the approach that Victor favors, he notes that the most difficult thing to establish will be ambitious commitments to which nations are held accountable. While Victor has illuminated some of the key barriers to Kyoto's effectiveness, and some of the very useful things that will arise out of fragmented, non-universal approaches, he is not able to push us past the central conundrum of any climate policy. He notes that "binding treaties negotiated through processes dominated by diplomats do not offer good prospects for serious cooperation." This is true, but they may be the best prospect we have. The sum of efforts across individual nation-states and groups of nation-states will always be less than what would be necessary to achieve the efficient quantity of emission reduction, or likely any meaningful reductions at all.

While they are reasonable accommodations to the reality of collective action problems in a system of nation-states, the first two characteristics of Victor's proposed new architecture, nonuniversality and the absence of integration, may prevent any significant accomplishments from arising in the third category – ambitious commitments and accountability. That is, the whole of such a structure (or lack thereof) may be less than the sum of its parts, due to leakage, free riding, and other problems.

Key parameters of a post-Kyoto climate architecture

Any post-Kyoto international climate agreement must address three crucial questions: who, how, and when.[3] The roster of participants – and a means to ensure that key industrialized and developing countries participate – is the "who" element. Victor reminds us that the optimal "who" is not necessarily "everyone." Our list of countries that must participate in a meaningful agreement would likely be very similar, but Victor and I disagree on the second question – how these nations should be engaged in GHG emission reductions and other climate change mitigation activities.

The incorporation of market-based policy instruments, such as international emission trading, is essential to any international agreement on this issue. The United States, for one, is unlikely to participate meaningfully in a regime that does not rely on market mechanisms. The possibility of generous permit allocations for developing countries (paired with growth targets that allow for minimal emission con-

[3] This framework is developed in Olmstead and Stavins (2006).

straints early, increasing in stringency with wealth) provides an important "carrot" for their participation. Victor suggests an alternative carrot that may be equally, or even more, appealing to developing countries – direct investments in their climate-friendly energy infrastructure by industrialized countries. However, it is difficult to see why industrialized countries will willingly provide this carrot as, unlike compensation through permit allocations, it comes with no corresponding reduction in their own GHG emission abatement costs.

An important issue on which Victor's proposal does not focus is "when." Climate change is a long-term problem, as GHGs, once emitted, remain in the atmosphere for decades and even centuries. The time-path of mitigation efforts going forward will be a critical component of post-Kyoto climate policy architectures. Under a system of targets and timetables, targets should be moderate in the short term to avoid rendering large portions of the capital stock prematurely obsolete. Targets could then increase in stringency in the long term (with commitments, while flexible, mapped out in principle *ex ante*) to motivate technological change and bring costs down over time.

Victor's proposal eschews targets and timetables. It would be useful to think carefully about the dynamic aspects of Victor's proposed alternative structure – a more decentralized, bottom-up approach. The timing of commitments (both within the group of nations Victor envisions agreeing to climate change mitigation strategies, and between this group and the major developing countries receiving infrastructure investments) will have much to do with their eventual costs and benefits.

References

Aldy, J. E., S. Barrett, and R. N. Stavins (2003). "Thirteen Plus One: A Comparison of Global Climate Policy Architectures," *Climate Policy* 3(4): 373–397.

Brunnermeier, S. B., and A. Levinson (2004). "Examining the Evidence on Environmental Regulations and Industry Location," *Journal of Environment and Development* 13(1): 6–41.

Cooper, R. N. (1998). "Toward a Real Global Warming Treaty," *Foreign Affairs* 77(2): 66–79.

Ederington, J., A. Levinson, and J. Minier (2005). "Footloose and Pollution-Free," *Review of Economics and Statistics* 87(1): 92–99.

Edmonds, J., S. H. Kim, C. N. McCracken, R. D. Sands, and M. A. Wise (1997). *Return to 1990: The Cost of Mitigating United States Carbon*

Emissions in the Post-2000 Period, Washington, DC: Pacific Northwest National Laboratory, October.

Ellerman, A. D., P. L. Joskow, R. Schmalensee, J.-P. Montero, and E. M. Bailey (2000). *Markets for Clean Air: The US Acid Rain Program*, Cambridge: Cambridge University Press.

Frankel, J. A. (1999). "Greenhouse Gas Emissions," *Brookings Policy Brief*, No. 52 Brookings Institution, Washington, DC.

(2005). "You're Getting Warmer: The Most Feasible Path for Addressing Global Climate Change Does Run Through Kyoto," in J. Maxwell and R. Reuveny (eds.), *Trade and Environment Theory and Policy in the Context of EU Enlargement and Transition Economies*, Cheltenham: Edward Elgar, pp. 37–55.

Hahn, R. W. and R. N. Stavins (1992). "Economic Incentives for Environmental Protection: Integrating Theory and Practice," *American Economic Review: Papers and Proceedings* 82(2): 464–468.

(1999). *What Has the Kyoto Protocol Wrought? The Real Architecture of International Tradable Permit Markets*, Washington, DC: American Enterprise Institute Press.

Jaffe, A. B., S. R. Peterson, P. R. Portney, and R. N. Stavins (1995). "Environmental Regulation and the Competitiveness of US Manufacturing: What Does the Evidence Tell Us?" *Journal of Economic Literature* 33(March): 132–163.

Kolstad, C. D., and M. A. Toman (2005), "The Economics of Climate Policy," in K. G. Mäler and J. R. Vincent (eds.), *Handbook of Environmental Economics,* Vol. III: *Economywide and International Environmental Issues*, Amsterdam: Elsevier Science, North-Holland, pp. 1561–1618.

Levinson, A. (1996). "Environmental Regulations and Industry Location: International and Domestic Evidence," in J. N. Bhagwati and R. E. Hudec (eds.), *Fair Trade and Harmonization: Prerequisites for Free Trade*? Vol. I, Cambridge, MA: MIT Press, pp. 429–457.

(2003). "Environmental Regulatory Competition: A Status Report and Some New Evidence," *National Tax Journal* 56(1): 91–106.

McKibbin, W. J., and P. J. Wilcoxen (2002). "The Role of Economics in Climate Change Policy," *Journal of Economic Perspectives* 16(2): 107–129.

Newell, R. G., and W. A. Pizer (2003). "Regulating Stock Externalities under Uncertainty," *Journal of Environmental Economics and Management* 45(2S): 416–432.

Nordhaus, W. D. (2005). "Economic Analyses of Kyoto Protocol: Is There Life after Kyoto?" paper presented at Global Warming: Looking Beyond Kyoto, New Haven, CT, October 21–22.

Olmstead, S. M., and R. N. Stavins (2006). "An International Policy Architecture for the Post-Kyoto Era," *American Economic Review, Papers and Proceedings* 96(2): 35–38.

Pizer, W. A. (2002). "Combining Price and Quantity Controls to Mitigate Global Climate Change," *Journal of Public Economics* 85(3): 409–434.

Stavins, R. N. (2003). "Experience with Market-Based Environmental Policy Instruments," in K.-G. Mäler and J. R. Vincent (eds.), *Handbook of Environmental Economics Vol. III: Economywide and International Environmental Issues*, Amsterdam: Elsevier Science, North-Holland, pp. 355–435.

Victor, D. G. (2001). *The Collapse of the Kyoto Protocol and the Struggle to Slow Global Warming*, Princeton, NJ: Princeton University Press.

Victor, D. G., J. C. House, and S. Joy (2005). "A Madisonian Approach to Climate Policy," *Science* 309: 1820–1821.

Watson, Robert T. (2001). *Climate Change 2001: Synthesis Report: Third Assessment Report of the Intergovernmental Panel on Climate Change*, Cambridge: Cambridge University Press.

5 A credible foundation for long-term international cooperation on climate change

WARWICK J. MCKIBBIN AND
PETER J. WILCOXEN

THE first step toward meaningful progress on climate change is to be realistic about institutions – both about how existing institutions, such as national governments, can be brought to bear on the problem, and also about the prospects for creating powerful new international institutions. It is, in essence, a decision about whether it is more productive to bring existing tools, however imperfect, to bear on the problem or to design new and better tools at the international level. The latter course has attractions, but the risk is that the design process may go on indefinitely – with greenhouse gas emissions rising unchecked – without producing a viable new institution.[1] Such has been the case over the last decade as attention has focused on designing the Kyoto Protocol, an elaborate new international institution without any real precedent that may do nothing to slow emissions.[2]

In this chapter we argue that a better alternative would be to tackle climate change with simpler policies that can be carried out by national governments immediately. As David Victor noted in chapter 4, that process is happening by default already. We discuss key characteristics needed in an effective approach to climate change and argue that prospects for creating a powerful international institution to control greenhouse gas emissions are dim at best. We then outline one policy,

[1] An old joke once told by physicists about computer scientists illustrates the point vividly. The joke was that physicists solve tomorrow's problems with today's computers, while computer scientists do the opposite. Over the last decade, international negotiations seem to have taken the approach attributed to computer scientists: designing an ideal system to be implemented by future institutions while doing nothing about current emissions.

[2] David Victor has provided an outstanding discussion in Chapter 4 of the gulf between international climate negotiations and what national governments can actually carry out. Moreover, as noted by Jeffrey Frankel in Chapter 2, even if fully implemented, the Kyoto Protocol does nothing to address emissions after 2012.

an internationally coordinated system of national policies based on a hybrid tradable permit mechanism, that can be implemented with minimal development of new international institutions. It focuses on international cooperation and coordination, rather than on coercion. Moreover, the policy has a number of key strengths that make it a solid, long-term foundation for addressing climate change.

Credibility and climate policy

For a climate policy to be effective, it must satisfy three broad requirements: it must be widely adopted; it must remain in force indefinitely; and it must provide credible incentives for individuals and firms to make the investments that will be needed to reduce emissions. The third point is particularly important. Although international negotiations focus on commitments by governments to achieving particular emission targets, most governments have only indirect control over emissions within their borders. Emissions arise as a result of choices made by households and firms over energy technology and fuel consumption, not as a result of administrative decisions by government agencies. In contrast, other treaties often apply to actions that can be taken directly by governments themselves. International trade agreements are a good example: they restrict tariffs and other policies that are unambiguously under the control of participating governments. Even a government with the best of intentions on climate change will be unable to achieve much unless it can spur its citizens into action.

Moreover, the actions that individuals and firms will need to undertake in order to reduce emissions involve enormous investments in capital equipment and research and development, both with long payback periods. A climate policy will be unable to induce such investments unless it is clear that the policy is likely to be enforced and is unlikely to be repealed.[3] The single most important characteristic of a climate policy, in other words, is to provide a solid foundation for large, long-term investments by the private sector.

Although credibility is essential to an effective climate policy, it does not arise automatically. In a democracy, a policy does not become credible simply by being written into law. As Jeffrey Frankel noted in chapter 2, governments cannot bind their successors. Every subsequent

[3] This point is also discussed by William Pizer in Chapter 7.

legislature will have the authority to repeal the law, and subsequent administrations will be able, if they choose, to relax enforcement until the law is irrelevant.[4] A current government thus has little direct ability to constrain the actions of its successors. As a result, a policy will be credible only if it is clear that future governments – whether controlled by other political parties, or facing very different economic circumstances – will *want* to continue carrying it out. Structuring a policy to provide powerful incentives for continuing enforcement by future governments is a critical step in designing an effective climate change agreement.

At first, the problem of credibility might seem insurmountable. If a current government cannot adopt rules that future governments cannot reverse, what else could it do? The answer is straightforward, but it has profound implications for the structure of a climate policy: it must create a constituency with a strong financial interest in perpetuation of the policy. Bluntly put, it must create a powerful lobby group that will vigorously resist any attempt at backsliding by future governments.

Before turning to a policy that would build such a constituency, it is instructive to consider a policy that would have exactly the opposite effect: a carbon tax. From an economic perspective, a carbon tax would be an ideal instrument for addressing climate change. It would be efficient given the uncertainties surrounding climate change,[5] and it would definitely work: high energy prices in the 1970s stabilized US emissions for nearly twenty years.[6] However, a carbon tax creates precisely the wrong constituency. No group in the private sector would have a large financial stake in seeing the policy continue, and all future users of fossil fuels would be motivated to lobby against it. Apart from satisfying the terms of an international agreement, the only incentive a government would have to keep the tax in place is the revenue it generates. However, that incentive may not be very strong: recent history has shown that governments may be willing to run large deficits for long periods of time in order to reduce taxes.

[4] A case in point is the Bush administration's decision to abandon the 1972 Antiballistic Missile Treaty.

[5] This is an application of Weitzman's seminal 1974 paper on prices vs. quantities; see McKibbin and Wilcoxen (2002) for a detailed discussion in the context of climate change.

[6] Jorgenson and Wilcoxen (1993).

The broader lesson is that an international agreement cannot succeed in the long run if it relies on pitting national governments against broad, highly motivated groups of their own citizens. Ultimately, international agreements are voluntary and a climate change treaty will be no exception. Faced with a choice between angering constituents by adhering to an unpopular treaty, or repudiating the treaty and angering the international community, few democratic governments would be able to take the former course year after year. To be successful, an international agreement must be designed from the start to enhance and coordinate the efforts of national governments, not to use them as instruments of enforcement that are subsidiary in authority to an international regime. In terms of the analogy at the beginning of the chapter, national governments may not be the ideal tools for controlling climate change, but they are by far the best tools available today.

Returning to the issue of credibility, building a national constituency with a financial stake in maintaining a climate change policy is possible if the policy involves long-lived tradable emissions permits. We discuss long-lived permits in more detail below but the key feature of such a permit would be to allow one ton of emissions every year for the life of the permit. A perpetual permit, for example, would allow one ton of emissions every year forever. Once long-lived permits have been distributed, permit owners will have a valuable financial asset whose price depends directly on the health of the policy. With scrupulous monitoring and enforcement of the policy, firms will pay high prices to emit carbon and the permits will be very valuable. However, if enforcement is lax, or if the policy is repealed, the value of the permits will drop to zero. Permit owners thus have a strong financial interest in supporting the policy. In essence, the permit system replaces the conflict that a carbon tax would cause between a government and energy users with conflict between two private sector groups: permit owners and energy users. It doesn't eliminate the difficulty of reducing emissions but it does even out the political landscape and reduces the pressure on future governments to repeal the policy.

Despite this advantage, an international climate policy based entirely on long-term permits is not a viable option. The reason is straightforward: to ratify such a policy, a government would have to be willing to agree to achieve a specified emission target by a given date, regardless of the cost of doing so. That approach would be appropriate if carbon dioxide were a threshold pollutant. Threshold pollutants cause little or no damage when emissions are low but cause substantial damages once a threshold is exceeded. As a result, keeping emissions below the threshold may indeed be imperative. Carbon dioxide, however, is a stock pollutant and *not* a threshold pollutant.[7] Excess emissions accumulate in the atmosphere and remain there for decades: current annual emissions are equal to only about 1 percent of the total anthropogenic carbon dioxide in the atmosphere. The risks associated with climate change result from the accumulated stocks of carbon dioxide and other greenhouse gases. Each additional ton of emissions increases the risks, although very slightly, and there is no threshold below which risks are zero.

This point is often misunderstood in the public debate because some of the consequences of climate change might occur suddenly, such as rapid melting of the Greenland ice sheet. However, a potentially sudden consequence does not necessarily indicate a distinct threshold in emissions of the underlying pollutant. One way to understand the distinction is by analogy to the effects of cigarette smoking. Each cigarette raises the risk of lung cancer slightly. If cancer occurs, however, it doesn't make sense to argue that a particular cigarette caused it. Doing so would be to argue that all previous cigarettes were insignificant in causing the cancer. In the same way, all emissions of carbon contribute to future climate risks. The damage caused by one ton of emissions is essentially the same as the damage caused by the next.

In the absence of a clear threshold, basing a climate policy on a rigid emissions target makes little sense: achieving the target does not eliminate the risk and exceeding the target does not cause consequences markedly different from achieving it. Put bluntly, when every ton of emissions contributes equally to the problem, it is impossible to justify any particular emission target, other than possibly no emissions at all. As a result, a rigid system of targets and timetables for emission reductions is not economically efficient.[8] Nor is it politically realistic: a climate policy that does not take costs into consideration will never be ratified by the US Senate and is likely to be rejected – or ratified but later repudiated – by many other governments as well.

[7] See Newell and Pizer (1998) for a discussion of the economic theory behind regulation of stock externalities.

[8] For a detailed discussion, see McKibbin and Wilcoxen (2002).

In summary, neither of the two main market-based mechanisms for pollution control are suitable for climate change. A carbon tax would be economically efficient but is not a credible long-term policy because of the conflict it would create between a government and its constituents. A permit system based on a fixed number of long-term permits is also unsuitable but has the opposite weaknesses: it would be credible, but it would be inefficient since carbon dioxide is not a threshold pollutant. Although both mechanisms have serious economic and political disadvantages when used alone, those problems can be overcome by a hybrid policy that combines the best elements of both.[9] For efficiency, the hybrid should act like an emission tax at the margin: it should provide incentives for abatement of all emissions that can be cleaned up at low cost while not requiring that a particular emission target be achieved. For long-term credibility, the hybrid should create a private sector constituency with a clear financial interest in seeing the policy maintained and enforced. The structure and operation of a hybrid policy for addressing climate change at the national level are discussed in the following section; a subsequent section will discuss the international implications of the policy.

A hybrid policy for controlling national emissions

A hybrid policy for climate change is discussed in detail in McKibbin and Wilcoxen (2002) and summarized briefly in Box 1. It combines a limited supply of long-term permits good for multiple years with a much more flexible supply of annual permits. Every year, firms would be required to hold a portfolio of permits equal to the amount of carbon they emit.[10] The portfolio could include any mix of long-term and annual permits. The long-term permits could be owned outright

[9] The economic theory behind hybrid regulatory policies is due to Roberts and Spence (1976). A hybrid approach to climate change was first proposed by McKibbin and Wilcoxen (1997a) and has subsequently been endorsed or promoted by a range of authors and institutions. Examples include Kopp, Morgenstern and Pizer (1997); Kopp, Morgenstern, Pizer and Toman (1999); Americans for Equitable Climate Solutions (2000); Aldy, Orszag and Stiglitz (2001); and Victor (2001).

[10] This approach is known as a downstream policy because it applies to fuel users. It would also be possible to apply the policy upstream by imposing limits on the carbon embodied in fuels when they are produced (e.g., at the mine mouth or wellhead).

Box 1: A hybrid policy for controlling national emissions

In its basic form, the hybrid policy allows each participating country to issue two kinds of emission permits: long-term permits that entitle the owner of the permit to emit one metric ton of carbon every year for a period specified by the permit, and annual permits that allow one ton of carbon to be emitted in a single, specified year. Key features of the policy are listed below:

Long-term permits:
- Limited quantity available, perhaps a specified fraction of 1990 emissions
- Distributed once, at the time when the policy is first enacted
- Could be bought, sold or leased within the country of issue without restriction
- Could only be used in the country of issue; no international trading
- Price will be determined by the market

Annual permits:
- Would be sold for a stipulated price, say $20 per ton of carbon
- Valid only in the year and country of issue
- No limit on the quantity that could be sold

by the firm, or they could be leased from other permit owners. In the sections below we discuss each type of permit in more detail.

Long-term permits

A country adopting the hybrid policy would create and distribute a set of long-term permits, each entitling the owner to emit a specified amount of carbon every year for the life of the permit. The simplest long-term permit would have no expiration date and would allow one ton of emissions every year forever. A more sophisticated alternative would be to issue long-term permits with a variety of expiration dates, much the way governments now issues bonds. For example, a country wishing to distribute 100 long-term permits might chose to issue 20 of them as perpetual permits, 40 as permits expiring in fifty years, and the remaining

40 as permits expiring in twenty years. In essence, this approach would create a family of assets with a term structure of expiration dates.[11]

Just as the expiration date of the permits can be varied, so could be the amount of emissions each permit allows at each point in time.[12] For reasons we will return to later in the chapter, governments might find it useful to have the amount of emissions allowed by a long-term permit decline over the permit's life. For example, a permit might allow 1 ton of emissions per year for the first 20 years after it is issued, 75 percent of a ton during years 21–40, 50 percent of a ton in years 41–60, 25 percent of a ton after that.[13] It would be analogous to distributing bundles of permits with varying expiration dates: an equivalent bundle would consist of four 0.25 ton permits: one valid for twenty years, one for forty years, one for sixty years, and one valid in perpetuity. Computing the market value of such permits would be slightly more complex than valuing permits allowing one ton per year. However, the added complications would be minimal as long as all long-term permits had the same issue and expiration dates, and hence allowed identical paths of future emissions.[14] Moreover, this approach has a very significant advantage relative to a system of one-ton permits with varying expiration dates: all long-term permits would be identical and would hence trade in a single market at a single price.

When initially distributed, the long-term permits could be given away, auctioned, or distributed in any other way the government of the country saw fit. One option would be to distribute them free to firms in proportion to their historical fuel use. For example, a firm might receive permits equal to 90 percent of its 1990 carbon emissions. Such an approach would be relatively transparent and would limit the incentives for lobbying by firms. It is important to note that although the allocation would be based on historical emissions, the policy would not have the disadvantages of a traditional grandfathering scheme.

[11] Nicholas Gruen and Geoff Francis have made similar suggestions to us along these lines.

[12] We are indebted to Rob Stavins for pointing out that long-term permits need not allow constant emissions over their lifetimes. As we will discuss later in the chapter, this feature would play a crucial role in the evolution of a hybrid system over time.

[13] This structure potentially addresses Frankel's suggestion in Chapter 2 that a climate policy should consist of a series of gradually tightening medium term targets.

[14] Varying *both* attributes of long term permits (the expiration date and the time path of allowed emissions) would be a mistake. It would create unnecessary transactions costs by fracturing the long term permit market into many submarkets.

Because the permits are completely tradable and are not tied in any way to the original recipient or any particular plant, they do not create differences in marginal costs across firms or plants. Moreover, the existence of annual permits limits the ability of incumbent firms to create entry barriers by keeping their long-term permits off the market: entrants could simply buy annual permits. Incumbent firms would benefit financially from the initial distribution of permits, but unless they were previously liquidity-constrained, they would not be able to use their gains to reduce competition.[15]

Another alternative would be to auction the permits, but few governments would find auctioning attractive. From the point of view of the energy industry, auctioned permits would be exactly like a carbon tax except with an added disadvantage: the industry would have to pay the entire present value of all future carbon taxes up front. To see why, suppose the number of long-term permits to be issued is small enough that at least a few annual permits would be sold in every year. The price of a permit during the auction would be bid up to the present value of a sequence of annual permit purchases. As far as the industry is concerned, the policy would be equivalent to a carbon tax set at the annual permit price, except that it would have to pay the entire present value of all future tax payments on the emissions allowed by the permit at the time of the auction.

Once distributed, the long-term permits could be traded among firms, or bought and retired by environmental groups. The permits would be very valuable because: (1) there would be fewer available than needed for current emissions, and (2) each permit allows annual emissions over a long period of time. As a consequence, the owners of long-term permits would form the private sector interest group needed for long-term credibility of the policy: they would have a clear financial interest in keeping the policy in place.

Annual permits

The other component of the policy, annual emission permits, would be straightforward: the government would agree to sell annual permits

[15] In passing, it's worth noting that anti-competitive behavior by the incumbents, while unlikely, would have an environmental benefit: it would reduce overall carbon emissions.

for a specified fee, say for $20 per ton of carbon. There would be no restriction on the number of annual permits sold, but each permit would be good only in the year it is issued. To put the fee in perspective, $20 per ton of carbon is equivalent to a tax of about $12 per ton of coal and $3 per barrel of crude oil; other things equal, the price of a $22 ton of coal would rise by about 50 percent and the price of a $60 barrel of oil would rise by about 3 percent. The annual permits give the policy the advantages of an emission tax: they provide clear financial incentives for emission reductions but do not require governments to agree to achieving any particular emission target regardless of cost.

Incentives for investment

Although the policy is more complex than an emission tax or conventional permit system, it would provide an excellent foundation for the large private sector investments in capital and research that will be needed to address climate change. To see why, consider the incentives faced by a firm after the policy has been established. Suppose it has the opportunity to invest in a new production process that would reduce its carbon emissions by one ton every year. If the firm is currently covering that ton by buying annual permits, the new process would save it $20 per year every year. If the firm can borrow at a 5 percent real rate of interest, it would be profitable to adopt the process if the cost of the innovation were $400 or lower. For example, if the cost of adoption were $300, the firm would be able to avoid buying a $20 annual permit every year for an interest cost of only $15; adopting the process, in other words, would eliminate a ton of emissions and raise profits by $5 per year.

Firms owning long-term permits would face similar incentives to reduce emissions because doing so would allow them to sell their permits. Suppose a firm having exactly the number of long-term permits needed to cover its emissions faced the investment decision in the example above. Although the firm does not need to buy annual permits, the fact that it could sell or lease unneeded long-term permits provides it with a strong incentive to adopt the new process. To keep the calculation simple, suppose that the permits are perpetual and allow one ton of emissions per year. At a cost of adoption of $300, the firm could earn an extra $5 per year by borrowing money to adopt the process, paying an interest cost of $15 per year, and leasing the permit it would no longer need for $20 per year.

The investment incentive created by a hybrid policy rises in proportion to the annual permit fee as long as the fee is low enough to be binding – that is, low enough that at least a few annual permits are sold. For example, raising the fee from $20 to $30 raises the investment incentive from $400 to $600. That makes sense: if emitting a ton of carbon becomes 50 percent more expensive every year, the amount a firm would pay to avoid that cost should rise by 50 percent as well. Raising the annual fee even further would continue to increase the incentive in proportion, provided that the policy remains credible: a $40 fee generates a $800 investment incentive; a $50 fee generates a $1,000 incentive; and so on.

The upper limit on incentives created by the annual fee is the market-clearing rental price of a long-term permit in a pure tradable permit system. Above that price, there would be enough long-term permits in circulation to satisfy demand and no annual permits would be sold. For example, if long-term permits would rent for $90 a year under a pure permit system, the maximum price of an annual permit under the hybrid will be $90.

Credibility and incentives

The critical importance of credibility becomes apparent when considering what would happen to these incentives if firms are not sure the policy will remain in force. If the policy were to lapse at some point in the future, emission permits would no longer be needed. At that point, any investments made by a firm to reduce its emissions would no longer earn a return. The effect of uncertainty about the policy's prospects is thus to make the investments it seeks to encourage more risky. Firms will take that risk into account when evaluating climate-related investments and will be willing to pay far less to undertake them as a result. The decline in incentives is surprisingly large. Consider the same investment that would save a firm $20 a year if the policy is in force, but now suppose the firm believes that there is a 10 percent chance each year that the policy will be repealed. That may sound like a small erosion of credibility, but it can be shown that it reduces the maximum amount the firm would be willing to pay for the innovation from $400 to only $133. The drop in credibility – from 100 percent confidence in continuation of the policy to 90 percent – reduces the incentive for investment by two-thirds.

Since the incentives created by the policy increase with the price of an annual permit, a government might try to compensate for low credibility by imposing higher annual fees. For example, suppose a government would like a climate policy to generate a $400 incentive for investment but firms believe that there is a 10 percent chance the policy will be abandoned each year. For the policy to generate the desired incentive, the annual permit price would have to be $60 rather than $20. That is, the stringency of the policy (as measured by the annual permit fee) must *triple* in order to offset the two-thirds decline in the incentives arising from the policy's lack of credibility. In practice, the situation is probably even worse. Increasing the policy's stringency is likely to reduce its credibility further, requiring even larger increases in the annual fee. For example, suppose that investors believe that the probability the government will abandon the policy rises by 1 percent for each $20 increase in the annual fee. In that case, maintaining a $400 investment incentive would require an annual fee of $70 rather than $60, which would be accompanied by an increase in the perceived likelihood of the policy being abandoned from 10 percent to 12.5 percent. The general lesson is clear and vitally important to the development of an effective climate policy: a modest but highly certain policy generates the same incentives for action as a policy that is much more stringent, but also less certain. A hybrid policy with a modest annual permit price would generate larger investment incentives than a more draconian, but less credible, emission target imposed by a system of targets and timetables.

Analogous institutions

Our discussion of the hybrid policy so far has been somewhat abstract, which might give the impression that it is a complicated and unfamiliar mechanism. In fact, nothing could be further from the truth. Stripped to its bare essentials, the main effect of the hybrid is to create a new asset – a long-term permit – that behaves very much like a conventional form of capital. Individuals face analogous decisions every day. The similarity can be seen by comparing the way a firm would use its permits with the way a household uses its automobiles. When a firm needed to emit carbon dioxide, it would compare the prospective emissions against its stock of long-term permits. If it had too few permits, it would have four choices: it could reduce its emissions until they matched its stock of permits; it could buy more long-term permits; it could lease additional long-term permits from other permit owners; or it could buy annual permits from the government. Someone planning transportation for household begins by comparing the number of passengers to the capacity of the available vehicle. If the number of passengers exceeds the capacity, the alternatives are very similar to those faced by the firm: reduce the number of passengers; buy another vehicle; rent or lease an additional vehicle; or send some of the passengers in a taxi (the option equivalent to annual emission permits). Although it may sound exotic, a hybrid policy is no more complex than other decisions involving capital goods routinely made by individuals and firms.

The analogy between permits and vehicles also helps clarify the role of annual permits. Someone with a predictable need to transport a large number of people would usually find it best to own several vehicles, while someone whose transportation needs were less predictable would find it better to own a smaller number of vehicles and use rental cars or taxis to cover peak periods. Similarly, firms with a predictable need for a large number of permits would generally find it best to own a large number of long-term permits, while firms whose emissions fluctuate a lot from year to year would find it profitable to own a smaller number of long-term permits and cover peak periods using leased or annual permits. Annual permits would thus play a valuable role in helping firms manage short-term fluctuations.

Summary

A hybrid policy combining a fixed supply of tradable long-term emissions permits with an elastic supply of annual permits would be a viable and efficient long-term climate policy at the national level. It would be more credible than many alternatives, especially a carbon tax, because it builds a political constituency with a large financial stake in preventing backsliding by future governments. It thus addresses the inherent difficulty that a democratic government faces in binding future governments to continue carrying out the policy. At the same time, the provision for annual permits allows the hybrid to avoid the inefficiencies and political hurdles that would arise with a conventional system of permits, which would impose a rigid cap on emissions. Thus, it would provide a strong foundation for investment decisions by the private sector because it creates credible, long-term returns for reducing greenhouse gas emissions. It combines the best features of a permit system and an emissions tax, as shown in Table 5.1.

Table 5.1 *Comparing key attributes of market-based climate policies*

Attribute	*International permits*	*Carbon tax*	*Hybrid policy*
Attributes in common			
Minimizes abatement costs within each country	✓	✓	✓
Encourages energy conservation and innovation	✓	✓	✓
Guarantees that benefits are greater than costs	no	no	no
Attributes of tax-based approaches			
Relies on national legal systems and institutions	no	✓	✓
Economically efficient response to uncertainties	no	✓	✓
Explicit upper bound on compliance costs	no	✓	✓
Avoids large international transfers of wealth	no	✓	✓
Provides incentives for domestic enforcement	no	✓	✓
Does not need strong international enforcement	no	✓	✓
Robust to accession or withdrawal of participants	no	✓	✓
Limits propagation of shocks across countries	no	✓	✓
Attributes of permit-based approaches			
Creates constituencies for enforcement	✓	no	✓
Flexibility in domestic distributional effects	✓	no	✓
Does not requires large transfers to the government	✓	no	✓
Easy to implement transition relief	✓	no	✓
Guarantees a given reduction in emissions	✓	no	no

International cooperation and harmonization

A key feature of the hybrid policy we propose is that emission permits would be valid only in the country of issue. They would not be tradable internationally – permits issued in one country could not be used to cover emissions in another country.[16] Each country would manage its own domestic hybrid policy using its own existing legal system and financial and regulatory institutions. There would be no need for complex international trading rules, or for the creation of a powerful new international institution, or for participating governments to cede a significant degree of sovereignty to an outside authority. As a result, a treaty built around the hybrid policy would be very simple and would focus primarily on harmonizing the price of annual permits across participating countries.[17] To join the agreement, a country would establish a hybrid permit system and agree to charge the price specified in the treaty for annual permits. Unlike an agreement focused on achieving a national emission target, governments would be making commitments that are within their direct control.

Easy accession is very important. To be effective in the long run, the agreement will eventually need to include all countries with significant greenhouse gas emissions. However, it is unlikely that all countries will choose to participate at the beginning. Developing countries, for example, have repeatedly pointed out that current greenhouse gas emissions are overwhelmingly caused by industrialized countries, and that those countries, therefore, should take the lead in reducing emissions. As a result, an international climate policy will need to cope with gradual accessions taking place over many years. Its design, in other words, must be suitable for use by a small group of initial participants,

[16] Strictly speaking, the term "country" is too narrow. The permits would be valid only within the political jurisdiction of issue. If the relevant jurisdiction is multinational – the EU, for example – permits could be traded between countries within the broader jurisdiction.

[17] Because the core of the treaty would be the price of annual permits, it would be relatively straightforward to negotiate: only one key number is really involved. That is not to say, however, that negotiations would be trivial: getting agreement on the annual price would require considerable diplomacy. It is interesting to note that a treaty of this form has a strong built-in incentive for countries to participate in the initial negotiations. Countries that participate will have a role in setting the annual price while those who remain on the sidelines will not. We are indebted to Jonathan Pershing for pointing this out.

a large group of participants many years in the future, and all levels in between. Because it is fundamentally a harmonized system of domestic policies, rather than a monolithic international policy, our hybrid proposal has exactly the flexibility needed. A country can participate by simply adopting the hybrid domestically, without any need for international negotiations.

Beyond specifying the price of annual permits, the treaty could provide a guideline for governments to use in determining the number of long-term permits to issue. It could, for example, suggest that signatories distribute no more long-term permits than their allotments under the Kyoto Protocol. However, governments wishing to tackle climate change more aggressively could choose to distribute fewer long-term permits.[18] Moreover, governments that for one reason or another would prefer a carbon tax could distribute no long-term permits at all.[19] The treaty does not need to specify rigid allocations of long-term permits because emissions will generally be controlled at the margin by the price of annual permits. As long as each country distributes few enough long-term permits that at least one annual permit is sold, the number of long-term permits only affects the distribution of permit revenue between the private sector and the government; it does not affect the country's total emissions. Distributing a small number of long-term permits means the government will earn a lot of revenue from annual permit sales, but it also means that the lobby group supporting the policy will be weak. Distributing a larger number means less government revenue and a stronger supporting lobby. In either case, one country's decision has little effect on other signatories.

Developing countries

One important role for the treaty's long-term permit guidelines would be to distinguish between developed and developing countries. Developing countries could be allowed to distribute more long-term permits than needed for their current carbon emissions. In that case, a

[18] As discussed by Victor in Chapter 4, countries have different degrees of concern about climate change and different abilities to implement climate policies. A coordinated system of hybrid policies provides participants with the ability to tailor the policy to their own circumstances.

[19] A government might prefer a carbon tax if it lacks the institutional and administrative mechanisms needed to operate a permit market.

country adopting the treaty would be committing itself to slowing carbon emissions in the future but would not need to reduce its emissions right away. As the country grows, its emissions will approach the number of long-term permits. The market price of long-term permits would gradually rise, and fuel users would face increasing incentives to reduce the growth of emissions.

A generous allotment of long-term permits would reduce the disincentives faced by developing countries, but that alone might not be enough to induce widespread participation. If stronger incentives are needed, it would be possible to augment the treaty with a system of foreign aid payments or with programs for technology transfer to participating developing countries. Alternatively, the infrastructure policies discussed by Victor in chapter 4 could be used as an inducement. In either case, the result would be more transparent and more attractive to developing countries than the Kyoto Protocol, which essentially requires that compensation payments from developed to developing countries be in-kind, in the form of improved energy technology.

Advantages of separate markets

Because the permit markets under this policy are separate between countries, shocks to one permit market do not propagate to others. For example, accession by a new participant has no effect on the permit markets operating in other countries.[20] Similarly, if a participating country withdraws from the agreement or fails to enforce its hybrid policy, permit markets in other countries are also unaffected.[21] Collapse of one or more national permit systems would be unfortunate in terms of emission control, but it would not cause permit markets in

[20] The mechanism Frankel proposes in Chapter 2 for accession would have the opposite property. New participants would be given permit quotas sufficient to guarantee that they could sell permits in their first few years of membership. As a result, permit prices in other countries would invariably drop every time a new country joined. The result would be good for energy users but bad for the owners of long term permits.

[21] As discussed by Scott Barrett in Chapter 6, an international permit system could be particularly difficult to enforce because of the links it creates between countries. Restricting sales of permits by non-complying countries, as would be required under the Kyoto Protocol, would harm the interests of compliant countries by raising permit prices. The international links between permit markets thus provide a strong incentive against enforcement of the agreement.

other countries to collapse as well. Separate permit markets are, in essence, a form of compartmentalization that lends stability to the international agreement. In contrast, under an international trading agreement, such as the Kyoto Protocol, shocks in one country – ineffective enforcement, or withdrawal from the agreement, for example – would cause changes in permit prices around the world. Permit owners would receive windfall gains or losses and permit users would be faced with volatile and unpredictable permit prices. From the perspective of both permit owners and permit users, investments in emission reductions would be more risky.

Compartmentalization is especially important for a climate change agreement, which must endure for many, many years. Not only must it be able to survive noncompliance by some of its members, it must be able to survive through economic booms and busts; through wars and pandemics; and through times of low concern about the environment as well as in times of high concern. Moreover, because of the uncertainties surrounding climate change, it must also survive through intervals where warming seems to be proceeding more slowly than expected and there could be political pressure to abandon the agreement on the grounds that it is not necessary. Such intervals could arise because of random fluctuations in global temperatures from year to year, or because the policy is actually succeeding in reducing the problem. The latter point is worth emphasizing: if a climate regime is successful at reducing warming and preventing significant damages, it will be easy for complacency to arise: many people may interpret the absence of disasters to mean that the risks of climate change were overstated.

Another advantage of multiple national permit markets, rather than a single international one, is that individual governments would have little incentive to monitor and enforce an international market within their borders. It is easy to see why: monitoring polluters is expensive, and punishing violators would impose costs on domestic residents in exchange for benefits that will accrue largely to foreigners. There would be a strong temptation for governments to look the other way when firms exceed their emission permits. For a treaty based on a single international market to be effective, therefore, it will need to include a strong international mechanism for monitoring compliance and penalizing violations. National permit markets reduce the problem substantially because monitoring and enforcement becomes a matter of enforcing the property rights of a group of domestic residents – the owners of long-term permits – in domestic markets.

Potential disadvantages of separate markets

One possible disadvantage of separate permit markets is that the prices of long-term permits might differ between countries. If so, the overall policy would not be minimizing the cost of abatement: it would be possible to lower overall abatement costs by doing more abatement in countries where permit prices are low and doing less abatement in countries where prices are high. However, it is unlikely that permit prices would differ significantly in practice.[22] As long as each country's stock of long-term permits is small enough that at least one annual permit is sold, long-term permit prices in all participating countries will be equal to the present value of buying a stream of annual permits. With annual permit prices harmonized across countries, permit prices will therefore be equal.

Separate markets have other liabilities as well. Because each country's trading rules will be different, it will be harder for multinational firms and brokers to operate across borders than it would be with a single international market. In addition, some markets will be small, which will mean that the liquidity of long-term permits will be lower in those countries. Also, small markets have a greater risk of anti-competitive behavior on the part of locally large buyers or sellers.

Balancing the pros and cons

Overall, the advantages of an internationally coordinated system of national hybrid policies outweigh the potential disadvantages of separate permit markets. The policy would be credible and efficient, thereby providing a solid foundation for investments by individuals and firms to reduce emissions. It would be implemented almost entirely via national governments and other existing institutions without the need for a powerful new international agency. It would require little sacrifice of sovereignty by participants. Accession would be straightforward and would not disturb existing permit markets. It would be robust,

[22] In Chapter 7, Pizer also argues that inefficiencies across countries are a relatively small problem.

because adverse shocks in one permit market would not propagate to others. Finally, it would eliminate the disincentives national governments would face in monitoring and enforcing an international trading regime. It might not minimize costs completely, but that outcome occurs only in situations that are unlikely to arise in practice. Moreover, the potential loss of efficiency is likely to be insignificant when compared to the administrative gains achieved by using existing institutions.

Evolution over time

Over time, more information will become available about climate change, its effects, and about the costs of reducing emissions. Revising the agreement in light of new information is straightforward: if it becomes clear that emissions should be reduced more aggressively, the price of annual permits can be raised. The political prospects for an increase would be helped by the fact that raising the price of annual permits would produce a windfall gain for owners of long-term permits, since the market value of long-term permit prices would rise as well.[23]

If new information indicates that emissions should drop below the number allowed by long-term permits, raising the price of annual permits would need to be augmented by a reduction in the stock of long-term permits. One option would be for each government to buy and retire some of the long-term permits it issued. However, the costs involved could make that approach infeasible. A better option would be to design the original permits to expire gradually. Of the options discussed earlier in the chapter, we favor an approach with perpetual permits whose allowed emissions gradually diminish over time. Such a system would establish long-term property rights and strong, credible incentives but would still allow future governments room to maneuver. In addition, the gradual reduction in emission entitlements will make the policy more attractive to environmental groups.

[23] Although long term permit owners would welcome an increase in the annual price, there is little risk that they would be able to drive prices up on their own. Given that other energy users provide countervailing pressure to keep energy prices low, it is hard to imagine that permit owners would be able to push a government into adopting an inefficiently high price and excessively stringent emissions policy.

Building on the foundation

The agreement outlined above – an internationally coordinated system of national hybrid policies for controlling carbon emissions – provides a solid foundation for private sector investments to reduce emissions of carbon dioxide. It provides clear and credible financial incentives for developing and deploying new innovations that reduce fossil fuel use, or capture and sequester carbon emissions. However, it need not be the only policy adopted and could easily be integrated with other actions taken at the national level. In this section, we discuss the advantages and disadvantages of including gases other than carbon dioxide; including sinks; and including measures focused on technology, such as product standards or subsidies for research and development.

Including other greenhouse gases in a unified hybrid policy would be possible using a system of weights at which emissions of different gases could be traded.[24] However, doing so would add considerably to the complexity and cost of monitoring and enforcement. Monitoring the use of fossil fuels is relatively easy because they are produced in a narrow segment of the economy. In contrast, monitoring emissions of methane from agriculture and landfills, for example, would be much more difficult. Because carbon dioxide accounts for most of the greenhouse gas burden, starting with a system focused on it would be an attractive approach. Other gases could be controlled later via separate hybrid policies for each gas. The policies would be coordinated via appropriate annual permit prices that were equal on a carbon-equivalent basis.

Credit for sinks could also be included in a hybrid policy. In fact, an important advantage of the hybrid approach is that the decision on whether or not to allow credit for sinks could be left to the discretion of individual governments and would not need to be a formal part of the international agreement. Including sinks would be straightforward. Individuals and firms carrying out sink-enhancement projects would apply to their own governments for certification of the number of tons of carbon sequestered by their projects each year. The owners of the sinks would then be allowed to sell an equivalent number of annual permits. In effect, a government granting credit for sinks would

[24] See Reilly, Babiker and Mayer (2001) for a discussion of the complexities of designing an economically appropriate system of weights.

subsidize the activity by shifting revenue that it might otherwise have earned to the owners of the sinks.

Finally, the hybrid policy could also be combined with a wide range of measures focused on energy technology, including product standards, informational campaigns, demand-side management, subsidies for investment in non-fossil energy sources, or research and development subsidies. Although each of these could be combined with the hybrid, none of them could replace it. Without the clear, credible incentives for investment provided by the hybrid, individuals and firms will be slow to adopt new technologies to reduce emissions. In fact, without a price-based instrument like the hybrid, many of these policies would be counterproductive. Subsidized research and development, in particular, would have the effect of *reducing* energy prices, thus tending to increase energy consumption and greenhouse gas emissions. Using the hybrid policy in combination with a research subsidy would offset this effect.

Conclusion

If an international agreement is to succeed in reducing global carbon dioxide emissions, it should build on existing institutions to establish credible long-term incentives for major investments in physical capital and in research and development. In particular, it should focus on fostering collaboration and coordination among national governments, rather than on attempting to create a new international organization that would be likely to place national governments in the position of imposing unpopular international policies on their constituents.

At the national level, a hybrid policy mixing long- and short-term emission permits has many features that would help provide credible incentives. It would create a powerful interest group – the owners of long-term permits – with a financial stake in the existence and enforcement of the policy. Creating such a group does not guarantee that the policy will be carried out indefinitely: a future government could still abandon the policy or reduce its stringency. However, it certainly improves the odds: a policy supported by a vested interest group clearly has a better chance of survival than one having only moral support.

In addition, the flexibility provided by annual permits allows the policy to be adopted without the need for a government to agree to achieve a rigid emission target regardless of the cost. Tradability of the permits also provides the usual benefit of market-based environmental policies: it ensures that within each country, emission reductions will be achieved at minimum cost.

References

Aldy, Joseph E., Peter R. Orszag, and Joseph E. Stiglitz (2001). "Climate Change: An Agenda for Global Collective Action," paper presented at workshop on the Timing of Climate Change Policies, Pew Center on Global Climate Change, Washington, DC, October 2001, Brookings Joint Center on Regulatory Studies Publication.

Americans for Equitable Climate Solutions (2000). *Sky Trust Initiative: Economy-wide Proposal to Reduce US Carbon Emissions*, Washington, DC: Americans for Equitable Climate Solutions.

Cline, William R. (1992). *The Economics of Global Warming*, Washington, DC: Institute for International Economics.

Coase, Ronald H. (1960). "The Problem of Social Cost," *Journal of Law and Economics* 3: 1–44.

Energy Journal (1999). "Special Issue: The Costs of the Kyoto Protocol: A Multi-Model Evaluation".

IPCC (1990). *Scientific Assessment of Climate Change*, Cambridge: Cambridge University Press.

(1996). *Climate Change 1995*, 3 vols., Cambridge: Cambridge University Press.

(2000). *Emissions Scenarios*, Cambridge: Cambridge University Press.

(2001a). *Climate Change 2001: The Scientific Basis*, Cambridge: Cambridge University Press.

(2001b). *Climate Change 2001: Impacts, Adaptation, and Vulnerability*, Cambridge: Cambridge University Press.

(2001c). *Climate Change 2001: Mitigation*, Cambridge: Cambridge University Press.

Jiang, Tingsong (2001). "Economic Instruments of Pollution Control in an Imperfect World: Theory, and Implications for Carbon Dioxide Emissions Control in China," Ph.D. thesis, Australian National University, Canberra.

Jorgenson, Dale W., and Peter J. Wilcoxen (1993). "Energy Prices, Productivity and Economic Growth," in R. H. Socolow, D. Anderson, and J. Harte (eds.), *Annual Review of Energy and the Environment*, Vol. 18, Palo Alto, CA: Annual Reviews Inc., pp. 343–395.

Kopp, Raymond, Richard Morgenstern, and William A. Pizer (1997). "Something for Everyone: A Climate Policy that Both Environmentalists and Industry Can Live With," *Weathervane*, Resources for the Future, Washington, DC, September 29.

Kopp, Raymond, Richard Morgenstern, William A. Pizer and Michael A. Toman (1999). "A Proposal for Credible Early Action in US Climate

Change Policy," *Weathervane*, Resources for the Future, Washington, DC, February 16.

McKibbin, Warwick J., and Peter J. Wilcoxen (1997a). "A Better Way to Slow Global Climate Change" *Brookings Policy Brief*, No. 17, June, Brookings Institution, Washington, DC.

(1997b). "Salvaging the Kyoto Climate Change Negotiations" *Brookings Policy Brief*, No. 27, November, Brookings Institution, Washington, DC.

(2002). *Climate Change after Kyoto: Blueprint for a Realistic Approach*, Washington, DC: Brookings Institution.

McKibbin, Warwick J., Robert Shackleton, and Peter J. Wilcoxen (1999). "What to Expect from an International System of Tradable Permits for Carbon Emissions," *Resource and Energy Economics* 21(3–4): 319–346.

Newell, Richard G., and William A. Pizer (1998). "Regulating Stock Externalities Under Uncertainty," Resources for the Future Discussion Paper 99-10, Washington, DC.

Nordhaus, William D. (1991). "The Cost of Slowing Climate Change: A Survey," *Energy Journal* 12(1): 37–65.

(1992). "The DICE Model: Background and Structure of a Dynamic Integrated Climate-Economy Model of the Economics of Global Warming," Cowles Foundation Discussion Paper No. 1009, New Haven, CT, Cowles Foundation for Research in Economics, Yale University, February.

(1993). "Reflections on the Economics of Climate Change," *Journal of Economic Perspectives* 7(4): 11–25.

(1994). *Managing the Global Commons*, Cambridge, MA: MIT Press.

Nordhaus, William D., and Joseph G. Boyer (1999). "Requiem for Kyoto," *Energy Journal* (Special Issue): 93–130.

Reilly, John, Mustafa Babiker, and Monika Mayer (2001). "Comparing Greenhouse Gases," MIT Joint Program on the Science and Policy of Global Change Report 77, Cambridge, MA.

Roberts, Marc J., and A. Michael Spence (1976). "Effluent Charges and Licenses under Uncertainty," *Journal of Public Economics* 5: 193–208.

Schelling, T. C. (1992). "Some Economics of Global Warming," *American Economic Review* 82(1): March, 1–14.

(1997). "The Cost of Combating Global Warming: Facing the Tradeoffs," *Foreign Affairs* 76(6): Nov./Dec., 8–14.

Tol, Richard S. J. (1999). "Kyoto, Efficiency, and Cost-Effectiveness: Applications of FUND," *Energy Journal* (Special Issue): 131–156.

Victor, David (2001). *The Collapse of the Kyoto Protocol and the Struggle to Slow Global Warming*, Princeton, NJ: Princeton University Press.

Weitzman, Martin L. (1974). "Prices vs. Quantities," *Review of Economic Studies* 41: 477–91.

Commentaries on McKibbin and Wilcoxen

5.1 *The case for greater flexibility in an international climate change agreement*

RICHARD D. MORGENSTERN

Introduction

Warwick McKibbin and Peter Wilcoxen, authors of "A Credible Foundation for Long-Term International Cooperation on Climate Change," challenge the Kyoto orthodoxy, rejecting a top-down international structure in favor of what they term "coordinated national approaches" which involve only minimal intrusion on national sovereignty. They articulate the key goals of an international climate policy and, critically, they highlight the basic limitations of government in implementing the policies needed to achieve long-term reductions of carbon dioxide and other greenhouse gases. Recognizing the inherently voluntary nature of an international agreement, they develop a proposed architecture that is robust with respect to the withdrawal of individual participants from the agreement and, at least in principle, attempts to align the incentives for protection of the atmosphere with economic realities. However, given the complexity of the problem they are addressing, it is not surprising that neither the political economy nor the pragmatic aspects of their approach are fully convincing. In the first part of my comments, I briefly summarize the McKibbin–Wilcoxen assessment of the Kyoto-style top-down approach. The second section focuses on their proposed policy architecture. Third, I consider how the McKibbin–Wilcoxen proposal differs from an approach previously advanced by myself and several of my colleagues. The fourth section examines the political economy issues associated with the McKibbin–Wilcoxen proposal. Finally, I offer some practical suggestions on how to move ahead on the difficult issue of international climate change policy.

The problem definition

McKibbin and Wilcoxen eschew radically new, international institutions as unworkable and seek, instead, to address climate change via policies that can be implemented by national governments, largely via existing arrangements, and with a minimum of international coordination. They identify three key elements of a successful policy: it must be widely adopted; it must remain in force indefinitely; and it must provide credible incentives for individuals and firms to make large-scale, long-term investments to reduce emissions, including support for research and development. They also stress the importance of creating incentives for individuals and firms as opposed to governments to make the needed investments since, unlike most treaties which apply to actions taken directly by governments, the private sector is the key actor in reducing greenhouse gas emissions.

McKibbin and Wilcoxen recognize that new policies do not become credible simply by being written into law, since subsequent legislation can repeal a law, and subsequent administrative actions can weaken enforcement. To push back against the likely pressures for policy relaxation that might occur in the event of an unanticipated technological breakthrough or other major change, McKibbin and Wilcoxen attempt to create a constituency with strong financial interests in perpetuating current climate policies. Such a constituency, they argue, will lobby now and in the future to maintain or increase the stringency of the current policies. At a minimum, this new constituency will serve as a counterweight to those seeking policy relaxation. In that sense, their approach at least partially addresses the potential for dynamic inconsistency in climate policy.

As an example of the incentives for policy relaxation in most climate policy proposals, they note the instability of a carbon tax, economists' favored approach. Despite the obvious efficiency advantages of such an instrument, and the potential for a large revenue stream, imposition of a carbon tax would not create a major private sector constituency for perpetuating the policy over the long term, especially in the face of a technology breakthrough. On the contrary, the very existence of the tax would likely motivate powerful interests to seek rate reduction or total abolition. As they note, even if these interests fail to achieve their goal of policy relaxation, the very act of lobbying against the carbon tax increases the uncertainties associated with long-term investments in research and development and in direct mitigation, and thereby reduces the level of such activities. Thus, despite the economic efficiency of a carbon tax, it is not a viable long-term instrument because it creates inherent conflicts between governments and their constituents. What to do?

The elegant solution

McKibbin and Wilcoxen's proposed solution is quite elegant, marrying the efficiency of a carbon tax with the credibility of long-term permits. Like the late David Bradford (2002), they identify property rights issues as critical and emphasize the importance of defining such rights upfront. Specifically, McKibbin and Wilcoxen propose a hybrid system involving two types of emission permits: (1) long-lived (perpetual) permits which allow one ton of emissions per year for the life of the permit; and (2) short-term (annual) permits, valid for a single ton of emissions, that act at the margin like a carbon tax. The price of the short-term permits would be controlled by the use of a safety valve, presumably one with an escalating price cap. Once allocated, the permits could be traded or leased without limitation. They also recognize that other policies to encourage development and adoption of new technologies, including product standards, demand-side management, direct subsidies, etc. could be combined with their hybrid permit scheme.

Noting that greenhouse gases represent a classic example of a stock pollutant, wherein damages are a function of the total accumulation of these gases rather than short-term emission spikes, McKibbin and Wilcoxen argue that strict adherence to a rigid system of targets and timetables is less important environmentally than the long-term effort to reduce emissions. Because their approach would not oblige a nation to meet its short-run target irrespective of the cost, on efficiency grounds it is clearly superior to one based solely on a fixed quantity system. At the same time, they note that a system based on fixed quantity targets is unlikely to be ratified by the US Senate. Neither is it likely to be a viable long-term option for most other nations. Thus, adherence to a short-term cap is neither economically efficient nor politically realistic.

A key feature of the proposed policy is that the emission permits would be valid only in the country of issue. Thus, in principle, each nation would be able to manage its own domestic activities. To join the agreement, a country would agree to establish a comparable hybrid system with both perpetual and annual permits *and* to coordinate the ceiling price for the annual permits via an (unspecified) international process. Although only limited international negotiations would be required upfront, to be effective over the long term, the agreement would (eventually) need to include all countries with significant greenhouse gas emissions. At the outset, the signatories would be the industrial nations. In the future, other nations would be expected to join, perhaps according to some predetermined formula based on per capita income, size, or other factors.

Because the system would not involve international trading, the withdrawal of a participant would not directly affect compliance strategies or costs of other nations. This is in contrast to the Kyoto Protocol where, as Bohringer (2002) and others have shown, the withdrawal of a large nation like the United States substantially reduced the mitigation costs for the remaining signatories.

Over time, as knowledge about climate science and mitigation costs increases, the signatories might want to further reduce the long-term emission path. In that case, McKibbin and Wilcoxen would oblige the national governments to re-purchase some of the previously issued perpetual permits. The individuals or firms gaining the most from the government buy-back would represent a natural constituency for the more stringent policy. Of course, the converse is also true. That is, if the new scientific and economic knowledge indicated that emissions did not need to be reduced as much as previously believed, the existing permit holders would be expected to lobby against a relaxation.

To hedge against some of the extremes in either direction, McKibbin and Wilcoxen would issue some long-lived permits with a fixed term, say forty or fifty years, rather than giving them all an infinite life. Thus, there would be a term structure of permit lives, much like the current range of bond maturities in financial markets. Interestingly, their proposal is somewhat similar to the architecture used in individual fishing quotas (IFQs), although these quotas are typically allocated for a relatively short period (seven years) (Sanchirico *et al.*, 2006). Despite the current practice of re-issuing the IFQs *gratis* to the existing permit holders, the requirement for new decision creates some uncertainty about the future. In all likelihood, adoption of a similar mechanism for climate change would create much greater uncertainty about future policies, thereby reducing incentives for investment in research and development or direct mitigation. Thus, adoption of the IFQ-type mechanism would undermine one of the principal rationales for the McKibbin–Wilcoxen approach – namely, to increase long-term investment in carbon mitigation.

Comparison with Kopp *et al.*

In 1997, around the time McKibbin and Wilcoxen first introduced the policy framework that has evolved into their present proposal, Kopp, Morgenstern, and Pizer (1997) advanced a policy framework with many of the same characteristics.[1] While a number of the specifics of both proposals have evolved over the years, the key elements remain.[2] This section briefly outlines some of the key similarities and differences between the two approaches.

Like McKibbin and Wilcoxen, Kopp and his colleagues take a long-term, price-based perspective and require governments to frame commitments in terms of both the level of effort required (the safety-valve price) and the emission reductions they seek to achieve (a quantity target). Unlike McKibbin and Wilcoxen, however, who propose to make the key decisions upfront about the level of control and the allocation of permits decisions, Kopp and his colleagues seek greater flexibility. Specifically, they are more focused on the possibility of changing emission targets over time in response to new information about climate science and economics. Whereas McKibbin and Wilcoxen would force the government to repurchase allowances in the event the target needs to be reduced, Kopp and his colleagues would have the long-term property rights reside in the public sector, thereby reducing the government's cost of such a decision. Regarding the allocation of permits, McKibbin and Wilcoxen would make an upfront decision to allocate the permits freely, thus foreclosing options for redistribution to new or different sources, to heavily impacted consumer or regional groups or, more generally, to the possibility of an auction. In their writings, Kopp, Morgenstern, and Pizer are more concerned about the equity and efficiency aspects of the upfront allocation, while McKibbin and Wilcoxen

[1] While the two proposals were developed independently, the Kopp *et al.* paper was published several months after McKibbin and Wilcoxen.

[2] See also Kopp *et al.* 1999 and Kopp *et al.* 2004 for their updated work.

focus on building a political constituency for perpetuation of the policies (see next section for further discussion).

A second difference between the two approaches involves support of research and development (R & D). Whereas McKibbin and Wilcoxen are quite open to such a program, they do not advance a specific funding mechanism to support it. In contrast, Kopp *et al.* (2004) explicitly earmark some of the revenues from the sale of safety-valve permits to finance an R & D program. Third, although McKibbin and Wilcoxen oppose a Kyoto-style agreement, they do advocate a specific international framework to be negotiated upfront. Kopp *et al.* (2004) focus on a bottom-up strategy, preferring to lead by example, building on one or more successful domestic programs before locking into a specific international architecture.

Political economy and pragmatism

Despite its elegance, the McKibbin–Wilcoxen approach faces numerous hurdles in both the domestic and the international spheres. Even though it is somewhat less restrictive than the Kyoto Protocol, it would impose a one-size-fits-all domestic architecture on participating nations. Gaining broad international support for such a system would be a daunting task, especially in light of the range of policy preferences already manifest by Annex I nations. A second issue involves the allocation of permits. Given the political difficulties associated with allocating annual permits, e.g., in the European Union, one can only imagine the problems associated with allocating perpetual permits. Third, there are concerns about both the fairness and efficiency of grandfathering as the sole mechanism for permit allocation. Fourth, there are serious questions about the ability of the system to respond to new scientific information. Finally, it is unclear whether McKibbin and Wilcoxen's approach to developing nations is workable.

Notwithstanding William Pizer's observation (Chapter 7 of this volume) that current permit prices are clustered in the relatively narrow range of \$10–25 per ton of CO_2, the variety of domestic policy designs currently in use or planned suggests considerable disagreement among nations about the preferred approaches for domestic policies. The European Union has embraced a fixed cap-and-trade system on heavy industry (including electric generators), involving about one-half of total CO_2 emissions. Japan has considered a modest carbon tax but rejected it in favor of a voluntary program for large firms combined with efficiency standards and aggressive use of the Clean Development Mechanism. Canada initially announced a cap-and-trade program on heavy industry (including electric generators), based on intensity targets and including a safety valve. However, in March 2006, the newly elected government indicated its intention to step back from this approach (details are not yet available). In New Zealand, a modest carbon tax is apparently the policy of choice, although a number of large exemptions have been proposed, including for energy-intensive industries and major segments of the agricultural sector.

With such a wide array of designs, the prospects do not seem bright for the adoption of a common domestic policy framework along the lines of McKibbin and Wilcoxen. Not surprisingly, each country seems to have crafted its domestic policies to meet it own national circumstances and needs. While it might be plausible to integrate the Canadian cap-and-trade with a safety valve and the New Zealand carbon tax approaches, the EU fixed cap system reflects quite a different philosophy. Japan has yet to embrace a transparent market-based system of any kind.

A subsidiary issue involves the sectoral coverage of each of the national systems. Among the individual EU nations, there are small differences in coverage. Among all Annex I countries, the differences are more pronounced. Thus, even though they reject new, top-down international agreements, McKibbin and Wilcoxen must first persuade a number of major nations to alter the arrangements currently established within their own borders. While this may not be an insurmountable task, it is not trivial either, especially in the near term.

A second concern with the McKibbin–Wilcoxen approach involves allowance allocation, especially for the long-lived permits. In the United States, the allocation of annual SO_2 and other non-carbon allowances has often proven to be quite contentious. For carbon, the situation is likely to be more difficult. Reportedly, serious conflicts arose in the EU ETS. Recent Senate hearings on the Bingaman–Domenici proposal highlight the diversity of views on the appropriate allocation in the United States. Powerful interests, including energy producers, large industrial users, consumer groups, and others are lining up to fight for these potentially valuable assets. Imagine if the stakes were raised to include the allocation of *perpetual* permits where the present value of the allowances was ten to fifteen times greater. McKibbin and Wilcoxen are completely silent on this point.

A third issue concerns both the fairness and efficiency of the McKibbin–Wilcoxen architecture. Grandfathering perpetual permits worth tens or hundreds of billions of dollars clearly raises equity concerns among individual consumers, and across industries and regions. New (industrial) entrants face special problems. As is well known, many public goods, including radio spectrum, are not freely allocated in the United States or in other developed nations. At the same time, as demonstrated by Burtraw *et al.* (2002), there are efficiency gains from auctioning rather than freely allocating permits. Inherent in the McKibbin–Wilcoxen argument, however, is the need to create a constituency that does not have to pay for its emissions, either now or in the future. This dependence on grandfathering seems inconsistent with the long-term goal of achieving greater fairness and efficiency via auctioning of permits. This point has also been emphasized by Kopp *et al.* (1999).

A fourth issue concerns the implications for changing the long-term target in response to new scientific or economic information. Imagine a situation where new scientific or economic analysis suggests that further tightening of the long-term targets is appropriate. Under the McKibbin–Wilcoxen approach this could only be accomplished by having national governments purchase perpetual permits in the open market. Yet, as Jorgenson and Yun (2001) have demonstrated, the burden to the economy of raising taxes in excess of the additional tax revenue collected is of the order of 27 cents per dollar of extra tax revenue raised (based on the 1996 tax law). Thus, to finance the purchase of these perpetual permits, the nation would bear a considerable cost beyond the funds actually paid to the permit holders. Combined with the political difficulty of raising taxes in most nations, this creates a strong disincentive – despite the lobbying efforts of the existing permit holders – to increase stringency of the policies in response to new scientific or economic information.

The converse situation may also be problematic. Imagine that new scientific or economic information suggested that loosening rather than tightening was the appropriate policy. In that case, the Jorgenson and Yun analysis would support the sale of additional permits and use of the proceeds to reduce existing taxes. The average citizen, if convinced of the scientific merits of the case, presumably would also favor the tax cuts. But in this instance the owners of existing permits would tend to resist what economists would see as a welfare-enhancing decision.

Thus, the McKibbin–Wilcoxen approach may alter but it would not eliminate the inherent conflicts between governments and their constituents.

A final issue concerns the application of the McKibbin–Wilcoxen approach to developing nations. Under their proposal, developing countries would be expected (eventually) to establish a hybrid permit system of the type proposed for developed nations and coordinate permit prices internationally. Since this is the only means available under the McKibbin–Wilcoxen approach to equalize marginal abatement costs across nations, it is essential to achieving cost-effective reductions on a global scale. But what happens if some nations, most likely developing countries, are unable or unwilling to establish the necessary institutional structures to support a McKibbin–Wilcoxen type system? Although there are some limited exceptions, the evidence so far indicates that developing nations have generally been quite slow to embrace market-oriented approaches for environmental protection. The Clean Development Mechanism, which involves carrots but no sticks, is not really a relevant precedent for a hybrid trading system. In the future, with greater market reforms, it is possible that hybrid-type trading systems would become viable in significant parts of the developing world. In the near term, however, the institutional arrangements in many of these nations could not support such an approach. Ironically, if the institutional structure needed to implement a hybrid system was interpreted to be a threshold for the adoption of meaningful commitments in the developing world, the McKibbin–Wilcoxen mechanism might actually hinder near-term progress on softer, less-demanding mitigation measures.

Conclusions

McKibbin and Wilcoxen correctly identify the need for long-term incentives to accelerate the development and adoption of new technologies in the context of a widely accepted international agreement. Overall, however, despite the elegance of their approach, there are many impediments to wrapping the diverse national approaches now in operation or planned into a coordinated, one-size-fits-all national system – even one with flexible timing of entry and without international trading. As noted, there are also concerns about the fairness and adaptability of their proposed system, as well as its efficiency. In many developing nations, the proposal simply may be a non-starter.

A more pragmatic approach would try to build on existing efforts in a bottom-up fashion, reflecting national circumstances and needs – à la Kopp *et al.* (2004). Toward that end, the international community might step back from either a top-down Kyoto or McKibbin–Wilcoxen type approach and seek, instead, to encourage meaningful action by individual nations without an overly rigid structure. Thus, emission trading (with or without a safety valve), carbon taxes, regulatory approaches, and even voluntary programs could all be considered acceptable approaches, at least for now. For poorer countries, more development-friendly options are clearly in order.

The prospects for future international progress would certainly be enhanced if one could point to genuine success in the United States or other large nation with relatively high rates of GDP growth. Even though international negotiations on climate change have been under way for almost two decades, to date no major nation has yet demonstrated a viable domestic architecture suitable for achieving large-scale emission reductions and none, except for special cases like the United Kingdom, which experienced large changes in its resource base, or Germany, which benefited from economic restructuring, has made substantial progress in actually reducing its emissions. We need to walk before we run.

Although my intention is not to endorse a specific option, I would make the case for a more flexible, bottom-up alternative to either a Kyoto-style agreement or the McKibbin–Wilcoxen approach. As Bodansky (2004) and others have noted, the literature on such options is substantial and it is growing. What these alternatives share is the deference they give to national circumstances, especially at the early stages of the process.

The key to success for any international system is to establish a credible evaluation process which encourages institutional learning at the scale of the nation-state. Such a process should include specific performance metrics involving transparent, internationally acceptable, measures of effort and/or actual environmental outcomes. In the context of a formal international process to review the progress made by individual nations, these metrics would provide feedback to individual countries on their own progress and, simultaneously, serve as a prod to further action.

Over time, once nations have gained experience with their own policies and have had the opportunity to compare their experiences with others, the system would certainly evolve. Strong leadership by one or more large nations would certainly accelerate the process. While some may criticize the lack of elegance or speed of the walk-before-you-run approach, in my view it is the only realistic way forward.

References

Bodansky, Daniel (2004). *International Climate Efforts Beyond 2012: A Survey of Approaches*. Arlington, VA: Pew Center on Global Climate Change.

Bohringer, Christoph (2002). "Climate Policies from Kyoto to Bonn: From Little to Nothing," *Energy Journal* 23(2): 61–71.

Bradford, David (2002). "Improving on Kyoto: Greenhouse Gas Control as the Purchase of a Public Good," Dept. of Economics Working Paper, Princeton University.

Burtraw, Dallas, Karen Palmer, Ranjit Bharvirkar, and Anthony Paul (2002). "The Effect on Asset Values of the Allocation of Carbon Dioxide Emission Allowances," *Electricity Journal* 15(5): 51–62.

Jorgenson, Dale W., and Kun-Young Yun (2001). *Investment Volume 3: Lifting the Burden*, Cambridge, MA: MIT Press.

Kopp, Raymond J., Richard D. Morgenstern, and William A. Pizer (1997). "Something for Everyone: A Climate Policy that Both Environmentalists and Industry Can Live With," *Weathervane*, Resources for the Future, Washington, DC, September 29.

Kopp, Raymond J., Richard D. Morgenstern, William A. Pizer, and Michael Toman (1999). "A Proposal for Credible Early Action in US Climate Change Policy," *Weathervane*, Resources for the Future, Washington, DC, February 16.

Kopp, Raymond J., Richard D. Morgenstern, Richard G. Newell, and William A. Pizer (2004). "Stimulating Technology to Slow Climate Change," in Richard D. Morgenstern and Paul R. Portney (eds), *New Approaches on Energy and the Environment*, Washington, DC: Resources for the Future Press.

Sanchirico, James N., Daniel Holland, Kathryn Quigley, and Mark Fina (2006). "Catch-quota Balancing in Multispecies Individual Fishing Quotas," *Marine Policy* 30(6): 767–785.

5.2 *Using the development agenda to build climate mitigation support*

JONATHAN PERSHING

McKibbin and Wilcoxen, in their "A Credible Foundation for Long-Term International Cooperation on Climate Change," have articulated a powerful argument for a new structure for climate mitigation efforts.

In particular, they highlight the importance of two elements for success:

(1) the long-term nature of the problem – and thus, the requirement for long-term solutions. Solutions that are implemented erratically, with varying levels of stringency over time, and with periods of non-enforcement will not likely yield success.

(2) The requirement that any regime be built around national systems for compliance and enforcement, not predicated on what may likely be weak international systems where there is no guarantee for comparability of rigor and action.

There is no question that both elements are indeed central to any successful climate mitigation solution. However, it behooves us to look closely at some of the implications of the McKibbin–Wilcoxen analysis before committing to the path they suggest.

Political buy-in: a critical element?

There is no question that climate change itself is a long-term phenomenon, and that mitigation efforts are therefore likely to be needed over a period of many decades. McKibbin and Wilcoxen argue that the political will to enact such long-term policies can best be achieved through allocation of property rights (in the form of long-term or "perpetual" emission permits); holders of such permits, in order to maintain their value, will support continued government action on climate change. The greater the stringency of the government's intervention, the greater the value of the permit – and hence the greater the incentive for the permit holder to support policies.

This argument is built on a number of assumptions:

(1) An emission trading regime is an effective mechanism to address climate change (hence, allocating and allowing trade in permits is a legitimate part of a successful solution);

(2) Holders of permits will be supportive of stringent policies – as such policies will enhance the value of their assets; and

(3) It is possible to determine an allocation that is large enough to ensure political support, but still small enough to assure that appropriate levels of climate mitigation are achieved.

With respect to the first, analyses by multiple authors suggest that a cap-and-trade emission trading program is indeed an effective approach to addressing climate change. However, these same analyses suggest that not all sectors are necessarily amenable to a cap-and-trade approach. Thus, while the power sector and heavy industry may be well served through such a regime, residential and commercial space, forestry and agriculture may not be. At the global level, emissions of the six major greenhouse gases from electricity and heat, industry, and fugitive emissions (those usually thought best suited to a trading regime) account for less than 45 percent of total emissions (see World Resources Institute Climate Analysis Indicators Tool, WRI/CAIT, a database with information culled from more than a dozen different sources covering greenhouse gas emissions and other climate change relevant indicators from 175 countries around the world). While tradable permit programs have been suggested for transport (which would bring another nearly 14 percent of total emissions into a system), few analysts think significant emission reductions in the transport sector are likely unless permit prices climb quite high. Furthermore, while a number of suggestions have been offered as to how to incorporate the transport sector into the cap-and-trade regime, no program has yet applied any of these models. Of course, this does not inherently obviate the potential advantages of employing an emission trading program for the sector (after all, the very nature of a trading regime allows any actor to purchase emission reductions from the least-cost seller – and transport sector participants in the market would thus not be required to implement costly reductions internally but could take advantage of lower-cost options elsewhere). Notwithstanding the economic argument, a political one may still hold: there may be strenuous resistance against a regime that serves only as a transfer mechanism from one sector to another. Given the limited options for low-cost reductions within the sector, individual drivers are likely to see the permits passed through as increases in vehicle or fuel prices, making the trading approach very like a tax. While a similar pass-through may also occur in other sectors (e.g., electricity), low-cost emission reduction opportunities within other sectors may lead to a lower-level pass-through than in the case of transport.[1]

This conclusion has implications for whether the allocation can be large enough to attract political buy-in. If we begin with the Intergovernmental Panel on Climate Change (IPCC) contention that emissions must be reduced by 60 to 80 percent[2] from current levels in order to stabilize global GHG concentrations, we have an initial boundary condition for permit allocations: conservatively, no more than 20 percent of total current emission levels can be allowed over the long term. However, using the analysis above (suggesting that a permit system might only be applicable for about a half of total emissions), it may only be permissible to allocate permits amounting to about 10 percent of current national emissions. Even a 10 percent figure may be too high: as McKibbin and Wilcoxen note, emissions are not currently distributed evenly across countries, and even less so on a per capita basis (on a per capita basis, developing countries emit only about one-third the total of developed countries, see WRI/CAIT). Correcting for inequities between developed and developing countries, and seeking to move toward convergence on per capita emissions, a rough estimate suggests that about two-thirds of the allowable long-term emissions should go to developing countries, and only one-third to developed countries. Doing so would thus allow for long-term permits to be

[1] The author is indebted to Rob Stavins and Joe Aldy for pointing out that a cap-and-trade system looks very like a tax to end-use energy consumers across the economy, not just in transportation. They in particular note that this consumer-felt price increase has been problematic in electricity and heating, especially given the run-up in electricity, natural gas, and heating oil prices over the past year, and suggest (in the author's view accurately) that some of the opposition to the EU ETS over the past year, especially when permit prices exceed 30 euros/ton, reflected the surprise among consumers that their electricity prices moved with the carbon permit prices.

[2] It might be noted that while this level of reduction is required to meet any level of stabilization between 400 and 700 ppm, the timing will vary depending on the final desired concentration. According to the IPCC Third Assessment Report, global peaking and subsequent declines in greenhouse gas emissions will need to occur by about 2015 if we wish to limit concentrations to 450 ppm, by 2030 if limits of 550 are set and 2045 if limits of 650 are to be allowed.

issued only in the amount of about 3–4 percent of current national emissions for developed nations. It is not clear whether allocating such a limited quantity of permits would generate the kind of political support McKibbin and Wilcoxen seek.

If a larger portion of permits were to be allocated, it is not clear that the total reductions sought would be achieved. Similarly, if a higher proportion of permits were allocated to developed nation emitters, developing countries, citing equity concerns, may be difficult to bring into the system. In any event, the allocation levels are likely to be substantially below the "90 percent of 1990 carbon emissions" level suggested by McKibbin and Wilcoxen.[3]

One solution proposed by McKibbin and Wilcoxen for this problem is that the allocation of very long-term property rights be limited – they suggest variously twenty, forty, or seventy-five years. While this may theoretically alleviate the problem, there are potential conflicts. The timetable for long-term reductions has been discussed in a number of recent meetings and papers (see, for example, the UK government-sponsored meeting in Exeter in 2005, "Avoiding Dangerous Climate Change"). Building on a variety of statistical and scientific analyses, both governments and individual researchers have argued that achieving stabilization at levels that are not dangerous requires temperature increases to be held below 2°C – and hence, that concentrations be limited to no more than 450 ppm (e.g., according to Meinshausen 2005, at 450 ppm, there is only a 50 percent chance that temperatures will be held below 2°C). Modeling work suggests that to do so will require global reduction of between 30 and 50 percent below current levels by 2050 – less than forty-five years out.[4] Thus, the McKibbin–Wilcoxen proposal for expiry dates would fit – but would require the preponderance of permits to expire in the relatively nearer term – or, alternatively, that a higher proportion of the allocations be in the form of annual permits.

[3] It might be noted that McKibbin and Wilcoxen's analysis is based on assumptions that the intent is to meet the Kyoto targets. However, virtually all analysis is clear on this: meeting Kyoto level targets will have at best a very limited effect on global emissions, and virtually none on changing the long-term concentrations required to solve the climate change problem.

[4] The reader should note that there is no inconsistency between a reduction of 50 percent by 2050 and a long term need for 60–80 percent reductions; Meinshausen (and other undertaking similar modeling work) assumes emissions continue to decline after 2050 to meet targeted stabilizations levels.

As allowances are fully transferable, initial allocations in an emission trading system do not affect the marginal cost of abatement. However, this does not obviate the political issue of allocation. McKibbin and Wilcoxen argue that political expediency dictates that allocation of permits be based on historical emissions: those who have emitted significant quantities in the past should be given allowances to continue to do so. Even in the case of allocations of annual permits, this may be unacceptable – but if permits are allocated in perpetuity, this may sit even less well with the public.

Currently (due to factors well beyond climate policy), electricity prices are rising rapidly. Allocating additional (and permanent) entitlements to companies already collecting record profits may be a difficult political choice. A similar objection may be directed at the oil and gas sector: US efforts to increase subsidies for oil and gas production have been faced with increasing opposition. While this objection may be relevant to any form of allocation for emission permits, it is exacerbated when those permits are assigned in perpetuity.

Another aspect of concern on permanent allocations is that those to whom permits are given may not, over the longer term, need them for compliance. For example, if permits are allocated to the power sector, but over the next fifty years, power production becomes GHG-neutral, power companies will hold an asset and reap rewards from an entitlement with historic but no ongoing rationale. While McKibbin and Wilcoxen rely on the value of such long-term holdings to generate political pressure, it is precisely in such cases that government may choose to renege on its early commitments.

This would not be without precedent: there are many cases of the government providing property rights which they later take away. Examples of takings of physical property are not uncommon. Tax incentives and benefits are even more frequently given and removed. Given that no session of Congress can pass a law that another cannot revise, the certainty of "permanence" may be somewhat fragile. It is not clear what kind of discount rate a holder of a theoretically permanent permit would apply to a holding, but it is clearly higher than zero. Furthermore, in determining the political value of maintaining the permit, holders would also need to calculate the cost of engaging in the political debate to retain their rights – a cost that may not be easily passed on to consumers of the emissions generated by their permits.

Creating a GHG market: problems with price caps

The system suggested by McKibbin and Wilcoxen is not limited to a long-term element; it also includes, for the remainder of the annual obligation, a short-term component. McKibbin and Wilcoxen assume that "annual" permits would be sold by the government at a fixed price; they suggest, initially, a price of $20 per ton of carbon. If emission reductions cost more than $20 per ton of carbon, permit sales would perform as taxes (the permit price acts as a cap on costs); if reduction costs were lower, they would provide investor certainty that compliance costs would not rise.

A number of concerns have been raised with regard to the barriers and consequences of government seeking to set and maintain market prices. We do have some experience with such efforts: local electricity markets, where government or quasi-government public utility commissions set prices, or internationally, with OPEC price bands. In both cases, there have been huge complaints. In the electricity case, public utility commissions have been accused variously of bowing to pressure from power producers (who argue they need high prices to allow for investment in new production), or from the public (which always prefers low prices whether or not these provide an adequate return on investment). In the international oil price case, when lifting costs are less than $10 per barrel of oil, OPEC price setting is largely about rent transfers – and many governments find it an unacceptable basis for international relations.

Essentially, picking a price is (albeit in a less constrained way) a variant of picking a specific technology – and governments have been historically poor at picking specific technologies to solve problems. Technologies that may today have a price higher than the cap are less likely to be developed. If instead the desirable emission level is picked and price not specified, companies have a considerably greater incentive to seek least-cost near- and long-term compliance opportunities rather than working only on options that are below the price cap level.

Another rationale for supporting price caps is that political objections to high prices could lead to efforts to dismantle the entire system. To date, we have a rather mixed record. For example, Figure 5.2.1 shows the prices of the US NO_X trading program. In its early history, prices spiked at nearly $7,000 per ton, nearly five times the longer-term

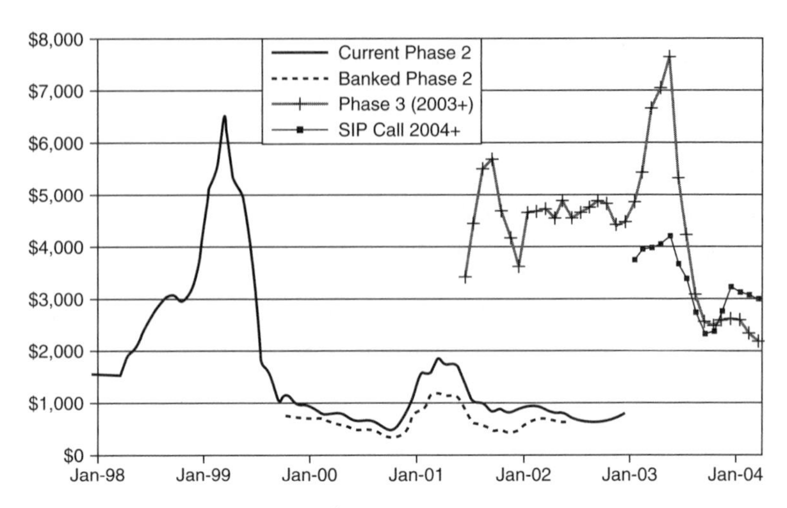

Figure 5.2.1 US OTC NO_X market prices

average costs. Yet, the program was not halted – and in fact, investments continued and the overall environmental effectiveness of the program was very high: emissions were well below allocated amounts in all years of the program (see Aulisi *et al.* 2005).[5]

A similar example in emission trading under the European Emission Trading Scheme (EU ETS) suggests that a price cap may not be a necessary component. Prices in the EU market rose steadily over the first year of operation (from under €20 to nearly €30 before falling sharply to their mid-2006 levels of about €15/ton (see Figure 5.2.2). A number of rationales for the price collapse have been offered, including efforts at market manipulation, inadequacy of reporting, and expectations that the market would be short. However, had a price cap been set and had the government issued additional permits once the cap had been breached, it seems likely that an even greater price collapse would have been observed as oversupply would have left little if any reduction

[5] The author is indebted to Joe Aldy and Rob Stavins for noting a different NOX case as a contrast: California's South Coast Air Quality Management District's Regional Clean Air Incentive Market (RECLAIM). In this case, emissions caps were not entirely met, and companies paid a non-compliance penalty. This may have had an effect similar to that of a price cap – although some authors (see for example, Kolstad and Wolak 2003) argue that the entire price spike was the result of an effort by power generators to manipulate their market control over electricity prices.

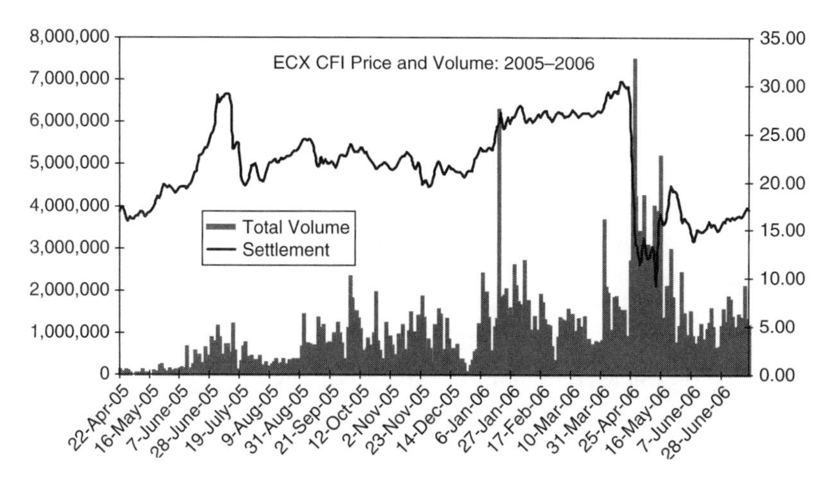

Figure 5.2.2 European emissions market, volumes and settlement prices

Source: European Climate Exchange

requirement for firms with allocations. With demand near zero, prices would have been extremely low and would have generated little if any incentive for changes in performance.

The European market offers another lesson: one related to political support for the system. During the course of the negotiations on permit allocations, some of the most vocal opposition came from the electricity utilities, which argued for generous allocations in order to limit the emission reductions they would need to make. When allocations were made, in many cases these same companies benefited from the asset value they received, incorporating them into their corporate valuations. As long as ETS permit prices were rising, so did corporate value – but as prices began to fall, these same companies again petitioned for redress, claiming the entire emission trading system had been mismanaged. This appears to be leading, in many cases, to a weakening of the EU ETS regime in its second period – quite the opposite effect to that suggested to occur by McKibbin and Wilcoxen.

Conversely, the deregulation of the California energy market led to a significant spike in prices (a function of poor market design, price gouging by monopolistic producers, and scarcity of generation capacity), and a re-regulation by the California public utility commission following not only numerous blackouts but essentially a voter uprising leading to a change in government (see Bushnell 2003). However, even in this case it is not clear that a price cap would have led to a fundamentally different outcome: the utilities argued that the design constraints made it impossible for them to invest in new capacity – and a price cap may not have fixed this problem.

Perhaps the most problematic aspect of establishing price caps is that they constrain the cost of compliance by voiding the environmental outcome desired. Given the relatively limited literature on the costs of climate change, it is difficult to make an assessment of how much we "should" pay for emission reductions. Mid-range estimates using IPCC projections suggest global GDP losses from climate change of several percent (for a new computational analysis showing an impact of up to 3 percent see Nordhaus 2006); more optimistic numbers suggest that at a low level of impact, small benefits might even accrue in certain countries (e.g., see Jorgenson *et al.* 2004). However, these analyses do not take into account abrupt change and acknowledge that uncertainties around adaptive capacity are critical factors. Consequences for the United States, with high capacity, may be much more limited than those globally – and certainly are likely to be lower than those in developing countries.

Given that a significant share of climate change will be locked in over the next few decades, it is difficult to judge how high (or low) an initial cap price should be. Technology optimists suggest that very low prices (and hence low cap levels) might drive adequate change over long time periods, although most acknowledge they will provide little change in the near term. Conversely, higher prices or caps would yield greater near-term reductions but are likely to be politically unpalatable. With the difficulty of establishing precise cost information, the debate is likely to be politically charged. In this context, it seems likely that the relative greater level of political power of those objecting to any price, much less a high price, will win the debate. Thus, we are likely to emerge with low prices, and concomitantly relatively limited action being undertaken. Conversely, if environmental benefits rather than costs were the basis for debate, a different constituency may be brought to the table, likely leading to more-aggressive near-term efforts.

There is one final issue that must be incorporated into any system design: negotiating complexity. While, as discussed above, there are both merits and disadvantages to the use of both long-term property rights and short-term permits when taken individually, together they pose a significant additional problem: they require the negotiation of a rather complex system. Given the political difficulties inherent in complex negotiations, combining them may prove an overwhelming obstacle.

Political buy-in must be found not only for a permit system (currently attracting rather limited support in the developing world), but also for what looks very much like a tax – to which, historically, there has been even stronger opposition. Up to the point of the final allocation, those who might be expected to support the regime (i.e., companies being given permanent rights) cannot be guaranteed those permits and thus are likely to be lukewarm to the idea. Conversely, as with any permit or tax system, those expecting to need to pay for permits are likely to oppose action. In this sense, the system, while possibly generating long-term support once implemented, faces some of the same problems of any new regime at the point of origination.

Limiting international trade: an inefficient approach

McKibbin and Wilcoxen argue that, for reasons of maintaining robustness and integrity in their permits, trade between countries should be prohibited. This, they rightly argue, would have the benefit of eliminating a possibly costly requirement for an international system of compliance and regulatory oversight. They also suggest it would also prevent market dynamics in one country from affecting the markets in other countries. Instead, they argue that the international agreement should focus on establishing the price for annual permits.

However, this approach seems to gloss over several important concerns. While McKibbin and Wilcoxen note that such a system would be difficult to negotiate, they assume that it could quite readily be done. This does not seem plausible. It is interesting to note, however, that an approach calling for globally harmonized taxes *was* proposed during the climate change negotiations and failed to garner support even from within the coalition suggesting it (in this case, the European Community).

The rationale behind the failure to adopt such a system may equally apply in this case. Countries are not persuaded that they should all pay the same price for emission allowances. Some, including most developing countries, believe that they should have no obligation, as they were not responsible for the current problem. For them, any price is too high. Others, seeking to use prices to force technology development, believe too low a price will fail to generate an acceptable outcome. The result is more likely to be an impasse than an agreement.

McKibbin and Wilcoxen suggest that one solution to this is to assign more permits to developing than to developed countries. However, as suggested above, there may not be that many permits to allocate, and providing more to developing countries means fewer in the developed world if stringent emission reductions are to be achieved (and consequently lower levels of political support).

Initial participation in a price-setting regime could also be problematic. Presumably, the initial discussion would occur between countries prepared to act. Thus, additional countries could only accede by adopting the already agreed price. In many cases, it might be supposed that countries would only choose to join in a second or third round – at which point prices are likely to have risen above initial levels (otherwise, emissions would not be reduced). In such a case, new entrants would be faced with a considerable hurdle; the analog is one where the carbon cost, instead of steadily ratcheting up, were to be imposed at the highest level in the first instance. Under such circumstances, it may be impossible for new accession (and relatively poorer) countries to participate.

Finally, the idea of an autarchic approach is not likely to be a least-cost solution. McKibbin and Wilcoxen suggest that differences in national efficiencies are small. This clearly is not true. According to GHG and economic statistics (see WRI/CAIT), the GHG intensity of economies varies by a factor of at least three. Such differences – likely substantially greater when it is individual companies that are considered – are certain to make trading much less efficient if limited to a single country than if it were open to global markets.

This suggests that the system envisaged by McKibbin and Wilcoxen is not likely to be implemented, or if implemented is likely to be ignored. One possible outcome is that it could lead to trade distortions as some countries seek to compensate for differences in costs imposed by permit prices. This, in turn, could lead to tariff and non-tariff barriers to trade as competitiveness was affected. Alternatively, it could lead to increased levels of corporate relocation, as well as to accounting manipulation to manage corporate accounts across borders.

An alternative suggestion

McKibbin and Wilcoxen offer critical insights into the demands placed on a new system. The need to affect long-term change in policy *does* require creating supportive constituencies. However, they perhaps overemphasize this point at the expense of ignoring other critical problems facing the development of a new climate regime. These include the lack of political will to negotiate, the need to broaden participation to include developing countries (and the United States), and the requirement to make fundamental changes in the structure of economies if the scale of emission reductions (60–80 percent) is to be achieved.

It is clear that no single solution is likely to be adequate to addressing all of these barriers. Rather, it is likely that a hybrid, employing multiple elements will be used, and that the choice of the elements will differ from country to country.

However, if we look for synergies between the barriers suggested by McKibbin and Wilcoxen (the need for sustained long-term buy-in), and those on developing country engagement, there may be a different approach that offers promise: that of "sustainable development policies and measures" (SD-PAMs). First identified by Winkler *et al.* (2002), and elaborated by Bradley and Baumert (2005), this approach seeks to promote specific development approaches that are compatible with GHG reductions. Thus, for example, it looks at issues such as providing electricity to the rural poor – a matter that garners considerable public support – and examines options for electrification that incorporate low or zero-emitting forms of energy. It looks at energy security (particularly relevant to the modern world economy as oil prices soar) and considers options expanding public transport, or for development of biofuels. Essentially, it incorporates climate mitigation measures into development priorities that are of enormous import already. Such an approach – more simply than that of McKibbin and Wilcoxen's permanent allocations – addresses the issue of constituencies and political buy-in and, as all SD-PAMs inherently have long-term value, creates incentives for similarly long-term policy support. This approach also has the merit of engaging developing countries: in international climate change negotiations over the past decade and more, this bloc has stated that development issues are their first priorities.

There may be some synergies with the McKibbin–Wilcoxen approach and that of SD-PAMs. In particular, capital costs are a critical issue in the SD-PAMs debate – and a trading regime in which governments auction permits may provide some of the revenues required. Such revenues could be derived from any auctioned permit system – and need not be tied to one with permanent property rights. Revenues could also be derived (although perhaps in smaller amounts) from project-based offsets (currently not discussed in the McKibbin–Wilcoxen proposal) that could be attached to any tradable permit system (and currently used both in the EU ETS as well as in the Kyoto Protocol framework).

Ultimately, new designs for international agreements must accurately identify the questions they seek to resolve. In the McKibbin–Wilcoxen proposal, the focus is on the long-term need for consistency and continuity in program implementation. While other issues are clearly of equal (or greater) concern in the near term, McKibbin and Wilcoxen's proposal does open the door to different ideas. As such, it usefully pushes us to think about options that build long-term engagement rather than only developing a policy that lasts a few years before collapsing for lack of political buy-in.

References

Aulisi, Andrew, A. Farrell, J. Pershing, and S. Vandeveer (2005). *Greenhouse Gas Emissions Trading in US States: Observations and Lessons from the OTC NO_X Budget Program*. Washington, DC: World Resources Institute.

Bradley, R., and Baumert, K., (eds.) (2005). *Growing in the Greenhouse: Protecting the Climate by Putting Development First*. Washington, DC: World Resources Institute.

Bushnell, James (2003). *California's Electricity Crisis: A Market Apart?* Center for the Study of Energy Markets (CSEM) Working Paper 119, Berkeley, University of California Energy Institute.

Jorgenson, D. W., R. Goettle, B. Hurd, and J. Smith (2004). *US Market Consequences of Global Climate Change*. Pew Center on Global Climate Change, Arlington, VA. Available at www.pewclimate.org/.

Kolstad, J., and F. A. Wolak (2003). *Using Environmental Emissions Permit Prices to Raise Electricity Prices: Evidence from the California Electricity Market*. Center for the Study of Energy Markets (CSEM) Working Paper 113, Berkeley, University of California Energy Institute.

Meinshausen, M. (2005). *On the Risk of Overshooting 2°C*. UK Conference on Avoiding Dangerous Climate Change, February 1–3, 2005. Available at www.stabilisation2005.com/14_Malte_ Meinshausen.pdf.

Nordhaus, William D. (2006). "Geography and Macroeconomics: New Data and New Findings," *Proceedings of the National Academy of Sciences* 103(10): 3510–3517.

Winkler, H., R. Spalding-Fecher, S. Mwakasonda, and O. Davidson (2002). "Building on the Kyoto Protocol: Options for protecting the climate," in K. Baumert, O. Blanchard, S. Llosa, and J. F. Perkhaus (eds.), *Sustainable Development Policies and Measures: Starting from Development to Tackle Climate Change*, Washington, DC: World Resources Institute, pp. 61–87.

World Resources Institute (2006). *Climate Analysis Indicators Tool (CAIT)*, Washington, DC: World Resources Institute. Available at www.cait.wri.org/.

PART III
Coordinated and unilateral policies

6 *A multitrack climate treaty system*

SCOTT BARRETT

Introduction

THE global regime for climate change consists of two agreements, the United Nations Framework Convention on Climate Change (UNFCCC) and the Kyoto Protocol, neither of which adequately addresses the real challenge. Certainly, neither agreement has had much effect so far. Atmospheric concentrations of carbon dioxide have increased every year since the UNFCCC was negotiated in 1992. The trend did not change after the Kyoto Protocol was negotiated in 1997. Although Kyoto is still some way from being implemented, its shortcomings are already apparent. A new approach is needed.

In this paper I sketch out an alternative treaty system – a new "architecture," in the language of this book. Taking the objective to be sustainable development that reduces climate change risk, this new system consists of four parts: first, protocols that promote research and development (R & D) into new energy, air capture, and geoengineering technologies; second, protocols that encourage the development and diffusion of new mitigation technologies emerging from this R & D; third, cooperation in financing investments that will make the poorest countries less vulnerable to climate change (an example being investments that reduce malaria prevalence or that improve malaria treatment); and, fourth, agreements on the deployment of geoengineering technologies, particularly in response to the first warning signs of abrupt or catastrophic climate change.

This is a fundamentally different approach, and to understand the reason for it, it is best to begin by outlining the limitations of the current treaty arrangement.

I am grateful to Joseph Aldy and Robert Stavins for inviting me to rewrite the first draft of this paper, and for providing me with the comments that helped me do it. Jake Jacoby also provided helpful comments on an earlier version of this paper.

Problems with the current regime

The UNFCCC has all the appearances of an effective treaty. It has 188 parties (the only non-parties are Andorra, Brunei, Holy See, Iraq, and Somalia) and it aims to achieve something of seemingly undeniable value – to stabilize "greenhouse gas concentrations in the atmosphere at a level that would prevent dangerous anthropogenic interference with the climate system." So, what is wrong with this agreement? There are three problems.

First, the goal implies a discontinuity, with concentrations above some critical level being "dangerous," and concentrations below it (presumably) being "safe." Although abrupt climate change is possible, we do not know the critical concentration level(s) that would trigger it (because the climate follows concentrations with a lag, even the current level of concentrations could prove to be "dangerous"), and we cannot simply choose a "conservative" level arbitrarily (by appealing, for example, to the "precautionary principle"), because to meet that would entail costs having other consequences and require actions that would introduce different kinds of dangers (for example, it is inconceivable that emissions could be reduced substantially in the medium term without a major shift towards nuclear power worldwide, raising problems for waste disposal and proliferation).

Let me be more specific. The European Union has set the goal of limiting climate change to 2°C, and to meet this has recommended a concentrations goal of 550 ppm (parts per million). However, climate sensitivity means that, at 550 ppm, mean global temperature could rise from 1.5° to 4.5°C. Put differently, the 2°C goal could with some confidence be met by limiting concentrations to between 380 and 700 ppm (Caldeira, Jain, and Hoffert 2003: 2052). The midpoint of this range is approximately 550 ppm. But why choose the midpoint? If it is absolutely essential to limit climate change to 2°C, shouldn't concentrations be limited to the current level, 380 ppm? One reason that possibility is not being contemplated may be that it would be virtually impossible to achieve stabilization at this level, at least in the near term. To do this would require not only a substantial cut in global emissions. It would also require a huge effort to suck CO_2 out of the air.

Moreover, it is not clear that a temperature change greater than 2°C is "dangerous" or that a change less than 2°C is "safe." O'Neill and Oppenheimer (2002) have identified three discontinuous changes that should concern us: the destruction of large-scale coral reef ecosystems; the disintegration of the West Antarctic Ice Sheet; and the collapse of the thermohaline circulation that warms the air over western Europe. They then suggest that all three changes can probably be avoided by limiting long-term warming to 1°C; that the last two can probably be avoided by limiting change to 2°C; and that the last one can probably be avoided by limiting change to 3°C above 1990 global mean temperature. Their analysis thus demonstrates that there are substantial uncertainties, and that even the 2°C limit is unlikely to avoid at least *some* abrupt change. But why then should this particular target be the focus of attention? To sum up, while there may exist discontinuities in the climate system, we do not know the temperature change, let alone the concentration level, that would trigger these. Moreover, not all discontinuities are worth spending any amount of resources to avoid.

Here, in my view, is a better way to frame the problem: there is reason to believe that climate change may be abrupt, and that the probability of triggering a threshold increases as mean global temperature (and the concentration level that causes temperature to change) rises. So a "premium" should be added to the damages associated with "gradual" climate change – a premium that is positive today and that increases as the concentration level increases. (Note the difference: a discontinuity implies a zero premium for concentrations below the threshold concentration level and an infinite premium above it.) The magnitude of the premium should reflect climate change risk – the probability of abrupt climate change occurring, and the expected damage associated with that change (destruction of the coral reefs is to be distinguished from disintegration of the West Antarctic Ice Sheet, which would increase sea level by as much as five to six meters).

The second problem with the UNFCCC is that, even if a threshold concentration level could be identified, how would it be met? Presumably, global emissions would need to be restrained, to ensure that the threshold was not breached. But the international system is not well suited to limiting global emissions. Treaties can only enforce *individual* obligations, not collective responsibilities.

Finally, a focus on concentrations ignores two other policies. Climate change is almost certain to happen no matter what is done to limit atmospheric concentrations, and so the world will need to adapt to climate change. Focus on attainment of a global concentration level does not address this need. Moreover, stabilization of concentrations is not the only way to limit climate change. It is also possible to engineer climate stabilization. Indeed, geoengineering is the only intervention that could prevent abrupt climate change after the warning signs first appeared (because of lags in the climate system, the effects of mitigation are delayed). As I explain later in this paper, both of these policies need to be incorporated within a new climate treaty system.

The Kyoto Protocol was meant to be a first step towards achieving the UNFCCC's mitigation goal, but by design it will make little difference. It aims to cap the emissions of 38 industrialized countries, while imposing no limits on the world's more than 150 other countries, including large, fast-growing countries like China and India. Moreover, 13 of the 38 countries facing emission limits – all from central and eastern Europe – face caps that far exceed their actual emissions (the difference is popularly known as "hot air"). The emissions of these countries are therefore unconstrained. Finally, of the countries facing real limits, the largest, the United States, withdrew support for the treaty, as did Australia. This means that Kyoto really only constrains the emissions of about 23 countries.

After the United States withdrew support for Kyoto, the emission limits for Canada, Japan, and Russia were eased (by a technical renegotiation that made a generous allowance for carbon "sinks") to make their ratification more attractive. In the end, Russia's support was needed to bring the treaty into force, but Russia only acceded to the agreement after being given yet another carrot – the assurance that Europe would endorse its accession to the World Trade Organization. At around this same time, the rules for trading in emission entitlements were liberated. While free trade in emission entitlements will reduce compliance costs, it will also reduce the benefits of the treaty, because countries with binding emission constraints can then comply by buying "hot air," reducing the overall volume of emission cuts. Indeed, if the opportunities for trade are fully exploited, Kyoto may not reduce global emissions at all (Buchner, Carraro, and Cersosimo, 2001).

If Kyoto fails in terms of breadth (country participation) and depth (real emission cuts) it also fails in terms of time horizon. Kyoto limits emissions only over the period 2008–2012. It therefore provides no incentive for countries to make investments that will pay off (in terms of emissions reductions) in the more distant future.

Compliance may prove another problem, as the example of Canada illustrates. An official document of the Government of Canada (2005) says that Canada's emissions in 2010 will be at least 45 percent above the country's (already relaxed) target. To reduce emissions by that much, in a short period of time (by 2012), would be very costly – and perhaps hard to justify, given that Canada's big neighbor and main trading partner was doing virtually nothing to limit its emissions. Canada could still comply with the treaty, by purchasing "hot air." But that is likely to be politically difficult. Why should Canada pay Russia, say, for its surplus allowances, just to comply with the agreement, when doing so would have no environmental benefit? Finally, Article 18 of the agreement says that, "any procedures and mechanism . . . entailing binding consequences shall be adopted by means of an amendment to this Protocol." As no such amendment has been agreed, there is no serious penalty for noncompliance – and so little incentive for Canada to shoulder a huge cost in the effort to comply. Making matters worse, many of the other twenty-three parties subject to binding emission limits are also in danger of missing their targets, some by huge margins. Being the only country to fail to comply would be awkward. Being one among a crowd would be very different. My guess is that, as the prospects for full compliance appear less likely, attention within the agreement will shift. Countries will focus less on their compliance (particularly if the compliance of many parties is in jeopardy) and more on whether they are acting responsibly, by taking actions that are comparable with those taken by other parties.

As the implementation period nears, the broader picture is also likely to come into view. Attention will shift to the emissions of the countries unconstrained by Kyoto – like China, which is building a new coal-fired power plant every week to ten days, and India, which is also expanding coal consumption dramatically. Finally, the whole world will be waiting to see if the United States is prepared to take a leadership role. So far, it has not.

Attention has already begun to turn to the future. At the meetings held in Montreal in late 2005, a two-track negotiation process was initiated, with the Kyoto parties now bound by emission ceilings pledging to negotiate new ceilings (presumably for the 2013–2017 period), and with the parties to the UNFCCC (including the United States) agreeing to hold a nonbinding "dialogue on long-term cooperative action." This "dialogue" (the United States refused to call it a "negotiation") is likely to be influenced by a new model for international cooperation that is just beginning to take shape. The Asia-Pacific

Partnership, consisting of China, India, Japan, and the Republic of Korea (all parties to the Kyoto Protocol) and Australia and the United States (non-parties), was established "to promote and create an enabling environment for the development, diffusion, deployment and transfer of existing and emerging cost-effective, cleaner technologies and practices, through concrete and substantial cooperation so as to achieve practical results."[1] It represents a fundamentally different approach, focusing not on emission limits but on technologies. It offers an alternative to Kyoto, and – though not intended for this purpose – could attract the attentions of more countries disenchanted with emission caps (Canada has expressed an interest in joining). What this alternative model lacks is ambition and a strategy. As currently designed, it is unlikely to make much of a difference. The proposal I develop later in this paper also emphasizes the importance of technologies, but it offers an alternative structure that could achieve much more. This alternative could also embrace a modified version of Kyoto.

Overview of the real challenge

Leaving aside the existing arrangements, what should be done about global climate change? Generically, the problem is not only to limit concentrations; it is to reduce climate change risk: to make climate change, especially abrupt and catastrophic climate change, less likely; and to make the consequences of climate change less harmful. It is also to recognize that the actions taken to mitigate or otherwise reduce climate change entail costs and introduce new risks, and that what we do about climate change cannot be separated from the broader needs for sustainable development.

Achieving this goal will require four different kinds of intervention:

1. *mitigation*, including reductions in emissions from "business-as-usual" and reductions in atmospheric concentrations by air capture (that is, sequestration by biomass or other means);
2. *R & D* into technologies that can lower mitigation costs;
3. *adaptation* to climate change; and
4. *geoengineering*, or deliberate climate modification.

The existing climate change regime concentrates attention on mitigation. It addresses R & D timidly, and adaptation ineffectively. It

[1] See www.state.gov/g/oes/rls/fs/50335.htm.

completely ignores the possibility of engineering the climate. As it stands, it is an incomplete model.

Mitigation

There are three ways to reduce emissions: by reducing gross domestic product (GDP, an economic measure of output); by reducing the energy intensity of GDP (conservation); and by reducing the emission intensity of energy (fuel switching and carbon capture and storage). Reductions in GDP, at least on a significant scale, will not be tolerated – certainly not by the poorest countries. Indeed, GDP can be expected to rise substantially over time. The burden for reducing emissions must therefore fall on the other two terms. This burden will only increase over time.

What is this burden? If the concern were with "gradual" climate change, or abrupt and catastrophic climate change only with low probability and with global society having a fairly mild aversion toward risk, then the burden, overall and for the present time, appears not to be very great (Nordhaus and Boyer 2000; Pearce 2005). Under these circumstances, economic analysis suggests that radical changes in the near term are unwarranted; the current climate regime, though falling short of the effort needed, would thus not fail us badly. Over time, however, even from this perspective, much greater action will be required than Kyoto demands at present.

Preventing abrupt or catastrophic climate change will require substantially more effort than that. As already mentioned, no one knows whether any particular concentration level is especially "dangerous," let alone what this level may be (this is from an *ex ante* perspective; of course, the actual consequences of climate change will be revealed to us in time). Indeed, if the worst fears about climate change are proved right, it may already be too late to avoid abrupt climate change over the next century – at least by mitigation alone. Even stabilizing concentrations at levels much higher than 550 ppm will require a fundamental change in technology later this century. These technologies do not now exist, even as pilot models (Hoffert *et al.* 2002).

Concentrations can also be reduced by technologies that capture CO_2 and sequester it somewhere – in trees, depleted oil and gas fields, plankton, the oceans, or mineral carbonates. This approach has a tremendous political economy advantage: it allows us to continue to burn fossil fuels. Like the other options, however, it introduces new risks, including the risk of CO_2, stored in, say, deep coal seams, being suddenly released (a local risk), and the risks of CO_2 stored in the deep ocean changing ocean chemistry and being released slowly over time (both global risks); see Parson and Keith (1998). To develop these technologies will also require substantial R & D (Hoffert *et al.* 2002).

R & D

As just mentioned, R & D is needed to lower the future costs of mitigation (and so to make mitigation more attractive in the future), and to make very large reductions in concentrations feasible, without undermining development (a capability that will be needed to pursue ambitious mitigation goals, whether embraced today or in the future). R & D should have been the centerpiece of the global climate regime, but it is barely mentioned in either of the two climate treaties.

Of course, Kyoto was meant to stimulate R & D *indirectly*. If Kyoto succeeded, it would have put a "shadow price" on emissions, and so created a market for emission-saving technologies and an incentive for these technologies to be developed and brought to the marketplace. But so long as Kyoto fails to constrain emissions, this "pull" incentive for R & D will have little strength; and the very short-term nature of this agreement guarantees that Kyoto will not promote the scale of R & D needed to change technologies fundamentally.

Like mitigation itself, the knowledge of how to reduce atmospheric concentrations using new technologies is a global public good. It requires basic research, knowledge that cannot be patented. Even if Kyoto worked as intended, additional "push" measures would be needed to promote R & D.

Adaptation

Adaptation does not reduce atmospheric concentrations. It does not even reduce climate change; it reduces the *effects* of climate change. Like mitigation, however, adaptation reduces climate change risk.

Most adaptation will be market-driven (the development of new seed varieties that perform better in an altered climate being an example). Some adaptation will involve local public goods (reinforcement of the Thames Barrier protecting London from flooding). These responses do not need to be addressed by a treaty.

Poor countries, however, are a special case. They are especially vulnerable to climate change, mainly because they lack the resources, institutions, and technologies that could shield them from climate change. For these same reasons, they also are the least able to adapt to climate change.

So, how should these countries be helped, by reducing climate change or by making their economies and societies less vulnerable to climate change? To take a concrete example, consider the possibility that climate change might increase the range of both endemic and epidemic malaria. To limit this risk, we could reduce atmospheric concentrations, so that climate change – and the attendant change in malaria – deviated little from the business-as-usual baseline over time. But we could also supply bed nets, carry out more environmental controls, prevent resistance to antimalarials, and invest in the research needed to develop a malaria vaccine. The former approach would prevent malaria from getting a lot worse than it already is because of climate change. The latter approach would reduce malaria more generally (currently, about one to two million children die every year of malaria). Which approach should we choose? Of course, we need to do both. However, the resources spent reducing greenhouse gas emissions cannot also be spent addressing the malaria threat directly. As Thomas Schelling (2002) has reminded us, choices have to be made. And they need to be made now. Sea walls can be built after sea levels begin to rise, but the kind of adaptation needed in the poorest countries requires sustained, long-term development.

Kyoto does establish an adaptation fund for developing countries, but this is to be financed by taxing the so-called Clean Development Mechanism (CDM), and the money raised by this means is entirely unrelated to the need for adaptation assistance. As presently structured, Kyoto fails to address this fundamental resource allocation problem.

Geoengineering

Geoengineering would not change atmospheric concentrations, but it would change the climate, and so reduce climate change risk. For example, aerosols might be blasted into the stratosphere to scatter solar radiation back into space, counteracting the effect of rising atmospheric concentrations.

The idea of deliberately altering the climate sounds bizarre upon a first reading, but we are already doing this unwittingly. The aerosols released by burning coal have the same effect; the difference is that this approach to geoengineering is inefficient (it would be better to put the aerosols in the stratosphere, and possibly to use engineered particles) and unintentional. Geoengineering is also not as unnatural as it may seem. The eruption of Mount Pinatubo in the Philippines in 1991 injected huge quantities of SO_2 into the stratosphere, lowering global temperatures for several years. An analogy might also be helpful: to counteract the effects of acid rain, ground limestone is routinely added to Sweden's pH-sensitive lakes and soils. Though only reductions in acidic emissions can *prevent* acid rain (just as only reductions in greenhouse gas emissions can mitigate human-induced climate change), liming preserves the pH balance of these environments (just as geoengineering would maintain a more stable climate).

Unfortunately, geoengineering would also introduce new risks. For example, stratospheric aerosols might destroy ozone (the aerosols released by Mount Pinatubo did so; see Robock 2002). As already mentioned, however, trade-offs are inherent in every measure we might take to reduce climate change risk. And geoengineeering has the advantage of being able to alter temperature in fairly short order. Indeed, it is the only intervention that could halt abrupt climate change, after the first signs of rapid change first appeared. For this same reason, of course, geoengineering (unlike mitigation, R & D, and adaptation assistance) need not be undertaken now. However, research on geonengineering is needed now. We should be prepared to deploy this technology, should the need arise, and should we choose to do so, in the future.

Coordination

All four of these categories are interrelated and so need to be addressed jointly. Mitigation and R & D into new energy technologies are complements: R & D, if it succeeds in reducing mitigation costs, makes mitigation more attractive, while a more stringent mitigation goal increases the returns to R & D investment. Adaptation and mitigation, by contrast, are substitutes: if mitigation fails, the returns to adaptation will increase, while if adaptation can be relied upon to reduce climate damages, mitigation need not be as robust. If the optimists are right, and geoengineering can regulate the climate without serious adverse effects, this measure would substitute for all of the others.

International climate policy must therefore coordinate the use of these different measures, and this means we need a set of interrelated protocols, rather than a linear progression of a mitigation-only protocol – the original Kyoto model. In the section below I develop this multitrack model in more detail.

A new framework

The alternative model consists of a differently specified objective, coupled with a multitrack system of protocols.

Appropriate measures

Rather than aim to satisfy a particular concentration goal, countries should be encouraged to take "appropriate measures" to reduce the risk of climate change within a framework of sustainable development.

To enjoin countries to take "appropriate measures" is to focus on actions, rather than on targets, on inputs rather than outputs. Of course, climate change will depend on the output of atmospheric concentrations (leaving aside geoengineering), but this cannot be determined directly in an international setting. A focus on output is only appropriate when a collective response can be enforced, and international law prevents easy enforcement of a treaty. Participation in a treaty is voluntary, and treaties must create their own incentives for compliance. Indeed, Kyoto also focused on inputs. The emission limits in this agreement apply only to individual countries, and because of trade leakage, global emissions are likely to fall by less than the sum of the reductions specified by the treaty. The bigger problem with this agreement, however, is that its success depends on a capability for enforcement that is lacking. As we shall see, a focus on "appropriate measures" imposes a lesser burden on enforcement, although the constraint of sovereignty cannot be lifted entirely. In some cases, it probably cannot be eased at all.

The objective of taking "appropriate measures" has other advantages. It allows consideration to be given to the costs and benefits of action, whereas the output goal of limiting concentrations looks at only one side of this equation (the benefits of action). It also provides a more appropriate framework for addressing uncertainties. Though we cannot identify the concentration level that will avoid dangerous interference with the climate system, we know that the probability of abrupt and catastrophic climate change increases in the concentration level, and we therefore know that the measures taken to address climate change should be tightened up as concentrations increase. Finally, "appropriate measures" can, and should, embrace a wider range of actions than mitigation. As already explained, a climate change treaty system should also address R & D, adaptation, and geoengineering.

Though my concern in this paper lies with the need to take very substantial action, and to tie mitigation to new technologies, there is of course a need also to reduce emissions in the short term. The Kyoto Protocol already provides a vehicle for doing this, though it suffers by being unable to support the enforcement needed to achieve even its modest ambition. As I suggested earlier, it is likely that the efforts of the parties to this agreement will essentially find their own level, one in which there is some kind of comparability in terms of actions but not in terms of emission reductions or compliance. A version of this agreement could be retained: an agreement in which countries pledged to take certain actions (these could include the setting of emission limits), but in which the pretense of strong international enforcement was dropped. This would see the agreement as providing a kind of tote board for action (see Levy 1993), with the burden on compliance being borne more by domestic institutions and the informal international mechanism of naming and shaming.

My suggestion to focus on the need to take "appropriate measures" was not chosen arbitrarily. The Vienna Convention for the Protection of the Ozone Layer, which launched the remarkably successful Montreal Protocol, established the same goal. Though ozone depletion and climate change are alike in many ways, the Vienna Convention did not specify a limit on ozone concentrations in the stratosphere; it directed parties to "take appropriate measures . . . to protect human health and the environment against adverse effects resulting or likely to result from human activities which modify or are likely to modify the ozone layer." If this approach worked for the ozone layer, perhaps it could be made to work for the climate.

R & D

As noted before, the knowledge resulting from basic research is in part a global public good. The incentive for individual countries to undertake research unilaterally is often powerful (think of the medical research undertaken by the National Institutes of Health in the United States), but for climate change, no country has a strong enough incentive to invest the substantial sums that will be required; the benefits to any single country are simply not big enough; international cooperation is needed. Examples of multilateral R & D collaboration include nuclear fusion research through the International Thermonuclear Experimental Reactor (ITER), the new particle collider being built by the European Organization for Nuclear Research (CERN); and the International Space Station. In many ways, however, a climate R & D agreement (or set of agreements) would have to be more ambitious than any of these current efforts. It may also need to be structured differently. One possibility is that separate protocols could be negotiated for different R & D projects – one for clean coal, another for electricity distribution networks, a third for emission-free hydrogen production, and so on. Another possibility is that a single protocol would guide all decision making, both about funding and project selection. Decisions must also be made about the organization of the R & D. Is it to be undertaken by a single organization (such as the one at CERN) created for this purpose? Is it to involve collaboration among different entities? Is it to contract out the research? Is it to offer prizes for research success? There may not exist a generally preferred model. A variety of approaches may be needed.

How much R & D funding is required? This is a hard question to answer, not least because an assumption needs to be made about how R & D expenditure translates into technological success. David Popp has made a first attempt to estimate this value. In a model in which the "optimal" carbon tax is about $10 per ton of carbon in 2005 (and rising after that), he finds that, today, a little more than $13 billion (in 1990 dollars) should be spent on R & D that improves energy efficiency generally, and just over $1 billion on R & D that lowers the cost of the backstop technology (Popp 2005: 15). Of course, a more focused concern on avoiding abrupt and catastrophic climate change would warrant even greater investment.

Can we expect that the money needed will be provided? Perhaps surprisingly, R & D financing is unlikely to be undermined by free-riding concerns. R & D, for the purpose of gaining discrete pieces of knowledge, such as how to generate net power from nuclear fusion, is fundamentally different from reducing emissions. Emission reductions are a continuous choice; we can reduce emissions at a given source by

5 percent, or 60 percent, or 60.125 percent, and so on. Research, by contrast, is often of a discrete nature. To produce net power from nuclear fusion, for example, requires a massive effort and fixed inputs. Half of an experimental reactor is of no value, and the scale of a reactor cannot be chosen arbitrarily.

This feature of R & D makes financing more attractive, for the simple reason that, as more countries contribute to a project requiring a fixed sum of money, the costs of financing the remaining balance – and, therefore, bringing the project to the point where it can deliver benefits to all countries – shrinks. Financing discrete projects is essentially a coordination problem. Provided other countries contribute enough, it will be in the interests of each country to contribute the balance (Barrett 2006b).

Coordination is a lot easier than cooperation, but it can fail. My analysis of another discrete project – the financing of the effort to eradicate smallpox – shows that other considerations affect financing, including expectations about ultimate success, whether each country perceives that others have paid their "fair share," and domestic politics (is there a domestic constituency willing to lobby for financing?); see Barrett (2006c). Still, the problem is unlikely to be insurmountable (the experimental fusion project, for example, is going ahead), and there are ways to make financing more reliable (such as negotiating financing shares up front, perhaps by relying on the focal point of the United Nations scale of assessments).

The bigger problem for financing is likely to be of a different nature. Although R & D must be undertaken before technologies are developed and diffused, the incentives to undertake R & D depend on the prospects of those technologies being developed and diffused. The demand for R & D – the willingness to pay for the knowledge gained from R & D – is derived from the expectations that the R & D will ultimately cause atmospheric concentrations to fall (or, in the case of R & D into geoengineering, that it will cause temperatures to stabilize without serious adverse effects); see Barrett (2006b).

Mitigation and technology diffusion

How might R & D aid mitigation? There are three possibilities.

First, it is possible, though very unlikely, that R & D will discover a "silver bullet" technology that produces climate-friendly energy at lower cost than fossil fuels. This is the best outcome that can be hoped for. For reasons of cost only, substitution would be universally attractive. There would be no need for enforcement.

Second, and much more likely, R & D might lower the cost of reducing emissions. This would not improve the prospects for international cooperation, but it would increase the incentive for countries to reduce emissions unilaterally. Without improving the prospects for cooperation, however, the effect is likely to be "incremental." Indeed, precisely because the R & D would have little effect, it may not even be funded (Barrett 2006b).

The third possibility offers an exception to this gloomy outlook. If R & D leads to development of a technology that has certain key features – such as economies of scale, network externalities, and domestic-related benefits – then that technology might spread *even without enforcement*. This is how the catalytic converter coupled with the use of unleaded gasoline became a global standard. The standard spread because the production and adoption of this new technology exhibited increasing returns.

The tendency to standardize implies a threshold in adoption. If every other country uses technology-energy combination A, it may not be in the interests of any country to adopt alternative B. However, if enough other countries use B, then it may pay every other country to adopt B.

Here, now, is the idea: suppose initially that every country adopts A, and that A is cheaper than B, but that B is more climate-friendly. Without the tendency to standardize, B would not be adopted, even if all countries collectively would be better off using B rather than A. The tendency to standardize, however, can be exploited for the purpose of getting countries to switch to B. What is likely to be needed is a threshold level of adoption of B. A treaty would thus be needed, but only to identify the appropriate technology, and to provide the assurance to every country that "enough" other countries will adopt B. This is another coordination problem (in Barrett 2005, I call this a "tipping treaty").

So long as the costs of B exceed those of A, another problem arises. Even if it would pay developing countries to adopt B, given that enough industrialized countries adopted B, should the developing countries be expected to pay the full costs of switching to B? This is a question of equity, and it seems that, since the need to switch is due to the historical buildup of concentrations by industrialized countries since the beginning of the industrial age, compensation may be warranted. As well, the incentives for developing countries to switch to B on their own may be weak. Developing countries, after all, are likely to attach a lower value to local environmental improvements, and to the extent that they are less well integrated with the rest of the world, the network externalities of switching may be weaker for them also (though note that sub-Saharan Africa, for all its poverty, civil wars, political oppression, and HIV/AIDS cases, is phasing out leaded gasoline).

So, international financing may be needed to promote the use of new technologies in developing countries. Will this financing be vulnerable to free riding? It is certainly not the same problem as financing incremental emission reductions. Getting developing countries to adopt a technology, such as a new form of automobile coupled with a new fuel source, or a new means for power generation, is a discrete activity, entailing a fixed (net) cost. So long as the benefit to industrialized countries of a switch by developing countries exceeds the compensation that developing countries require to make the switch, the financing should be forthcoming (Barrett 2001 and 2005: Chapter 13). The logic is similar to the reason why financing of R & D should not be vulnerable to free riding.

Is there a technology–energy combination that exhibits the required degree of increasing returns? A shift to a breakthrough automobile technology such as hydrogen is very likely to be characterized by increasing returns, because of knowledge spillovers, economies of scale, and especially the need to combine a new automobile technology with a supporting energy infrastructure. However, increasing returns are likely to be less important for the production of the hydrogen fuel.

Increasing returns also do not feature large in electricity generation. The light water standard may dominate nuclear reactor design (Cowan 1990), but nuclear power is not generally favored over alternative sources of generation. However, there may exist other possibilities for this sector. Hoffert *et al.* (2002), for example, explain that renewable energy would become economically more attractive if the electricity transmission grid were re-engineered so that power could be redistributed between continents and time zones. Caldeira *et al.* (2005) claim that development of high-temperature superconductor or carbon nanotube cables or even wireless power transmission could make creation of a global electricity grid feasible sometime in the future.

So there does not exist a technology today that has the desired characteristics for global diffusion, but this only means that R & D needs to be directed. R & D should be devoted to developing technologies that not only produce energy without greenhouse gas emissions but can be diffused globally without the need for enforcement. As mentioned before, R & D and mitigation technologies must be chosen jointly.

To this point, I have considered only the collective action challenge of mitigation. There is another perspective that can be important: domestic politics. Carbon capture and storage has an obvious disadvantage as regards collective action: it is an add-on cost and so will not be diffused easily without enforcement. However, this technology does have an important political-economy advantage: it allows fossil fuels to be burned even while greenhouse gas emissions are cut. If proven effective, this technology would quiet the coal lobby. As explained earlier in the context of R & D financing, overcoming obstacles in domestic politics could be as important as defeating international free riding.

Agreements establishing technology standards may also have related political economy advantages. Technology standards create trade restrictions, which are legal and easy to enforce. Moreover, they allow the transfer of technologies with little in the way of transactions costs (in contrast to the CDM). Finally, if the technologies are produced at home, domestic politics will favor financing their use abroad.

The need to transfer technology is manifest. Poor countries like China and India are growing very rapidly, and it is important that the investment underlying this new growth be climate-friendly. Indeed, from this perspective also Kyoto got the design exactly wrong. Rather than have the poor countries grow like the rich countries and *then* transition onto a new development path, relying on a different kind of technology, it would have been much better for the agreement to have focused on getting these countries onto a new, more climate-friendly development path (financed by the rich countries) as a matter of urgency. The rich countries could then transition to this same path more gradually, as their own capital stock was retired.

Adaptation assistance

How should adaptation assistance be incorporated within a climate treaty? The answer is not obvious. Even if adaptation required taking steps only to reduce climate change damages, judgment would be needed to determine which changes were due to human influence and which were natural. In addition, the knowledge that adaptation costs would be paid for by others might create a moral hazard problem. Finally, and as noted before, adaptation requires more than investments that respond to climate change. It requires investments in development that make countries less vulnerable to climate change and that offer a higher return compared with the alternative of mitigation.

Investments in global public goods for development – investments that will benefit developing countries even if climate change turns out to be gradual, and that will protect them from the worst possible outcomes should climate change be abrupt or catastrophic – would seem especially appropriate. A prime example was mentioned previously: investments in malaria control and prevention. One such investment would reduce substantially the likelihood of the deadly malaria parasite becoming resistant to the new artemisinin-based antimalarials. Combination drugs are much less prone to resistance but are also more expensive. Currently, monotherapies are being used, despite the global risks. To prevent resistance from developing and spreading, an international standard should be created for this drug (monotherapies should be banned) and subsidies made available to encourage the adoption of the new combination therapy (Arrow, Panosian, and Gelband 2004). Another example would be to invest in R & D into a malaria vaccine or a similar innovation that could reduce malaria prevalence.

Geoengineering

The idea of engineering the global climate has a number of attractions:

First, and as noted previously, it can be undertaken as needed. In contrast to mitigation, which affects the climate only after a lag, geoengineering can alter the climate relatively quickly, as the example of Mount Pinatubo's eruption demonstrated.

Second, it is cheap. By one estimate, scattering back into space the sunlight needed to offset the warming effect of rising greenhouse gas concentrations by the year 2100 would cost just $1 billion per year (Teller *et al.* 2003: 5).[2]

[2] Nordhaus and Boyer's analysis of the economics of climate change policies assumes that geoengineering would be costless (Nordhaus and Boyer 2000: 127).

Third, geoengineering essentially involves a large project. It is entirely different from mitigation, which requires constraining the activities of billions of people all over the world. In Thomas Schelling's (2006: 48) words, "The first thing to say about the economics of geo-engineering compared with CO_2 abatement is that probably it totally transforms the greenhouse issue from an exceedingly complicated regulatory regime to a simple – not necessarily easy, but simple – problem in international cost sharing." A large project, as Schelling suggests, simply needs to be paid for. I noted the need for cost-sharing earlier, as regards both R & D and technology transfer. Here, the matter is even simpler. If the costs are as low as the above figure suggests, then there essentially is no free-rider problem. Nordhaus and Boyer (2000: 131) calculate that climate change would cost the United States alone about $82 billion in present value terms. To avoid this loss by paying out just $1 billion per year would be in the self-interests of the United States, even if no other country contributed to the effort.[3]

Finally, geoengineering may offer environmental benefits, the main one being the blocking of harmful UV radiation. Indeed, Teller *et al.* (2003: 5–6) calculate that this health-related benefit, for the United States alone, would exceed the cost by more than an order of magnitude. If correct, the economics are even more favorable than suggested above. Geoengineering would also allow CO_2 concentrations to remain elevated, which would be beneficial for plant growth, including agriculture.

Balanced against these advantages are some disadvantages:

First, geoengineering would entail a large-scale experiment, very much like the experiment of rising concentrations of greenhouse gases. There is uncertainty about the full consequences – these, like climate change itself, will be revealed to us in time. Fortunately, however, the risks are bounded. Aerosols pumped into the stratosphere would survive only about five years before falling over the polar regions. Risks could also be reduced by carrying out experiments and related research on the effects of geoengineering.

Second, geoengineering may not alter the climate uniformly. Even if the Earth's average temperature were stabilized, the spatial distribution might be altered (greenhouse gases and sunlight scattering have different radiative forcings). Govindasamy and Caldeira (2000) and

[3] There is an analogy here with another single project, that of protecting the Earth from an asteroid collision; see Barrett (2006a).

Govindasamy, Caldeira, and Duffy (2003) have shown that this concern may be unfounded, but they also urge caution.

Third, geoengineering would not address the problem of ocean acidification. When CO_2 is pumped into the atmosphere, a portion is absorbed by the oceans, decreasing the pH level (as noted before, the portion added to the atmosphere might be a benefit). This is likely to change the process of calcification, endangering animals such as corals and clams. Limestone could be added to the oceans, just as we have added limestone to acid-sensitive lakes, but the scale required would raise a new set of problems. It may, however, be possible to lime certain sensitive coral areas (Royal Society 2005).

Finally, who is to decide whether geoengineering is to be applied? If the calculations developed above prove correct, and the United States (or some other country or group of countries) were willing to finance a geoengineering project all by itself, should it be allowed to do so? Could any country prevent it from doing so? These are serious questions. Some countries are expected to benefit from climate change, at least gradual climate change over the medium term; Nordhaus and Boyer (2000: 131) estimate that Russia, China, and Canada would all gain. Might they use geoengineering themselves to *absorb*, rather than to scatter, radiation, and so to counteract the effect of another country's intervention, or perhaps to make the planet warmer even than the greenhouse effect alone? Would they need to be compensated for damages resulting from a geoengineering intervention, even if that should go to plan? What about countries that have different attitudes towards risk, or that object to the concept of *deliberately* altering the climate. Should their views be heeded?

In a characteristically thoughtful essay on this subject, Daniel Bodansky (1996: 310) asks, would "the international community . . . recognize a right of unilateral action with respect to climate engineering? And is it likely that a country such as the United States would decide to proceed on its own?" His answer to both questions is no, and his assessment is that the international system is more inclined towards prohibitions than regulation. But is this correct? The US-led invasion of Iraq, undertaken without the approval of the United Nations Security Council (and after Bodansky wrote this article), suggests that the answer to the second question may, under the right circumstances, be yes.

A more illuminating example, however, is the decision of whether to destroy the last known stockpiles of smallpox virus, held by the United

States and Russia (it is not known whether clandestine stockpiles may exist elsewhere). Like geoengineering, retention of these stockpiles is of global interest. Should the virus escape into the environment – as a consequence of an accident, say – the entire world would be threatened. On the other hand, should the virus be retained for the purpose of developing a new antiviral drug or vaccine, the whole world would benefit (the risk of a bioterrorist attack would be reduced). The United States and Russia plainly have the upper hand; they possess the stockpiles. But they do not have a *free* hand to decide unilaterally what to do with their stockpiles. A decision by the World Health Assembly (in which virtually every country in the world receives one vote) struck a compromise between the countries wanting the stockpiles to be kept and those that wanted them to be destroyed. The resolution passed unanimously. Is a consensus on the use of geoengineering likely? That will probably depend on how climate change unfolds, but the process by which the smallpox decision was taken may serve as a kind of model for future decision making about geoengineering.

Conclusion

In this paper, I have pointed out the flaws in the existing arrangements and outlined a new proposal, or "architecture," that looks very different from the existing one. It focuses on promoting sustainable development by reducing climate change risk, rather than on meeting a particular concentration target; on the need for R & D and technologies, including mitigation and geoengineering technologies, rather than emission limits; and on the value of international investments in adaptation (development), rather than on mitigation alone. The proposal is an imperfect remedy to a complicated challenge, and its flaws derive from the same forces that make the existing treaty arrangements imperfect. My claim is modest: that the proposal outlined in this paper will *improve* on the existing approach. It is perhaps most appropriately thought of as the second-best next step in an evolving process that has a very long way to go yet.

References

Arrow, K. J., C. B. Panosian, and H. Gelband, eds. (2004). *Saving Lives, Buying Time: Economics of Malaria Drugs in an Age of Resistance*, Washington, DC: National Academies Press.

Barrett, S. (2001). "International Cooperation for Sale," *European Economic Review* 45: 1835–1850.

(2005). *Environment and Statecraft: The Strategy of Environmental Treaty-Making*, Oxford: Oxford University Press (paperback edition).

(2006a). "The Problem of Averting Global Catastrophe," *Chicago Journal of International Law* 6(2): 1–26.

(2006b). "Climate Treaties and 'Breakthrough' Technologies," *American Economic Review, Papers and Proceedings* 96(2): 22–25.

(2006c). "The Smallpox Eradication Game," *Public Choice* 130: 179–207.

Bodansky, D. (1996). "May We Engineer the Climate?" *Climatic Change* 33: 309–321.

Buchner, B., C. Carraro, and I. Cersosimo (2001). "On the Consequences of the US Withdrawal from the Kyoto/Bonn Protocol," Fondazione Eni Enrico Mattei, Venice, Italy.

Caldeira, K., A. K. Jain, and M. I. Hoffert (2003). "Climate Sensitivity Uncertainty and the Need for Energy without CO_2 Emission," *Science* 299: 2052–2054.

Caldeira, K., D. Day, W. Fulkerson, M. Hoffert, and L. Lane (2005). "Climate Change Technology Exploratory Research," Washington, DC: Climate Policy Center.

Cowan, R. (1990). "Nuclear Power Reactors: A Study in Technological Lock-in," *Journal of Economic History* 50(3): 541–567.

Government of Canada (2005). *Moving Forward on Climate Change: A Plan for Honouring our Kyoto Commitment*, available at www.climatechange.gc.ca/english/newsroom/2005/plan05.asp.

Govindasamy, B., and K. Caldeira (2000). "Geoengineering Earth's Radiation Balance to Mitigate CO_2-induced Climate Change," *Geophysical Research Letters* 27(14): 2141–2144.

Govindasamy, B., K. Caldeira, and P. B. Duffy (2003). "Geoengineering Earth's Radiation Balance to Mitigate Climate Change from a Quadrupling of CO_2," *Global and Planetary Change* 37: 157–168.

Hoffert, M. I. *et al.* (2002). "Advanced Technology Paths to Global Climate Stability: Energy for a Greenhouse Planet," *Science* 298: 981–987.

Levy, M. A. (1993). "European Acid Rain: The Power of Tote-Board Diplomacy," in P. M. Haas, R. O. Keohane, and M. A. Levy (eds.), *Institutions for the Earth: Sources of Effective International Environmental Protection*, Cambridge, MA: MIT Press, pp. 75–132.

Nordhaus, W. D., and J. Boyer (2000). *Warming the World: Economic Models of Global Warming*, Cambridge, MA: MIT Press.

O'Neill, Brian C., and Michael Oppenheimer (2002). "Dangerous Climate Impacts and the Kyoto Protocol," *Science* 296: 1971–1972.

Parson, E. A. and D. W. Keith (1998). "Climate Change: Fossil Fuels without CO_2 Emissions," *Science* 282: 1053–1054.

Pearce, D. W. (2005). "The Social Cost of Carbon," in D. Helm (ed.), *Climate-Change Policy*, Oxford: Oxford University Press, pp. 99–133.

Popp, D. (2005). "R & D Subsidies and Climate Policy: Is There a 'Free Lunch'?" Maxwell School, Syracuse University.

Robock, A. (2002). "The Climatic Aftermath," *Science* 295: 1242–1243.

Royal Society (2005). *Ocean Acidification Due to Increasing Atmospheric Carbon Dioxide*, London: Royal Society.

Schelling, T. C. (2002). "What Makes Greenhouse Sense?" *Foreign Affairs* 81(3): 2–9.

(2006). *Strategies of Commitment and Other Essays*, Cambridge, MA: Harvard University Press.

Teller, E., R. Hyde, M. Ishikawa, J. Nuckolls, and L. Wood (2003). "Active Stabilization of Climate: Inexpensive, Low-Risk, Near-Term Options for Preventing Global Warming and Ice Ages via Technologically Varied Solar Radiative Forcing," Lawrence Livermore National Library, 30 November.

Commentaries on Barrett

6.1 *Beyond Kyoto: learning from the Montreal Protocol*

DANIEL C. ESTY

Scott Barrett's "A Multitrack Climate Treaty System" provides a useful basis for thinking about how to realign global cooperation in response to the threat of climate change. In this commentary, I argue that Professor Barrett has made a strong diagnosis of the problems with the current international policy approach. In addition, he offers a solid description of what needs to be done to move the policy process forward. His prescription of how to reconstruct the climate change treaty regime is interesting but uneven and incomplete.

Rather than refocusing the global response to climate change around a series of new protocols, with added emphasis on geoengineering, I argue for a "Beyond Kyoto" approach that would restructure the current regime of international cooperation with an aim of creating a more comprehensive, effective, and efficient international policy architecture. In this regard, the model provided by the most successful global environmental policy structure, the Montreal Protocol and its affiliated agreements, deserves renewed attention.

Especially when compared to the success of the Montreal Protocol, which led to a phase-out of chlorofluorocarbons (CFCs) and other ozone-layer-damaging chemicals, there can be little doubt that the approach mandated by the Kyoto Protocol represents a flawed strategy for preventing global warming and other harms related to climate change. As Professor Barrett points out, the current regime has done little to stem the growth of greenhouse gas emissions. It also does too little to promote adaptation to climate change. Most seriously, the Kyoto Protocol imposes emission reduction obligations on only a few dozen countries – and fails to require any controls in the countries of the developing world, where emissions growth is rising most rapidly.

Professor Barrett is also correct in his assessment that the Protocol too narrowly focuses on a single five-year budget period from 2008 to 2012. It fails to establish a clear direction for policy in the decades beyond. Furthermore, the current international regime has little in the way of incentives for compliance and no mechanism for enforcement against those who either fail to uphold their obligations or do not sign on to binding commitments at all.

Professor Barrett forcefully argues for a multitrack climate change control structure. He suggests that we need enhanced efforts to promote greenhouse gas emissions control (mitigation). He would also like to see new protocols aimed at advancing research and development into technologies that can lower mitigation costs as well as more investment in strategies of adaptation to climate change. And perhaps most provocatively, he argues for a much greater emphasis on research into geoengineering in order to find ways to counteract the effects of a buildup of greenhouse gases in the earth's atmosphere.

Professor Barrett's argument for a multitrack approach is well reasoned and parallels the multifaceted nature of the challenge. Climate change is a complicated problem with multiple elements, sub-issues, and political implications. We should therefore expect that any serious response will need to be multidimensional.

It is increasingly clear that the Kyoto Protocol strategy of setting greenhouse gas reduction targets for a small set of industrialized countries is weak in its design and even more modest in its accomplishments (Victor 2004). If we are to blunt the risk from a buildup of greenhouse gases in the atmosphere to prevent "dangerous anthropogenic interference" with the planet's climate, then a much greater investment in mitigation across a wider swath of countries and industries must be undertaken. The critical issue in this regard, as Barrett notes, is the need to improve the structure of incentives for technology development. Indeed, there is substantial evidence that environmental gains are, almost always, a function of technological advances (Chertow and Esty 1997: 12).

Investments in new developing technologies such as alternative sources of energy or carbon sequestration are global public goods. As Barrett points out, those who make such investments cannot fully capture the benefits. Refining the incentives for investment in climate change mitigation technologies is therefore crucial. But the need for stronger inducements for technology innovation need not mean, as Barrett seems to suggest, that there must be much greater government involvement in research and development. To the contrary, government research efforts are less likely to produce good results than sharpened incentives for private sector action. In fact, with energy prices above $60/barrel, billions of dollars of venture capital are moving into

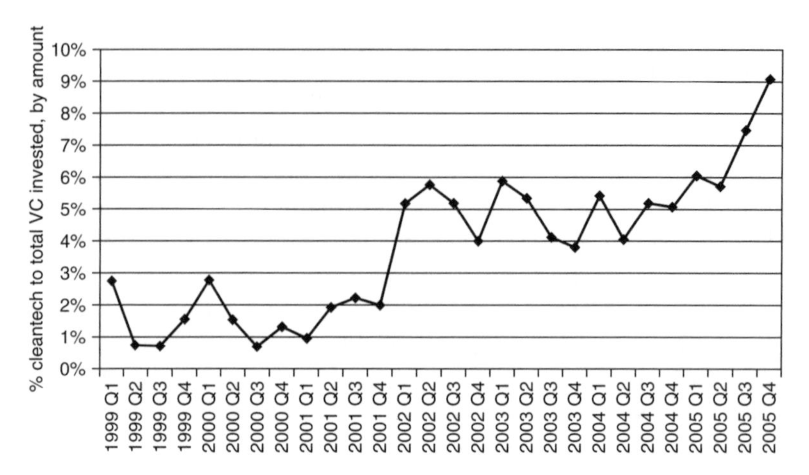

Figure 6.1.1 Growth of "cleantech" venture capital investment

Source: Parker and O'Rourke 2006.

the alternative energy marketplace as Figure 6.1.1 suggests (Esty and Winston 2006: 42).

In parallel with new technologies and mitigation efforts, investments are clearly also needed in adaptation. The current Kyoto Protocol regime places some focus on this policy arena, but not nearly enough. Barrett's call for more action in this realm makes good sense. Some countries, such as the Netherlands, are positioned to respond to rising sea levels and more severe windstorms and indeed are already adjusting land use strategies in preparation for climatic changes (Bressler, *et al.* 2005). Other countries that will be severely affected, including many of those in equatorial latitudes, such as Bangladesh, are much less well placed to address the harms they will likely face (Paavola and Adger 2006). Resources invested now may well help to avoid, or at least dampen, the threat of humanitarian crises and huge numbers of environmental refugees later.

What is somewhat curious is Professor Barrett's enthusiasm for geoengineering. While emphasis on trying to reflect solar radiation back into space is no longer considered a fringe approach to the climate change problem (Crutzen 2006), there remains a great deal of scientific doubt about whether geoengineering can be the central thrust of a global climate change strategy (Broad 2006). Barrett's focus on geoengineering does have an underlying logic. We all would like to be able to "fix" the climate if real problems emerge in the coming years.

But the geoengineering emphasis appears to rest on a series of assumptions that may be counterfactual. The analogy to adding limestone to lakes to reduce acidification due to acid rain cannot be seen as offering much of a promise of success. Liming lakes helps to re-establish a pH balance but it does not restore full ecosystem functionality.

The suggestion that technologies to reduce climate change impacts, such as aerosols pumped into the stratosphere to reflect sunlight, will be cheap and manageable is not shared by the bulk of the scientific community (Schmidt 2006). In addition, the proposition that geoengineering can be "undertaken as needed" implies a capacity for fine-tuned intervention that few scientists presently anticipate. And Barrett does not fully explore the downsides of dousing the atmosphere with reflective particles.

Finally, to the extent that Barrett's enthusiasm for geoengineering derives from his conclusions that, as a "discrete project," it will be easier to finance cooperatively, the argument seems too narrow. Why wouldn't a push for alternative energy supplies offer the same cooperation logic?

Thus, while Barrett's call for a new approach to reducing climate change risk provides a useful vision of the way forward by emphasizing multidimensional action on multiple tracks, his prescription for lowering the probability of big harms and reducing the impact of inevitable ones does not really provide an alternative policy architecture. Nor does Barrett develop in any detail how his proposed series of additional protocols would work. The centrality of technology innovation which he espouses cannot be disputed. But Barrett's call for a new treaty protocol that would promote research and development is not sufficiently explained. He does not say who would do the research. Nor does he spell out who in the global community would pay for the work undertaken.

Some of the assumptions that underpin Barrett's analysis seem quirky. For example, the suggestion that it is "very unlikely" that a silver bullet technology that produces climate-friendly energy at lower cost than fossil fuels will be found seems a bit strong. Indeed, there are many who believe that among a portfolio of alternative energy sources, it is quite likely that some will, at least over time, emerge as cost-competitive with fossil fuels and ultimately be widely adopted (Lovins *et al.* 2006).

More problematic is the assertion that technology breakthroughs that lower the cost of reducing greenhouse gas emissions "would not improve the prospects for international cooperation." There is, in fact, a fairly substantial body of literature that suggests that the degree of international cooperation that is possible around climate change is strongly shaped by the cost of participation in the system in question (Esty and Mendelsohn 1998).

Barrett's further assumption that the differences among nations with regard to how much they would be willing to invest in preventing climate change can be overcome more easily in a geoengineering context, as opposed to one focused on emissions control, appears a bit too facile. Differences in starting positions, economic capacity, and perceived threat from climate change have been a hallmark of the challenge of addressing this problem from the earliest days. The suggestion that a switch to geoengineering would somehow put these issues behind us cannot be sustained.

Finally, Barrett too quickly dismisses the possibility of constructing a treaty regime of greenhouse gas emissions targets and timetables that are enforced with a degree of rigor. In fact, Barrett fails to grapple with the fact that the status quo that he critiques has evolved considerably. While it seemed unlikely just a couple of years ago that cost-competitive alternative energy sources would emerge in the short- or even middle term, quite a different picture emerges when oil prices exceed $60 per barrel. At this much higher price, a series of technologies, including wind, solar, geothermal, and hydropower become potentially economically viable. Whether one accepts the argument that we are today at "peak oil" production with supplies likely to get even tighter, it cannot be disputed that worldwide energy demand will continue to rise. Driven by economic growth in China and India – and across a number of other developing countries – fossil fuel demand is poised to outstrip supply for years to come.

Perhaps more importantly, there is a growing sense that today's market price of oil understates the real cost. A "fully loaded" fossil fuel price would internalize a variety of "externalities" and include charges to reflect the geopolitical risk that accompanies the extraction of fossil fuels in volatile parts of the world such as the Middle East, local air pollution costs, and as well as a price for the carbon dioxide emissions associated with fossil fuel burning. With a public that increasingly sees these costs as *real* and wants policies that address these issues, a quite different energy future begins to come into focus.

So while the Kyoto Protocol does both too much (creating cost burdens that scare countries and major companies away) in the short run and too little in the long run, the existing treaty structure may still provide the architecture for a successful global response to climate change. What is needed is a "Beyond Kyoto" protocol that addresses the weaknesses of the current policy approach and sets out a better course for worldwide action.

In passing, Professor Barrett makes reference to the Montreal Protocol and the "appropriate measures" that it encouraged countries to undertake in order to reduce the breakdown of the earth's protective ozone layer. While Professor Barrett picks up on the objective of having countries take appropriate measures as individually defined, he might have more broadly focused on the elements of the Montreal Protocol that have worked to make it the world's most successful international environmental agreement (Speth and Haas 2006: 96–97). Indeed, I believe the Montreal Protocol provides the best model for a re-engineered architecture for the global response to climate change.

The Montreal Protocol has set the world on a course to restoring the Earth's protective ozone layer (Benedict 1991). Its targets and timetables for phasing out chlorofluorocarbons and other ozone-layer-threatening chemicals have proven to be relatively inexpensive to implement as well as effective. While the current international climate change regime has not engaged the developing world in the challenge of reducing greenhouse gas emissions through binding targets, the Montreal Protocol was a success precisely because it drew the developing nations into a worldwide effort to protect the ozone layer. Similarly, while policymaking on climate change has suffered from a lack of leadership (and even participation) by the United States, the United States led the effort to create the Vienna Convention, the Montreal Protocol, and the series of amendments to these core agreements that have united the world in response to the threat of a depleted ozone layer. Furthermore, while the climate change policy architecture is just beginning to translate obligations imposed on governments into incentives for action on the part of business, the Montreal Protocol quickly created incentives for industry to phase out of the use of CFCs and related chemicals.

What structural elements were critical to the success of the Montreal Protocol? First, the ozone layer protection regime imposed on all countries an obligation to take action. The central principle of "common but differentiated responsibility" meant that all countries were required to phase out the use of ozone-layer-damaging chemicals (Harris 1999). Those in the developing world were provided a longer timetable and financial support, but their obligation was ultimately the *common* one of fully phasing out CFCs.

Second, the Montreal Protocol provided a bundle of both carrots and sticks to promote ratification of the treaty and adherence to its terms (Esty 1994: 189–192). In particular, monies were made available through the Montreal Protocol Fund to subsidize the developing world's shift to CFC substitutes, and the developed world undertook a series of technology transfer initiatives. A parallel set of sticks were introduced to ensure compliance. Trade penalties were threatened against countries that either failed to become parties to the treaty or that fell down in the implementation of the treaty obligations.

Even more notably, special incentives were created to induce the most-critical developing countries into the ozone layer protection regime. Specifically, the Montreal Protocol Fund was initially capitalized with $160 million. But the United States and other countries committed to providing additional resources if and when both China and India ratified the agreement and began the implementation process. Getting the two developing countries with the greatest "democratic heft" into the treaty helped to ensure that other developing countries signed up as well (Esty 1999). The prospect of global markets for CFC substitutes intensified the incentives for technology progress.

The Montreal Protocol success demonstrates that institutional design matters. A carefully constructed treaty architecture that includes both positive inducements and an enforcement mechanism can create incentives for global participation and effective emission control. The commitment of countries to their treaty obligations can be translated into incentives for industry to develop strategies for mitigating emissions, generating alternative technologies, and disseminating breakthroughs across the world.

So what are the conclusions to be drawn? First, the fact that the Kyoto Protocol has failed to deliver the desired result does not mean that all structures of targets and timetables will be unsuccessful. To the contrary, a more thoughtfully designed climate change regime – that abandons the slapdash Kyoto approach – could be developed. Targets must be calibrated to what is achievable at reasonable cost – and stretched out over decades, not just five years. The key to success is a clear signal to the private sector that the energy future will be different from the past. Such a signal would allow those who are heavily invested in energy-intensive activities to write down their existing investments, develop more-efficient production processes, and prepare for a changed energy cost structure. Appropriate price signals would also help to spur investment, research and development, and entrepreneurial activity in pursuit of technology breakthroughs. A little "breathing room" in the form of a longer timetable would also help to get those who feel burdened by changing circumstances on the climate change bandwagon rather than feeling as though they have to try to derail the international policy process.

Second, universal participation, based on the principle of common but differentiated responsibility, must be insisted upon. Global problems require global solutions. Regardless of who is responsible for past emissions, all countries must accept responsibility for present and future harms. It does not work to give a "pass" to the developing world, particularly the emerging economic powerhouses of China and India. Simply put, those who enjoy the benefits of global integration (particularly trade liberalization) must share the burdens of addressing the costs of interdependence, including greenhouse gas emission control.

By moving toward a new Beyond Kyoto protocol, the nations of the world could simultaneously promote emission mitigation strategies by structuring incentives for adequate research and development into alternative energy sources, carbon sequestration, and geoengineering. In support of those least well positioned to help themselves, investments in varying forms of adaptation would also make sense.

Whether the world moves in this direction is by no means assured. A new climate change strategy and international policy architecture will not emerge without leadership. A successful restructuring of the international policy response to climate change requires, in particular, US participation and active support. Whether such leadership might be forthcoming after the 2008 presidential election remains uncertain. But there are clearly models to draw on and ways to move forward that might build on the United Nations Framework Convention on Climate Change of 1992.

As Barrett suggests, a multitrack structure will be needed. But the thought that the current platform is so deeply flawed that it cannot support the architecture that is required appears overstated. And placing geoengineering at the center of the global policy structure provides a shaky foundation on which to build.

References

Benedict, Richard (1991). *Ozone Diplomacy: New Directions in Safeguarding the Planet*, Cambridge, MA: Harvard University Press.

Bressler, A. H. M., M. M. Berk, G. J. Vanden Born, L. Van Bree, F. W. Van Gaalen, W. Ligtvoet, J. G. Van Minnen, and M. C. H. Witmer (2005). *The Effects of Climate Change in the Netherlands*, Bilthoven: Netherlands Environment Assessment Agency.

Broad, William J. (2006). "How to Cool a Planet (Maybe)," *New York Times*, Section F, Science Desk, p. F1, 27 June 2006.

Chertow, Marian R., and Daniel C. Esty (1997). *Thinking Ecologically: The Next Generation of Environmental Policy*, New Haven, CT: Yale University Press.

Crutzen, Paul J. (2006). "Albedo Enhancement by Stratospheric Sulfur Injections: A Contribution to Resolve a Policy Dilemma? *Climatic Change* 77 (3–4): 211–220.

Esty, Daniel C. (1994). *Greening the GATT*, Washington, DC: Institute for International Economics.

(1999). "Pivotal States and the Environment," in Robert S. Chase, Emily Hill, and Paul M. Kennedy (eds.), *The Pivotal States: A New Framework for US Policy in the Developing World*, New York: Norton, pp. 291–314.

Esty, Daniel C., and Robert Mendelsohn (1998). "Moving from National to International Environmental Policy," *Policy Sciences* 31 (3) September: 225–235.

Esty, Daniel C., and Andrew S. Winston (2006). *Green to Gold: How Smart Companies Use Environmental Strategies to Innovate, Create Value, and Build Competitive Advantage*, New Haven, CT: Yale University Press.

Harris, Paul G. (1999). "Common but Differentiated Responsibility: The Kyoto Protocol and United States Policy," *NYU Environmental Law Journal* 7 (1): 27–48.

Lovins, Amory, B., E. Kyle Dalta, Odd-Even Bustnes, Jonathan G. Koomey, and Nathan J. Glasgow (2006). *Winning the Oil Endgame: Innovation for Profits, Jobs and Security*, Snowmass, CO: Rocky Mountain Institute.

Paavola, Jouni, and W. Neil Adger (2006). "Fair Adaptation to Climate Change," *Ecological Economics* 56 (4): 594–609.

Parker, Nicholas, and Anastasia O'Rourke (2006). "The Cleantech Venture Capital Report 2006," Cleantech Venture Network, January.
Schmidt, Gavin A. (2006). "Geo-engineering in Vogue," Realclimate.org, June 28 2006, available at www.realclimate.org/index.php/archives/2006/06/geo-engineering-in-vogue/.
Speth, James Gustave, and Peter M. Haas (2006). *Global Environmental Governance*, Washington, DC: Island Press.
Victor, David G. (2004). *The Collapse of the Kyoto Protocol and the Struggle to Slow Global Warming*, Princeton, NJ: Princeton University Press.

6.2 Climate favela

HENRY D. JACOBY

The lure of a comprehensive climate "architecture"

From the earliest days of climate negotiations there have been danger signals about the pace and direction of the effort, particularly as regards participation by the United States but also for developing countries. The first Bush Administration was in the lead in insisting that nothing beyond a nonbinding "aim" for emission reduction be included in United Nations Framework Convention on Climate Change (UNFCCC) and followed up with little beyond voluntary measures. The Clinton Administration embraced the UNFCCC aim but also relied mainly on voluntary initiatives. And by its Byrd–Hagel Amendment, the Senate fired a shot across the bow of the US negotiators in the lead-up to Kyoto. In 2001 these problems were brought rudely to the surface as the second President Bush rejected the outcome of the negotiations carried out over the previous five years. Further, from the outset there was a distinction between Annex I parties and those in Non-Annex I, with the developing countries in the latter group consistently refusing to permit the agenda of the UNFCCC's Conference of Parties (COP) to include any discussion of ways they might enter into the emerging regime of national emission commitments.

As the negotiating process ran from Rio to Berlin to Kyoto and beyond, there emerged a body of reflection on the apparatus that was being created, and the metaphor that took hold was that of the "architecture" of the nascent regime (e.g., Schmalensee 1998). The concept has been stated in various ways, but a definition we have found useful is, "a unifying structure that restricts potential agreements in ways that both simplify negotiations and point them in desirable directions" (Jacoby, Schmalensee, and Sue Wing 1999). The Kyoto Protocol – a rolling system of legally binding national greenhouse gas (GHG) reduction targets carried out in the UN context – is an example. Aldy, Barrett, and Stavins (2003) and Bodansky (2004) explore the criteria for a satisfactory meeting of this loose definition and lay out literally dozens of proposals for the way forward.

The way this question has been approached, at least in the economics and public policy literature, highlights an important feature of the larger process that has drawn so much thought and analysis and so filled so many journal pages. It is the lure of the prospect of conceiving, designing, and implementing the comprehensive solution – not just a structure that points in useful directions but also promises to fulfill the ambitious goals of the UNFCCC. Consider the reasoning that led to the details of the Kyoto structure. Anthropogenic climate change is a global problem so (the diplomats decided) all nations should have a hand in creating the regime. Several gases contribute to radiative forcing and terrestrial sinks can absorb CO_2, so (as we economists have argued) the initial effort should be an all-gas agreement with sinks taken into account. Given a set of negotiated emission targets, costs will differ among countries, so (again as economists insisted) there should be provision from the start for permit trading to equalize marginal costs among parties. Developing nations should be somehow involved in reductions (as many argued) so add the Clean Development Mechanism (CDM) to allow crediting of actions carried out there. And of course there must be environmental improvement, so (as environmental advocates held) the agreement should seek the largest feasible emission reduction in the first accounting period.

The keystone of this structure, a compliance mechanism, proved elusive, but clearly what was sought was an architecture with the coherence and comprehensiveness of a Gothic cathedral.

Now the fact that an elegant and comprehensive architecture is so alluring is not to be disparaged. If we are to deal with the climate threat in the long run, such a structure for negotiation must emerge, and it is faint criticism to say the nations reached for it too quickly or that advocates pushed too hard. Similarly, to argue that Kyoto is not the basis of a long-term solution is not to demean all those analysts, advocates, and negotiators who contributed to the process, or to suggest that if other choices had been made in 1995 or 2001 we would be in a different position today. It is a fool's errand to argue, "Only if back then we had (fill in the blank), we wouldn't have this problem now." We are where we are, and despite much hard work the interplay of the politics of many nations and interests has led to a fragmented system – a process well elaborated by Victor in this volume. Moreover, the topic and organization of this volume suggests that we have reached a point where there is no clear, commonly accepted vision of how to proceed.

And thus, instead of the desired cathedral, we are headed into a period of construction more suggestive of the ramshackle neighborhoods or favelas that dot the hillsides of Rio de Janeiro. This is the world that underlies the Barrett paper, although the lure of the comprehensive solution – to be achieved by what he calls "a new framework" that will deal with the climate threat through "a multitrack system of protocols" – still motivates his analysis. As background for specific comments on what Barrett proposes in this context, it is useful to begin with an idiosyncratic view of our current situation.

Architectural efforts to date

The key element of the Kyoto Protocol was presaged by text of the UNFCCC which set negotiations on the path to national targets and timetables with its voluntary aim to return GHG emissions to 1990 levels by 2000, and locked in the two-tier system of negotiations with the clear division of expectations for Annex I and Non-Annex I parties. Also included was an overarching objective of atmospheric stabilization, though the level of the intended long-term goal has never been agreed, and that provision played no direct role in the negotiations that produced the Kyoto emission targets.

At the first meeting of the COP, held in Berlin in 1995, instructions were agreed for the negotiators to follow in their preparation of the first protocol under the Convention, to be agreed in Kyoto three years hence. Among the provisions of this so-called Berlin Mandate was a system of national targets and timetables that would be legally binding. (Incidentally, at the time there was a proposal to adopt a system of pledge and review rather than hard targets.) Importantly, it further locked in the division between industrialized and developing countries by stipulating that no emission commitments were to be introduced for parties outside Annex I. Also, negotiators were charged with agreeing to a set of common policies and measures (or PAMs in the jargon of the field) – that is, actions to take as well as targets to accept. Along the path from Berlin to Kyoto the system of targets and timetables was in fact converted from an "aim" to a legally binding commitment, whereas PAMs were dropped, largely at US insistence. Both PAMs and pledge and review now reappear in the Barrett proposals.

Much has been written about the flaws of the Protocol that resulted (e.g., Jacoby, Prinn, and Schmalensee 1998; Jacoby and Reiner 2005; Aldy, Barrett, and Stavins 2003; Barrett 2003, Victor, this volume). For purposes of this discussion, however, one circumstance stands out as a cause of the current fragmentation. Pressures on key political players led the negotiation process far out in front of the level of public concern in key nations about the seriousness of the climate threat, and thus ahead of any political commitment to take mitigation action. So, for example, there was no chance the US Senate would have ratified the Protocol whatever the outcome of the 2000 election.

This is not to say that there is no action on greenhouse gases in the United States, or among Non-Annex I members even apart from the CDM, or that no developing countries will accede to Kyoto commitments in a second commitment period. But extension of the system of binding targets and timetables to cover the major Non-Annex I parties like China, India, Brazil, and Indonesia is extremely unlikely. And even the adoption by the United States of substantial CO_2 reduction measures will not likely be accompanied by ratification of the Kyoto Protocol as it now stands.

So what happens next? Any scenario of future events must start with expectations for the institutions created thus far, and developments in the nations and nation groups that support them. Use of the favela metaphor, implying that there will be no universal, comprehensive architecture for some number of years, does not suggest that existing institutions will evaporate: efforts will be made to salvage, adjust, and correct. For example, a number of useful activities are organized under the UNFCCC including national communications, capacity building, and secretarial functions, and these will continue (Jacoby and Reiner 2005). The UNFCCC also may ultimately serve some role in the organization of additional multi-nation commitments, as can be seen, for example, in the participation of all parties in its Dialogue on Long-Term Cooperative Action to Address Climate Change by Enhancing Implementation of the Convention that first met in May 2006. The Kyoto Protocol likely will survive in some form as well. Annex I parties *ex* the United States and Australia will spend the next five and one-half years trying to make the best of their existing commitments, and some Kyoto-spurred developments like the European trading system and CDM are likely to become permanent features of economic life. Also, some form of second commitment period will be negotiated, perhaps with some smaller and richer developing countries taking on emission commitments, but with competitiveness pressures preventing substantial tightening of the first-period commitments. Too much effort has been spent to abandon the agreement and its bureaucracy altogether, but it will evolve into a softer, more pledge-and-review-type structure than originally intended.

Meanwhile, the search for a comprehensive approach continues in the COP, in other venues such as the Group of Eight (G8) summit, in an effort to convert the current Group of Twenty (G20) finance ministers into a group focused on climate (see Victor, this volume), and in bilateral and multilateral discussions. The result will be a period of regime construction with no global architect. Nations in and out of the Kyoto system are beginning to take action on mitigation and adaptation. Almost all have imposed voluntary schemes; regulations and technology standards are widely applied; subsidies to low CO_2 technologies appear in almost all countries; carbon prices are applied in various forms through taxes or cap-and-trade; and most of the richer parties are spending on R & D, commercial demonstration, and programs of technology transfer. Versions of the CDM outside the Kyoto agreement are being formulated, and industry-level agreements are already in place (e.g., in semiconductor manufacture) with others under consideration. Many of these activities involve bilateral or multilateral understandings and even explicit agreements. An ultimate comprehensive architecture, if ever reached, will be some integration of the favela of approaches developed in this period, including the Kyoto Protocol – a view explored at greater length by Pizer in this volume. In short, domestic actions will not follow international agreement but the other way around. What will matter most in the coming years will be those measures that domestic legislatures are willing to take even in the absence of any assurance that others are doing their share, and the ultimate structure for international negotiation in the long term will need to accommodate these developments.

The Barrett proposal

Barrett's proposal of a "new architecture" for this complicated setting has two key components: a shift from targets to actions and a focus on technology. He is led to these conclusions by the logic of his own prior analysis, which elaborates the difficulty of solving a global commons problem with sovereign states as parties (Barrett 2003). One can quibble with his focus on the atmospheric stabilization goal as the main characteristic of the UNFCCC/Kyoto structure when in fact it is the national emission reduction targets that are key. But either way his argument holds: incentives for participation in these kinds of targets are lacking and there is no means of enforcement among sovereign entities. He sees the Kyoto structure ultimately morphing from a legally binding agreement to a system of "shame and blame," or what was referred to earlier as pledge and review. In the near term this seems a reasonable forecast. Whether the gains to be had from such a system are as paltry as the paper suggests remains to be seen. If the major powers became committed to action domestically, such a system of shame and blame might be an important component of a long-term regime.

Barrett's suggested alternative, the negotiation of "appropriate measures," is a return to the PAMs requested of the negotiators by the Berlin Mandate, only now the measures are not limited to mitigation, as was the earlier focus, but also extend to adaptation and perhaps even to geoengineering (of which more at the end). The focus then is on the development of low-GHG technology and even more specifically on the Holy Grail of climate policy: the self-enforcing agreement. If Barrett turns out to be right about the limits to open-ended agreements to share the burdens of emission mitigation, then technology development with powerful incentives for adoption is the one place left to look for a solution. He finds such a prospect in technologies with economies of scale and positive network effects. Find such a technology and, with only minor agreement on common standards, the economics will "tip" so all nations have an incentive to adopt it freely. Barrett does not claim he knows of any such technologies that are actually out there to be found, or if a few did that they could make a significant dent in the climate risk – just that this is one approach that might actually work and lead parties beyond actions taken on a narrow national calculus.

It is a clever idea. Surely we should always be on the lookout for technical options with these characteristics. But is it an architecture for global agreement in the sense laid out above? An agreement on overall levels of R & D expenditure among nations capable of this work could be useful. But the suggestion of protocols on specific technologies targeted for their scale and network effects, or a treaty establishing a central body to allocate effort among options with this potential, is problematic to say the least. There are not one or a few technical advances relevant to the climate issue but many, and new technology often emerges not from a single focused project but as the combined effect of results gained in diverse areas. (An example, is the role of microprocessor technology in automotive power trains.) Network effects happen, but it is not clear that they can be predicted, and surely not by diplomats in protocol negotiations. Negotiation of common standards (e.g., auto mileage standards or refrigerator efficiency) may be possible, but these come only when the technology is reasonably well understood. Thus the proposed mitigation strategy boils down to agreements on the overall levels of expenditure on climate-friendly R & D and an increased focus on ways to agree on common policies and measures, with useful guidance as to where the most productive actions may lie.

On the issue of adaptation, Barrett suggests investments to raise the adaptive capacity of developing countries as a possible area of agreement, using malaria as an example. He does not, however, provide an analysis (as he does of emission mitigation) of whether under the calculus of national interest nations would be willing to invest in others in this way, nor does he suggest what types of agreement might prove constructive. And, importantly, for none of his four-track system of protocols is there any discussion of the institutional context. Are they to be pursued in the COP or elsewhere in the UN system? In the G8? In bilateral and multilateral agreements? All at the same time, or in separate negotiations perhaps coordinated in some way? The result, therefore, is the proposal of a set of promising directions for climate policy discussions constrained by a hard-edged view of the limits of international agreements – in effect a suggestion of additions to the existing fragmented structure. To point out the limited scope of the Barrett proposals is hardly a criticism but simply serves to highlight the fact that his broader and perceptive analysis of the limits of statecraft (Barrett 2003) tends to rule out proposals that range far from the current system of favela construction.

So is that it? From Barrett's analysis should we conclude that there is no path to a unifying structure, with strong enough incentives for participation and compliance to yield emission reductions that may prove to be needed? Hard as it is to separate hope from assessment, I think not. But the conditions are not yet in place for a serious discussion of the issue. On the view that international agreement follows domestic action, formulation of a comprehensive architecture must await direct action on CO_2 mitigation by the United States, taken because the activity is perceived to be in the national interest apart from what others are doing. And the same prerequisite holds for China. When these two giants have moved in their domestic policies serious discussion can begin with the EU and others operating under the current Kyoto regime. Some form of unifying structure could emerge out of the mix of approaches they are then taking. Until these conditions are met, uncoordinated construction of the kind actually implied by the Barrett proposals may be the best we can hope for.

Finally, Barrett goes where others have feared to tread (Summers and Schelling in this volume being conspicuous exceptions). For many years the "A" word could not be spoken in polite company, but once it was recognized that adaptation to some level of climate change is unavoidable the topic moved into the mainstream. At the margin now is the "G" word. But, as the difficulties of international agreement that Barrett so convincingly argues sink into public consciousness, the prospect of geoengineering – the ultimate technical fix – becomes an acceptable topic. In part, the earlier hesitance reflects our naiveté about the degree to which the Earth has already become a human-engineered system (Allenby 2000/2001), and it is not surprising that the coverage here mainly serves to reveal how far we have to go in integrating this option into policy discussion.

Barrett formulates geoengineering as a single "large project," using as his example the direct manipulation of the Earth's radiative balance by the injection of scattering aerosols into the atmosphere. In fact, this measure might not be one project but many, carried out in an uncoordinated fashion by diverse nations. Geoengineering also includes the possibility of selective manipulation of SO_X emissions from coal-fired power plants, relaxing controls where acid deposition is mainly over the oceans. Moreover, the purposeful geoengineering of climate includes modification of the carbon cycle, most prominently by iron fertilization of the oceans to increase the rate of CO_2 uptake. R & D on this latter option is well under way, and there are commercial firms setting up to provide this service as soon as some nation will credit the increased uptake in its GHG control regime.

Research to develop better understanding of these options, and their risks, is useful as Barrett suggests. But what of the larger question of collective action beyond R & D that motivates this volume? Most of the "architecture" discussion concerns the development of a structure to guide efforts to control the engineering of the planet that we are doing inadvertently, involving the need for some form of cooperation and/or coordination of effort. Here the problem is turned on its head: what is the structure that might lead to a constructive global outcome in a situation where any one nation (even a poor one if costs are as low as estimated by Barrett's source) could manipulate the solar insolation of the whole planet, or sets of nations could create crediting programs that unleashed private business to carry out these measures? And, as Barrett carefully notes, all this occurring in a situation where forecasts of the effects of these measures are at least as uncertain as those for GHG emissions and where there are unknown and perhaps significant distributional effects among regions and sectors.

The example cited of US unilateral military action nicely frames the issue: what architecture of climate negotiation will serve in the face of the power of individual nations to affect the global setting purposively? Discussion of that prospect needs further detailed work along the lines of Barrett (2003) before it can be thought of as a component of the integrated approach he suggests, or indeed of any serious international consideration.

References

Aldy, J., S. Barrett, and R. Stavins (2003). "Thirteen Plus One: A Comparison of Global Climate Policy Architectures, *Climate Policy* 3 (4): 373–397.

Allenby, B. (2000/2001). "Earth Systems Engineering and Management," *IEEE Technology and Society Magazine* (Winter): 10–24.

Barrett, S. (2003). *Environment and Statecraft: The Strategy of Environmental Treaty-Making*, Oxford: Oxford University Press.

Bodansky, D. (2004). *International Climate Efforts Beyond 2012: A Survey of Approaches*, Arlington, VA: Pew Center on Global Climate Change.

Jacoby, H., R. Prinn, and R. Schmalensee (1998). "Kyoto's Unfinished Business," *Foreign Affairs* 77(4): 54–66.

Jacoby, H., R. Schmalensee, and I. Sue Wing (1999). "Toward a Useful Architecture for Climate Change Negotiations," MIT Joint Program on the Science and Policy of Global Change, Report No. 49, May.

Jacoby, H., and D. Reiner (2005). "Getting Climate Policy on Track After The Hague: An Update," in R. Wilkinson (ed.), *The Global Governance Reader*, New York: Routledge, pp. 274–290.

Schmalensee, R. (1998). "Greenhouse Policy Architectures and Institutions," in W. D. Nordhaus (ed.), *Economics and Policy Issues in Climate Change*, Washington, DC: Resources for the Future Press, pp. 137–158.

7 *Practical global climate policy*

WILLIAM A. PIZER

Introduction

A MEANINGFUL discussion of international climate policy agreement needs to begin by asking the question, what is the goal of the international agreement? What defines success? One way to answer this question is to view the problem through the eyes of a stylized grouping of experts and stakeholders engaged in the issue: economists, environmental advocates, and technologists. Economists would likely describe the goal as maximizing welfare; that is, setting a global policy that balances expected costs and benefits of mitigation. Environmental advocates, and indeed the United Nations Framework Convention on Climate Change (UNFCCC), describe a goal of preventing dangerous interference with the climate system. Like the Clean Air Act in the United States, it suggests first consulting the science to establish a safe standard, then following up with a cost-effective (i.e., least-cost strategy) to achieve it. Alternatively, the goal might be described by technologists as the need to develop and deploy climate-friendly technology at a global level, without a heavy emphasis on near-term emission reductions.

Such stylized views typically lead to straw-man conclusions about the design of an agreement – agreements that are easily knocked down by practical critiques. Yet considering these straw men can help point us to an alternate, more practical goal. That is, the goal might be to encourage some balance of domestic mitigation and technology development across individual countries in the near term – embracing a fairly wide

I wish to thank Tom Heller, Juan-Pablo Montero, Rob Stavins, Joe Aldy, and participants in the May 2006 workshop "Architectures for Agreement: Addressing Global Climate Change in the Post-Kyoto World" at Harvard University, who all provided helpful comments on an earlier draft. Financial support from the Mistra-Funded Climate Policy Research Program is gratefully acknowledged.

range of efforts driven by domestic capacity in each nation – while laying the groundwork for a more coordinated international mitigation effort in the future. In other words, encourage at a global level more of what we are beginning to see happening, regionally and domestically, right now, in advance of stronger international efforts as institutions evolve.

While all of the architectures presented in this book seek to address practical concerns of one form or another, this chapter is more specifically focused on what we observe happening in the world right now and what those observations tell us about an international design that really might work, right now (and presumably accomplish something more than would occur in complete absence of an agreement). In this way, most of the analysis is positive, not normative; based on the observed suite of behavior over the last decade, what does it imply about what will and will not work? Foreshadowing the eventual conclusions, if the goal is to encourage action now, there needs to be much greater deference to domestic interests – whether it is concern about excessive reliance on natural gas in the United States or an overwhelming priority on economic development in countries like China and India. There needs to be a recognition of national differences in policy preferences. The agreement needs to consider technology development and investment activity (technology push), not just mitigation (demand pull). Efforts to engage developing countries need to proceed at all levels – project-based credits, sectoral or policy-based credits, and broader linkage with other issues such as energy security and trade. Finally, the emphasis needs to be more clearly on evaluating actions after the fact, rather than agreeing on targets and timetables in advance of any action.

If this seems relatively "squishy" compared to the elegance of legally binding commitments under the Kyoto Protocol, consider this: The Bonn and Marrakech agreements in 2001 literally renegotiated the Protocol targets four years after they were set (Russia, for example, received an additional 130 million tons in carbon dioxide sink credits[1]). The exit of the United States from the Protocol that same year further left the remaining participants with only a marginal aggregate commitment – if the Russian Federation and Ukraine sell their excess emission rights (aka "hot air") under the Protocol to EU member

[1] These credits under Article 3.3 were previously unspecified.

states, Japan, and Canada, those countries would be required to do very little. Finally, the Clean Development Mechanism (CDM) has the potential to flood the market with cheap credits – or not – depending on how the rules evolve. Arguably, the question is not whether an agreement is squishy, but whether it is explicit.

As we examine a "practical global climate policy," it will be important to recognize where the other goals lead. First and foremost, it is these other goals that bring many stakeholders to the table and give us an idea about what the longer-term effort might look like. Architecturally, it is the same as the "act-then-learn-then-act" framework discussed in Nordhaus (1994) except that rather than just learning about uncertain science and technologies, we also learn about how policies work and the capacity and commitment of other nations. We recognize the varied circumstances in different countries, as well as the difficulty in knowing, up front, how domestic policy processes will evolve, but we know the system needs to be prodded in certain directions. This is not an "anything goes" approach, even if the starting point is not that far from a non-cooperative solution.

What about the agreements suggested by other goals? For most economists, the defining features of global climate change are that it is a global environmental externality with uncertain mitigation costs and consequences. The global nature and the uncertainty have each spurred valuable lines of research. Economists have contributed to ideas about global cooperation and emission trading, the comparison of price and quantity controls, and option value and learning. At the end of the day, however, the basic message has been to use globally flexible market-based mechanisms – taxes, tradable quotas with banking, or some mixture – to best match the expected net benefits of mitigation, and to do so in the most globally cooperative way.

While much research has also suggested a preference for price-based approaches, this has not been the historical model applied to climate change. The basis of the UNFCCC, the structure of the Kyoto Protocol, and the EU Emission Trading Scheme (ETS) all center on quantitative limits. But even more than the quantity versus price distinction, the *level* of quantitative targets has typically ignored any assessment of benefits. This follows, in part, from a willingness in the economics community to focus attention on cost-effectiveness (e.g., minimizing costs) rather than an unyielding emphasis on the need to balance costs and benefits. This willingness also recognizes the inherent difficulties associated with valuing environmental amenities more generally and economists' comparative advantage on the cost side. Climate change has been no different; while numerous studies have studied costs, only a handful have examined benefits – to the point where it receives only a brief mention in periodic assessment reports of the Intergovernmental Panel on Climate Change (IPCC). As a consequence of economists being sidelined in the target-setting discussions, much of the debate has focused on quickly beginning to reduce emissions in absolute terms and then stabilize concentrations at levels that we will otherwise exceed in the next forty years.

The emphasis on quantitative limits is all the more remarkable because it is virtually impossible to tie something like the Kyoto Protocol emission limits directly to stabilization of greenhouse gases (GHGs) in the atmosphere at a particular level. That is, the short-lived nature of the commitments among industrialized countries and the lack of commitment among developing countries make extrapolation virtually a guess.[2] At the same time, calls for dramatic emission reductions have, in some ways, backfired in the United States. If such dramatic reductions are necessary, the reasoning goes, there is no way to provide enough of a price signal to the private sector; the solution must be government research and development – a Manhattan Project, or Apollo man-on-the-moon sort of effort – ironically the same response suggested by those who believe the science does not yet warrant mandatory controls.

Backing up for a moment, a number of assumptions underlie what was earlier described as the economist's, the environmentalist's, and the technologist's perspectives. Both the economist's and the environmentalist's perspective tend to assume that market mechanisms can be used to fully internalize an appropriate emission price. This assumption faces near-term opposition in countries like the United States and longer-term pressure more broadly if developing countries cannot be convinced (or do not have the institutional capacity) similarly to internalize the price. It also requires strong international institutions to

[2] The architecture proposed by Jeffrey Frankel would address these two criticisms – bringing in developing countries and creating a long-term trajectory. However, such an architecture requires assumptions about the ability both to entice developing countries into accepting targets as well as to encourage relevant domestic policies corresponding to those targets – assumptions that arguably may not be valid for some time.

enforce compliance – institutions that may or may not be achievable. Finally, the technologist makes nearly the opposite assumption, that government can instead fully incentivize not only technology development, but deployment and diffusion as well, without pricing emissions.

What if these assumptions do not hold? What if it proves impossible to encourage China, for example, to price emissions through mandatory policies? Or what if the kinds of prices required to achieve agreed environmental goals in an emissions tax or trading regime are higher than politically feasible? Conversely, what if the government cannot subsidize its way to a net-zero emission world? Finally, can we really expect world leaders to huddle together and negotiate a global plan, return home, and implement it?

If the arguably negative answer to all of those questions would seem to be cause for alarm, it need not be. Specifically, it suggests that rather than a centrally planned global solution of one flavor or another, we are likely to see a suite of domestic (or sub-domestic) responses that are gradually prodded and caroused into rough harmonization. And that is what we do see: the European Union issuing the directive for the EU ETS, and committing to a mandatory policy, before it is clear that the Kyoto Protocol will come into force. The northeast United States and California initiating mandatory policies in the absence of federal direction. Meanwhile, all the prices we have seen, from the EU ETS, to proposals in Canada, Japan, and New Zealand, are all in the \$10–25 per ton of CO_2 range. Autarkic prices upward of \$50 per ton of CO_2, predicted by most models in order to comply with the Kyoto Protocol, have not occurred – suggesting both that treaty commitments have been trumped by economic concerns *and* that the potential gains to trade are smaller. While this may not be the rapid start that many believe the problem demands, it represents the beginning of a practical start that is the most likely way forward.

Under these conditions, the value in an international agreement appears to be in prodding domestic action rather than elaborating mechanisms for global trading, tax harmonization, or enforcement. This prodding might come in the form of revisiting quantitative but nonbinding goals, or it might come in the form of evaluating current national action and providing consensus feedback. Technology development needs to be balanced with immediate efforts to reduce emissions. And, at the same time, industrialized countries are nudged toward action, developing country activities need to be encouraged in a variety of venues – project-based crediting, other financial incentives for climate-friendly policy reform, and linking climate policy to other issues including trade, development and security.

The remainder of this paper examines in more detail the lessons applied by economists and others to the problem of global climate change, and then revisits the underlying assumptions. The paper then considers new lessons to be drawn from observations and alternative assumptions and ends with a suggested architecture for moving forward.

Basic lessons from economics

It is impossible to look at the architecture of the Kyoto Protocol and, in many ways, not be proud of the influence economic thinking has had on the policy debate. The Clinton Administration worked hard to design a global agreement that had the flexibility to achieve its goals at the lowest possible cost (Aldy 2004). While many were quick to point out that the emission goals of the Protocol were unrealistic (e.g., Nordhaus and Boyer 1999), those critics also failed to appreciate the subtler assumptions presumed by negotiators: that forestry sinks would effectively relax the target and that cheap project-based credits would flow from developing countries. The first of these assumptions was born out in Marrakech (Babiker *et al.* 2002; Bohringer 2002); the second was potentially hinted at during the Conference of the Parties (COP) 11/Meeting of the Parties (MOP) 1 in Montreal. There, discussions emerged concerning both sector-based crediting as well as credit for avoided deforestation, both of which could lead to much larger – if not too large – credit flows (Aguilar *et al.* 2005).

But even in criticizing the targets, many economists would be relatively happy with the architecture. Countries are free to trade their emission commitments, just like firms in a domestic emission trading program. Commitments are fungible across all six greenhouse gases. Emission commitments are fully bankable and, effectively, borrowable at a 6 percent rate of interest (technically a country will be out of compliance, but the penalty is repayment in the next, five-year commitment period with a 30 percent "restoration" rate).[3] Project-based credits can be brought in from developing countries through the CDM.

[3] Here, 6 percent refers to a "real" rate of interest, denominated in terms of Assigned Amount Units (AAUs; allowances representing one ton of CO_2

The emphasis on flexibility can be traced to the strength of economic arguments. Studies around the time of the Protocol all described significant gains from trading among countries and over time (Wigley, Richels, and Edmonds 1996; Repetto and Austin 1997). Later studies emphasized the gains to trading across gases (Hayhoe *et al.* 1999; Reilly *et al.* 1999; Hyman *et al.* 2003). A particularly influential study by Weyant and Hill (1999) found that trading among countries with binding targets under the Protocol cut costs by more than half (versus autarkic policies that actually achieved their Kyoto targets). Access to developing country markets cut the costs in half again.

The question of whether the Kyoto target is, overall, too tough has been mooted by the withdrawal of the United States that, by far, had the most onerous reduction burden. Most estimates suggest that without the United States and with the negotiated sink allowances in Bonn and Marrakech, the effective target is barely binding when taken as a whole (Babiker *et al.* 2002; Bohringer 2002). That is, the net reduction requirements of Europe, Canada, Japan, and others with binding targets are basically offset by excess allowances held by Russia and Ukraine (also referred to as "hot air"). This highlights a rather central question: will countries accept large permit flows among countries as a mechanism either to achieve lower-cost compliance (or compliance at all) now, or to encourage participation later? The latter question we will take up momentarily, when we discuss cooperation. The former remains an open question. Currently, it is unclear whether the countries of the European Union, Japan, or Canada – that is, those countries facing the largest shortfalls – will buy excess credits from Russia, or whether Russia will even sell them (perhaps holding out for even higher value in future commitment periods). Other studies point out that emission trading and market-based climate policy more generally has significant effects on international trade flows, something that may diminish enthusiasm for significant market-based policies (McKibbin and Wilcoxen 1997; McKibbin *et al.* 1999).

equivalent in the Kyoto system). A country that is short of AAUs at the end of the first commitment period can instead retire 1.3 times the shortage at the end of the second period, five years later. Coupled with banking, this creates arbitrage opportunities that should keep first period AAUs priced somewhere higher than the price of second period AAUs (otherwise banking occurs), and somewhere less than 1.3 times the price of second period AAUs (otherwise borrowing occurs).

In January 2006, US Senators Domenici and Bingaman spurred a lively discussion in response to a White Paper question about whether a proposed US trading program should seek to link with other programs in the European Union and possibly Canada. Most responses indicated support for linking systems as a way to lower overall costs. However, such gains come from a disparity in prices, and some commenters also noted the potential problem of linking systems with different prices. Namely, while there is an aggregate gain in both countries, sellers in the high-price country and buyers in the low-price country both lose; there may also be concerns in the high-price country about whether the low-price country is doing enough.[4]

If the first lesson from the economics literature has been that flexibility reduces costs, the second has been that price-based approaches are likely preferable to rigid emission limits in terms of economic efficiency. The underlying intuition is that for a stock pollutant like CO_2 – that is, something that accumulates in the atmosphere over hundreds of years – the contribution in any one year is small. Therefore, the first ton emitted in a given year is likely to be just as damaging as the last. In other words, the *marginal* damages are relatively constant within a year and generally across several years or even a decade or two.[5] This is likely to be true even if we are unsure about the actual level of those marginal damages.

On the other hand, costs of mitigation can rise very steeply if emissions are required over short periods of time. And, importantly, random events like a cold winter, warm summer, dry season (limiting carbon-free hydroelectric supplies), or swings in economic activity, can shift the baseline emission level. Typically, we see emission fluctuations of 2–3 percent on an annual basis in the United States; other countries see fluctuations as high as 6–7 percent. That means that an otherwise modest emission cap – say a 1–2 percent reduction from expected levels – might suddenly become an exceedingly onerous 4–5 percent as the baseline shifts.

Under these circumstances, price-based mechanisms like an emission tax provide much-needed flexibility vis-à-vis an annual emission

[4] See www.energy.senate.gov/public/index.cfm?FuseAction=IssueItems.View&IssueItem_ID=38 for the full text of the comments.

[5] Economic growth and discounting generally make marginal damages grow over time, even as they remain constant at a given point in time for different emission levels.

trading program (Pizer 2002; Newell and Pizer 2003). Of course, most trading programs are not entirely annual in that most allow banking between periods. While arguments have been made that banking can provide similar flexibility (Jacoby and Ellerman 2004), the reality remains less certain. For banking to help, you need to either accumulate a sufficient bank, be given one to start, or have access to some amount of borrowing. For example in the NOx budget program in the United States, extra allowances (a "compliance supplement pool") were given out at the start of the program (US EPA 2004b). In the SO_2 program, a bank was quickly accumulated (US EPA 2004a). In the Kyoto Protocol, the use of five-year commitment periods was designed to facilitate a limited amount of borrowing within five-year blocks. Borrowing across commitment periods is also technically possible, as noted earlier, but it is associated with being out of compliance.

While banking and borrowing provisions may provide flexibility, only a price-based mechanism can provide cost certainty – that is, a fixed or maximum price per ton of CO_2. It is perhaps this feature, more than flexibility alone, that has led to price-based proposals in Canada, New Zealand, and most recently the United States. Of course, this cost certainty comes at a price: emissions *uncertainty*. Much as Heisenberg showed the trade-off between certainty about a particle's location and its momentum, the debate over prices and quantities shows that we cannot be certain about both a policy's cost and its environmental outcome. Economic efficiency, however, based on relatively constant marginal damages, argues for cost certainty over emissions certainty.[6]

A key element in the debate over prices and quantities has been the observation that you can introduce price-like features into an emission trading program via a safety valve.[7] That is, an otherwise ordinary emission trading program is complemented by an alternative compliance mechanism whereby firms can pay a fixed price per ton emitted, rather than surrendering an allowance. Such a mechanism was first proposed

[6] Climate threshold effects have the potential to reverse this result – but the threshold has to be well established. Uncertainty about the threshold tends to smear the effects and revert to a preference for prices. See discussion in Pizer (2002).

[7] It is also possible to design even more complex approaches to attempt to mimic the marginal benefit schedule or to make a permit system behave like a price policy. In the case of climate change, where there is strong evidence for flat marginal benefits, and where a safety-valve mechanism seems to be achievable, such additional complexity does not appear necessary.

in the international context immediately before the Kyoto meeting (Kopp, Morgenstern, and Pizer 1997; McKibbin and Wilcoxen 1997). Kopp, Morgenstern, and Pizer argued that this would combine the best of the price and quantity worlds – preserving the distribution of rents associated with a trading system but providing the cost-certainty of a fixed price. However, this proposal was roundly rejected by environmental advocates.[8] Eight years later, while the idea for a safety valve gains momentum in the United States, it continues to face opposition from environmental groups.[9]

If the economists' trading suggestions were broadly accepted, and the price-based approach is being hotly debated, the last area of economic research on climate change, into climate benefits, cooperation, and coalitions has been generally ignored in policy circles. While not as extensive as the work on mitigation costs, there have been a number of efforts to monetize mitigation benefits (see, for example, Smith and Tirpak 1989; Cline 1992; Mendelsohn, Nordhaus, and Shaw 1994; Nordhaus 1994; Fankhauser 1995; Tol 2005). Interestingly, while the IPCC devoted enormous attention to studies of mitigation costs (that is basically the task of the entire Working Group III report), estimates of mitigation benefits literally cover a few pages in the second, third, and fourth assessment reports.

Still, other economists often use these benefit assessments to examine why different countries and regions might join a climate change agreement. These economists typically consider two questions: based on the costs and benefits of mitigation, is an international agreement *profitable* relative to no cooperation, and is it *stable* in the sense that each participant has an incentive to remain in the agreement versus free riding (Carraro 1999)? Many analyses argued, for example, that the Kyoto Protocol was not profitable for the United States (Nordhaus and Boyer 1999). From a rational point of view, however, we need not only profitability but stability. Without this, individual parties to the agreement

[8] An October 8, 1997, letter to the President from seventeen environmental groups stated that "Such a mechanism would undermine the credibility of the entire treaty process, and . . . would weaken, if not eliminate, any incentive for private sector innovation and investment . . ."

[9] Bingaman's filed amendment S.A. 868 to the 2005 energy bill contained a safety valve mechanism based on NCEP (2004). Yet comments on the recent Bingaman–Domenici white paper (see footnote 4) from the Pew Center on Global Climate Change, Natural Resources Defense Council, and Environmental Defense, all argued against a safety valve.

will consistently have incentives to deviate from their commitment (and "free ride"). A key feature in international agreements, relative to domestic policies, is their inherently voluntary nature.

The message from this literature is that a self-enforcing agreement on mitigation is very hard. As an alternative, one might consider technology cooperation (Barrett 2001; Carraro and Buchner 2004). The arguable advantages are that, unlike mitigation where the benefits are non-excludable and everyone shares the improved climate, technology cooperation yields benefits that could be limited to participants; namely, appropriable knowledge. This provides greater stability in the agreement because deviating from the commitment means forgoing the technology benefits. The problem in the end, however, is that technology incentives alone are typically not enough to encourage mitigation (Fischer and Newell 2005).

Why has this work on cooperation, unlike the work on trading and instrument choice, been largely ignored in policy debates? To a large part, it is because this work is not particularly constructive – explaining why cooperation is so hard, not how to make it happen. The suggestion for a technology approach, while more rational, is also shown to have lower environmental benefits than even the current Kyoto approach in the absence of the United States and developing countries. Advocates and proponents of actions do not see this as a desirable direction.

Even if advocates find little value in economists' work on cooperation, the question remains, why have economic analyses of mitigation benefits not been used to justify action? The estimates have not only found a weak response in the IPCC, but are almost never used by advocates to argue for action. The answer, it seems, lies in how the problem is going to be framed.

Safe levels or marginal analysis

While economists are happy to put dollar values on virtually everything in order to make trade-offs, the rest of the world has yet to catch on. Environmental advocates, among others, have preferred to establish absolute standards for safety – standards below which consequences are acceptable and beyond which they are not. This is the basis for the 1970 Clean Air Act in the United States, and this is the basis for the UNFCCC. Most economists, in response, have tended to go along with the program, focusing their energy instead on achieving the standards at the lowest possible cost rather than battling over the level of the standards – or whether there should even be standards (e.g., versus prices).[10]

The 1970 Clean Air Act, for example, requires that air quality standards be set at levels sufficient to protect the public health. Legal challenges have elaborated that costs cannot enter into the standard setting process (Ross 2001). Regulations are then put into place that create progressively more-expensive control requirements depending on whether a region meets the standards or not. Various standards for new sources as well as retrofits for old sources enter the mix, although at this point the commercial availability of a technology becomes relevant. Sources are generally not asked to do something that is infeasible.

Thanks in large part to two decades of encouragement from economists, a significant improvement to this approach came with the 1990 Clean Air Act Amendments, which introduced emission trading. A national SO_2 cap (and regional NO_x cap) were used to achieve the standard everywhere, while providing considerable flexibility to sources. Required benefit–cost analyses of these regulations reveal that the benefits substantially outweigh the costs (US EPA 1999) – a somewhat amusing result given that the required benefit–cost analysis is not allowed to play a role in the initial setting of the policy-driving standard. Other government analyses suggest that marginal benefits continue to exceed costs (US Office of Management and Budget 2000). Even more interestingly, epidemiological evidence suggests that health and mortality continue to improve with lower and lower levels of fine particulates in almost a linear manner with no evidence of a "safe" threshold (Pope 2000). Despite our desire to ignore benefit–cost trade-offs in the law, nature may not cooperate.

Nature seems to be cooperating even less on climate change. With indications of climate change already occurring (IPCC 2001), it is even harder to argue that safe levels exist – on top of the economic arguments for stock pollutants presented earlier. Yet, the rhetoric and emphasis on safe levels persists in the public debate. Why?

[10] This crude generalization does not give sufficient credit to those economists who have argued for more emphasis in balancing costs and benefits but does reflect the relative weight in the published literature.

My guess is that arguing about the valuation of a long list of climate change consequences is less engaging than arguing about how clean, safe, or unspoiled we want the environment. For example, one of the best recent surveys of benefit estimates suggests a likely marginal benefit of about $15 per ton of CO_2 (Tol 2005). Within that best guess, however, there is uncertainty ranging from negative values (i.e., benefits), to prices many times higher. Does an advocate want to argue about the correct valuation being $100 per ton, or argue about whether atmospheric concentrations of 550 ppm (parts per million) is sufficiently safe versus 450 ppm? It is against this backdrop, where economists have had little success suggesting marginal valuations associated with incremental climate change, that cost-effectiveness analysis has flourished.

An important consequence of shifting the debate from marginal benefits to safe concentration targets, however, is that it spurs new debates about the timing of emission reductions. If new, inexpensive mitigation technologies are around the corner, would it not be better to wait and use them rather than more-expensive efforts now? Further, a concentration target indicates a zero-emission future, perhaps soon. Would we not be better off investing resources in zero-emission technologies rather than short-term mitigation efforts? For example, switching from coal to natural gas reduces emissions by half, but not to zero. Is that not a waste of resources compared to investment in zero-emission technologies that will eventually be cheaper? In economic terms, a price path that falls over time for an exhaustible resource cannot be optimal.[11]

Montgomery and Smith (forthcoming) consider an interesting extension of this line of thinking, namely, that zero-emission technology development can *not* be encouraged through market-based emission policies. They argue that while emission trading, for example, has been effective at finding the cheapest mitigation opportunities among existing options (across existing technologies and sources), as well as spurring innovations that have immediate cost reductions, its potential success at incentivizing innovation in the climate change context is unclear. In particular, they argue that the required innovations will take much, much longer, and, more importantly, that when the innovation

[11] The exception being arguments for learning-by-doing; see Edenhofer *et al.* (2006).

is realized, there will be strong pressure on the government to reduce allowance prices to the minimum necessary to marginally encourage technology adoption. This minimum price may be considerably less than the price required to spur innovation, thereby creating a dynamic inconsistency: they need to threaten the market with high emission allowance prices in the future in order to spur innovation, while everyone knows that when the innovation arrives the price will, in fact, not be high. In their discussion, they draw analogies to work on vaccine innovation by Kremer (2001), who argues that similar dynamic inconsistencies lead pharmaceutical companies to put few resources into vaccines for malaria and other Third World diseases.

While Montgomery and Smith are frequently interpreted as arguing against near-term mitigation, a careful read indicates they are arguing against mitigation beyond what is justified in a marginal benefit-cost analysis. Yet the basic message is, if you are really concerned about breaching a concentration threshold, focus on technology policy, not mitigation – ironically, the same message we hear from stakeholders generally viewed as *less* concerned about climate change. In any case, this is also where, we noted before, economists looking at stable coalitions end up: technology agreements. Viewed another way, Montgomery and Smith may not be that far off – concluding that emission pricing alone will likely be unsuccessful – but jump to the equally-likely-to-be-unsuccessful suggestion that technology policy alone is the answer.

Underlying assumptions and observations

The previous section tried to sketch out the goals stylistically associated with the economist, the environmentalist, and the technologist. That is, maximizing net benefits for the economist, avoiding dangerous levels of atmospheric greenhouse gases for the environmentalist, and developing and deploying zero-emission technologies for the technologist. It is now useful to separate out key assumptions underlying these perspectives, and test them – where possible – against what we observe.

For the economist, advocating a global carbon tax or emission trading program, two assumptions loom large: cooperation and capacity. Cooperation need not be global, but needs to include major emitters, such as China and India. As was evident in Montreal at the

COP/MOP meetings, developing countries are steering clear of commitments. There is a tendency among economists to reduce this to a question of incentives: It is just a matter of making China and India a good enough offer, in terms of extra allowances for growth or side payments in one form or another, to institute a binding policy. Bradford (2002), for example, calls for allocating all countries their business-as-usual emissions, and then having a global agency purchase allowances to create scarcity. Who could argue with that?

The problem is that developing countries may not see accepting a limit on their carbon dioxide emissions – essentially their use of fossil fuels – as a reasonable trade-off at any price.[12] More to the point, there is also a limit to the willingness of industrialized countries to pay a high price, perhaps even more so if it is paid in a very decentralized way (versus subsidizing technologies produced by the industrialized countries themselves). At the end of the day, if developing countries become sufficiently concerned about climate change, some arrangement should be possible. However, the question for economists is, what do we do in the mean time?

While the preceding point suggests that cooperation among all major emitters is unlikely in the near term, it is worth noting that many countries seem willing to take steps even without cooperation. The EU ETS was established prior to the Kyoto Protocol coming into force and without any comparable action (yet) by other countries (see Table 7.1). Within the United States, various states have begun establishing mandatory programs, despite inaction by other states and the Federal government. In other words, the argument that a binding international agreement is necessary for action, spurred by inaction under the original UNFCCC, appears to be missing something. Perhaps some regions collectively place a sufficiently high value on environmental amenities to justify unilateral action (or action they believe will inspire others to act). Or, perhaps they are sufficiently altruistic to internalize the externality

[12] There is a useful analogy to the plight of coal mines and mineworkers. Plenty of studies have shown that it would be relatively cheap to pay them to shut down (Bovenberg and Goulder 2001). Yet, in conversations with mining companies and mineworkers, they are less than enthusiastic about giving up their business and way of life in exchange for a government promise of its cash value. Similarly, developing countries may be reluctant to give up the tried and true approach to economic growth – freely burning fossil fuels – in exchange for industrialized country promises of allowance revenues or side payments.

of their actions. In any case, it is this observation – that mitigation efforts transpire despite the absence of a binding international agreement – that suggests that a global agreement designed in the style of domestic programs may not be necessary, at least as a first step. That is, an international agreement may be able to seek action from nations without, initially, establishing exactly what reciprocal actions by other nations will look like.

Distinct from the issue of whether countries want to cooperate is whether and how they can; that is, their capacity. Bell and Russell (2002) argue that the institutions necessary for market-based environmental policy – monitoring, enforcement, and effective legal systems – are unlikely to be up to the task in developing countries.[13] And, some countries may have strong preferences for certain policy designs – whether they are taxes, emissions trading, or (ideally tradable) standards. Trying to design a globally harmonized system using a single mechanism may not work.

Finally, it is worth noting that the economist approach also assumes we need an international agreement to equalize marginal costs. That is, we need either international emission trading or a negotiated price. An interesting observation is that among many of the domestic programs established or proposed so far – for example, the New Zealand tax, the Canadian safety valve, and the EU ETS – all are in the $15–25 per ton range (excluding an early 2006 run-up in the EU ETS price).[14] This is true despite the fact that, at those prices, none of these countries would expect to achieve their Kyoto commitments (without government offset

[13] International emission trading is particularly troublesome in developing countries because both firms and the government have an incentive to cheat (thereby sticking it to purchasers in industrialized countries who get bogus reductions). Harmonized taxes are better, in that the developing country government has an incentive to collect. Standards that can be enforced during production (e.g., energy efficiency) are much easier to enforce than trading or taxes that require continuous emission (or energy use) monitoring. In particular, they can be enforced if industrialized countries provide the technology themselves (e.g., Integrated Gasification Combined Cycle [IGCC] or other high-efficiency generation technologies).

[14] This interpretation of a similarity among observed prices contrasts with the interpretation David Victor offers in his chapter. The difference is that here the point of departure is the previous analyses (e.g., Weyant and Hill 1999) suggesting much higher autarkic prices to achieve Kyoto targets. Prices may be different, but not nearly as much as would be the case if nations were establishing domestic programs to achieve the Kyoto targets.

purchases of one form or another). This suggests that, even in the absence of an international agreement, there may be a natural tendency for policies to converge in marginal cost. The New Zealand program, in particular, specifically referenced allowance prices in the EU ETS as a guidepost for setting their domestic tax level.[15] In other words, countries care more about costs than the Kyoto architecture and commitments suggest.

Moving next to the environmentalist perspective, that goal – to limit atmospheric concentrations to safe levels – translates into emission limits, imposed one way or another. On the ground, we have seen the Kyoto Protocol, the EU ETS, and proposals in the United States, such as McCain–Lieberman and the Regional Greenhouse Gas Initiative (RGGI), all supported by the environmental community. Underlying this approach are assumptions that it is okay to move ahead, at least for now, with emission limits in some countries or regions only, that those emission limits will serve to both reduce emissions *and* spur needed technology development, and that perhaps these actions by some countries will spur actions by others.

Three concerns work against these assumptions. The first is leakage – that is, emission reductions mandated in one region "leaking" out into other, unregulated regions. In cases like RGGI, where there is a regional cap on the power sector, and power is freely traded with regions that are not regulated, leakage can be as high as 90 percent (ICF Consulting 2005). More moderate estimates for leakage rates under the Kyoto Protocol are around 5–20 percent, with the caveat that stricter targets tend to lead to higher leakage rates (IPCC 2001).

The second, related, concern is competitiveness. This was the explicit rationale behind the US Senate's Byrd–Hagel resolution (US Senate 1997) and led the National Commission on Energy Policy to recommend a price cap on any US emission trading program (NCEP 2004). The concern also motivated a recent analysis of the EU ETS, which found that while four of five industries studied did not face a competitiveness problem – and might even increase profitability – the one exception was aluminum smelting, where firms would be expected

[15] The New Zealand government announced in December 2005 that it was abandoning its plan for a carbon tax and would instead pursue a domestic emission trading program. Recent reports indicate that Canada may or may not continue with its plan for a domestic trading program with a safety valve (Ambrose 2006).

to exit the EU market (Grubb and Wilde 2004). Of course, concern over environmental regulation affecting firm production decisions is nothing new, and historical data has suggested only a small effect (Copeland and Taylor 2004). Nonetheless, it continues to be a contentious issue as manufacturing jobs shift overseas, even if driven by other factors.

Finally, connected with the preceding concerns as well, there is the issue of certainty with respect to future emission prices; for example, both the Kyoto Protocol and the EU ETS have an uncertain future post-2012. At the same time, energy sector investments last for long periods of time, and energy technology development even more so. Consequently, decisions in the face of an emissions pricing regime hinge not only on current prices, but on expectations about prices in the future. In the United States, for example, even the more conservative utility companies are making decisions about new power plants with a non-zero price of CO_2 assumed in the future despite the absence of any price now.[16] Meanwhile, carbon-saving energy investments in Europe face obstacles because of uncertainty about future prices (Gagnier 2005). The idea that emission limits will encourage new private sector technology development depends not only on the policies achievable now, but on the ability to give key stakeholders confidence that the policies will continue. Failing to address the earlier concerns about leakage and competitiveness erode that confidence.

All of these challenges suggest that the achievable price signal for technology development could likely be less than the desired price signal. Clearly, market-based mitigation policies like taxes and emission trading create an incentive to pursue new technologies – the question is whether it is enough and, if not now, whether an emissions pricing mechanism can be sufficiently ratcheted up over time given these concerns. To that, we add the further observation that there is, indeed, a second market failure that discourages technology development (Jaffe, Newell, and Stavins 2005). The private sector broadly underinvests in research and development because innovators typically are unable to fully appropriate the value of their inventions. A new technology that benefits society $100 may only benefit the innovator $20.

[16] Based on informal conversations with Southern Company and American Electric Power.

Taking this to an extreme position, we are led to the last perspective we have associated with the technologist. Unlike the preceding examples, that have presumed some form of mandatory controls to spur action, here the assumption is that government subsidies and incentives can lure companies to develop and deploy new carbon-saving technology. Montgomery and Smith (forthcoming) have emphasized this approach as a way to deal with potential time inconsistencies in the kind of mandatory policies necessary to spur technology development; it is also the approach embraced by the Bush Administration (White House 2002). More recently, it is the basis of the Asia-Pacific Partnership (US Department of State 2006).

Recent work suggests that emphasis on technology development is far less important than emphasis on technology use, in terms of providing cost-effective emission reductions (Fischer and Newell 2005). Subsidizing use is not only expensive, it also faces regular obstacles as governments must raise revenues to pay for the program (unlike emission trading programs or emission taxes that may *generate* revenue). The US experience with the production tax credit for renewable energy is a case in point: Congress renews the program for two years at a time; it has lapsed at times in the absence of funding; and, as a consequence, there has been a boom-and-bust cycle of technology deployment with the net effect being unclear. More specifically, given the magnitude of the challenge associated with climate change – eventually replacing conventional fossil energy with non-emitting alternatives – it seems unlikely that government technology programs alone will be sufficient, even if technology programs do play an important role.

Current national responses

One of the most interesting and revealing features surrounding climate change policy is the enormous range of domestic responses that have been implemented or proposed. Before considering the structure of an international agreement – which is inherently made up of a collection of domestic responses – it is useful to look at some of these responses in more detail. Table 7.1 provides an overview for all of the Annex B countries (e.g., countries with targets under the Kyoto Protocol). Five countries (or groups of countries), in particular, are worth looking at because of their variety of policies or proposed policies: the European Union, New Zealand, Canada, Japan, and the United States.

Table 7.1 *Summary of actions by Annex B countries*

Australia*	Voluntary programs; state actions
Austria	EU ETS
Belgium	EU ETS
Bulgaria	EU ETS (preparing to join)
Canada	Intensity-based trading program + safety valve; on hold in 2006
Croatia	EU candidate
Czech Republic	EU ETS
Denmark	EU ETS
Estonia	EU ETS
European Economic Community	EU ETS
Finland	EU ETS
France	EU ETS
Germany	EU ETS
Greece	EU ETS
Hungary	EU ETS
Iceland	EEA member**
Ireland	EU ETS
Italy	EU ETS
Japan	Voluntary programs, tax incentives; discussed taxes and trading
Latvia	EU ETS
Liechtenstein	EEA member**
Lithuania	EU ETS
Luxembourg	EU ETS
Monaco	
Netherlands	EU ETS
New Zealand	Dismissed taxes in 2005, considering trading in 2006
Norway	Emissions trading and taxes / EEA member**
Poland	EU ETS
Portugal	EU ETS
Romania	EU ETS (preparing to join)
Russian Federation	Exploring ways to channel AAU sales to "green investments"
Slovakia	EU ETS
Slovenia	EU ETS
Spain	EU ETS
Sweden	EU ETS

Table 7.1 ***Summary of actions by Annex B countries (continued)***

Switzerland	
Ukraine	
United Kingdom	EU ETS
United States of America*	Voluntary programs, tax incentives; state actions

Notes:
* not intending to ratify.
** European Economic Area. May join EU ETS under common directive.

With the exception of a few, limited, carbon tax programs in certain EU countries prior to 2003, the EU ETS is the first example of mandatory climate change mitigation policy in effect in the world. It stipulates an absolute cap on covered sources, which include the power sector and several energy-intensive industries (refining, paper, etc.). Like the NO_x program in the United States, member states in the EU are responsible for allocating allowances. Unlike the US program, however, member states are also responsible for setting their actual cap level. National allocation plans (NAPs) are proposed and then approved by the EU Commission. Importantly, NAPs must convey how limits on sources in the ETS, coupled with other national actions for non-ETS sources, will achieve the country's Kyoto commitment. So far, we have only seen plans that deviate slightly from business-as-usual. The real test will be in later in 2006, when member states submit plans for the actual Kyoto compliance period. Plans for the initial, warm-up phase presumably were subject to more lenient interpretations.[17]

At the other end of the spectrum of mandatory policies, New Zealand was on track until December 2005 to implement a CO_2 tax to have started in 2007. The government announced, in 2002, that they would implement an economy-wide carbon tax that would approximate the international price of emissions, but no more than NZ$25 (US$15) per ton CO_2. Energy-intensive industries that faced international competition would be allowed to enter agreements to avoid the tax, and agricultural methane and nitrous oxide (which accounts for

[17] More information can be found at www.europa.eu.int/comm/environment/climat/emission.htm.

more than half of total NZ GHG emissions) would be excluded entirely (Hodgson 2002, 2005). The initial level of the tax was to have been NZ$15 (US$9) per ton CO_2. Japan similarly considered a carbon tax during internal government discussions, at a level of ¥2,500–3,000 per ton of carbon (e.g., $6–7 per ton CO_2), but did not put it forward as a government position.[18]

In the middle sits Canada. Canada announced plans for a Large Final Emitter (LFE) trading program in April 2005 for the oil and gas, thermal electricity, mining and manufacturing sectors. The program is based on intensity targets; that is, the emission limit for firms is indexed to industry output. Further, the program has a C$15 (US$13) per ton CO_2 safety valve. That is, Canadian firms can always buy extra allowances in the domestic program at C$15 to meet the target, thereby providing a cost cap to firms. Of course, this does not comport well to the Kyoto Protocol, which includes neither an index to output, nor a safety valve. However, it does represent a compromise – perhaps a necessary compromise – between industry taking on a mandatory emission program while leaving the government responsible for meeting the specifics of the Kyoto Protocol. In any case, concerns about the LFE program comporting with Kyoto have been dwarfed by concerns that Canada will not even implement the LFE program. In March 2006, after the government changed parties, the environment minister indicated in a letter to a Toronto newspaper that emission trading *may* be part of an *eventual* strategy to reduce greenhouse gas emissions (Ambrose 2006).

And where is the United States? Aside from mandatory efforts at the state level, which are generally seen as vehicles to spur federal action rather than ends unto themselves, there has been a slew of proposals in the US Senate. These include economy-wide proposals put forward in the 109th session by Senators McCain and Lieberman and by Senator Feinstein, as well as proposals focused on power plants put forward in the 108th session by Senators Jeffords and Lieberman, Senator Carper, and Senators Smith, Brownback, and Voinovich. All of these feature the traditional cap-and-trade architecture, though with differences in coverage, the cap level, and opportunities to use offsets

[18] Instead, Japan has pursued a primarily voluntary/incentive approach based on voluntary efforts by the Keidanren (business associations), "top-runner" efficiency standards, and, more recently, a voluntary trading program and up-front payment for credits through the CDM (Pizer and Tamura 2005).

from different sources (e.g., agricultural projects, international projects, etc.). Perhaps most interesting, however, was a proposal put forward by Senator Bingaman, then ranking (Democratic) member of the Energy and Natural Resources Committee during a flurry of amendments to the 2005 energy bill. The proposal is interesting because it has garnered interest from Senator Domenici, then the chairman (highest ranking Republican) of the committee. While the proposal was not submitted for a vote in the 109th session of Congress, the two senators jointly held three hearings after the debate over the energy bill and their dialogue suggested the basic structure of the proposal could be correct.

What is that structure? Essentially, Senator Bingaman proposed a very modest emission trading program with a $7 per ton of CO_2 safety valve that escalates 5 percent per year starting in 2010 (US Senate 2005). The level of the cap is designed to be a growth cap for ten years, then flatten out in 2020. Rhetorically, the cap is defined in terms of intensity, but in practice it would be translated into an ordinary cap and the allocations would not change as output fluctuates.[19] An important part of the formula is a review, starting in 2015 and continuing every five years thereafter, to examine actions by other major emitters and trading partners and to make adjustments to the program based on that evaluation. These features, while riling environmentalists unhappy with the safety valve, and riling industry stakeholders unhappy with the notion of any mandatory program, actually proved relatively uncontroversial among senators like Domenici. The problem has been the further details surrounding point of regulation, allocation, and offsets (hence the White Paper questions referenced at the beginning of this paper).

Lessons and architecture

What can we learn from all of these actions and inactions? At least three things. First, a binding international agreement is neither necessary nor

[19] As specified in the filed amendment, the intensity target from 2010–2019 would be translated into an ordinary cap based on the GDP forecast of the Energy Information Agency (EIA) at the time of a rulemaking prior to 2010. That cap would not be adjusted as actual GDP varies from the original forecast. A new GDP forecast would be used prior to 2020 to compute the 2020–2024, and every five years thereafter.

sufficient for domestic actions. The EU initiated their trading program, via a 2003 directive, before it was clear Kyoto would come into force. While existence of the Kyoto Protocol was clearly an important influence on the EU decision, it is not clear that a binding commitment was the critical element. Perhaps a more convincing example of the potential for domestic action in the absence of international agreement is the United States, which seems poised to take some mandatory action in the near future without an international commitment. In fact, arguably the last thing that will motivate the United States is a binding international commitment *on the United States*. Meanwhile, countries inside Kyoto – Japan, Canada, and New Zealand – are all struggling to produce mandatory programs despite their commitments. Why this is the case, that at least initially some countries are willing to act unilaterally, is an interesting question but less relevant for the immediate design of an international program.

Second, whatever action a country takes, the form of that action is likely to be dictated by domestic features and forces. The form of a New Zealand policy is undoubtedly shaped by the relative share of agricultural emissions in their inventory. In the United States, comments on the Bingaman–Domenici White Paper were surprisingly favorable to an upstream program – something that has been eschewed in Europe.[20] Meanwhile, we have seen evidence that voluntary programs in some parts of the world – particularly vehicle efficiency standards in Europe and Japan – may work.[21]

Third, even without formal mechanisms to equalize marginal costs across countries – e.g., international trading or a single, agreed upon tax rate – various forces seem to keep those costs in line. For example, we saw the explicit linkage of the New Zealand tax rate to the international permit price. We have also seen that none of the domestic

[20] An upstream program would regulate *producers* of fossil fuels based on carbon content of those fuels versus a downstream program that regulates fossil energy *users* based on emissions. With adjustments for imports, exports, and sequestration, the upstream, carbon content-based approach is equally effective because the eventual emissions associated with a fossil fuel – coal, oil, or gas – are known once it is extracted. Even downstream programs can measure emissions based on the carbon content of the energy used (which is exactly what the EU ETS does).

[21] That is, manufacturers have voluntarily committed to improving fuel economy and are on target to achieve those commitments. This, of course, raises the question of whether the commitments were more than they were intending to do anyway.

prices – observed in the EU ETS or specified in the Canadian, New Zealand, or Japanese proposals – approach the high autarkic prices suggested by analyses of what would happen if countries sought to meet their Kyoto targets on their own. By extension, the expected gains to linking existing and proposed domestic systems are likely to be significantly less than those suggested by generic analyses of trading among Annex I parties under the Kyoto Protocol.

When we sit down to design an international climate agreement, therefore, we need to recognize that domestic circumstances and opportunities differ; that, at least right now, at the beginning of a perhaps century-long global effort to address climate change, binding emission limits, prices, or standards are unlikely to be helpful; and that formal mechanisms to equalize marginal costs (at this initial phase) are less important for efficiency than suggested in the literature. Instead, we should encourage countries to make *some* commitment to mandatory action, and focus our energy on a clear commitment to evaluate what actually happens. Bodansky *et al.* (2004) refer to this as a policy-and-measures approach (or sometimes, pledge and review). Here measures means both steps to be taken, as well as metrics for evaluating action.

An important point to realize is that the influence of this approach need not be any less than under a Kyoto-style targets-and-timetables approach. Whatever compliance and enforcement mechanisms one can dream up for the targets-and-timetables approach could be applied with equal vigor to a policy-and-measures approach. The difference is that all the work is on the front end for the timetables approach – picking the target – while all the work is on the back end for the policy-and-measures approach – evaluating action. Given experience at the COP/MOP in Montreal, where discussions of post-2012 targets are already hung-up on what to do about developing countries, it may even be the case that the possibility of again negotiating Kyoto-style targets, following national experiences with the first set of targets, is vanishingly small.

After working through this balancing act of prodding domestic action without either overreaching or achieving nothing, there are two more necessary elements in a global climate agreement that fall out of our discussion: attention to developing countries and technology. Developing countries need to be engaged on as many levels as possible. This follows from the observation that while there is a tendency of economists to assume developing countries can be cajoled into joining a global market-based solution through side payments or generous allowance allocation, that does not appear to be the case in practice at the current time (and even if they were cajoled, their capacity to implement market-based policies is suspect). It also follows from the observation that, with global trading, unquestionably the largest source of cheap reductions would be developing countries – meaning they cannot be ignored. So, until both their interest and capacity match that of the industrialized countries, we need to consider practical policies that will reduce emissions in developing countries as cost-effectively as possible.

The approach under the Kyoto Protocol involves the Clean Development Mechanism – in theory, a sensible market-based approach for projects in developing countries to earn credits that can be used to meet the Kyoto or EU ETS targets. However, there is considerable controversy about whether it is working. While slow to ramp up, as of April 2006 there were 161 registered projects, 4.5 million issued credits, and 340 million credits slated to be issued from registered projects through 2012 (UNFCCC 2006). There are more than a billion more credits associated with other projects in some phase of design. For reference, the surplus (e.g., extra allowances above what they need) expected in Russia and Ukraine is about 4 billion tons over the first commitment period (Babiker *et al.* 2002). A large supply, yes, but not large compared to the Russian surplus. What is also remarkable about the supply of CDM credits is the make-up: Roughly half are HFC23 projects; another sixth N_2O. That leaves about a third as energy-related projects (Victor 2006). Whether these metrics suggest modest success, or not, is somewhat in the eyes of the beholder.[22] Critics say this is too little action in the wrong sectors, or point to the inherent problem of establishing baselines for individual projects (suggesting that these reductions are not "real"); proponents say this is just the beginning.

My interpretation is that the CDM progress is fine so far – but that we need to consider more avenues to engage developing countries. Two proposals were discussed at the COP/MOP meetings: sector-based

[22] In a recent workshop, both perspectives were heard. See www.weathervane.rff.org/process_and_players/Policy_Collaboration/Understanding_Translatlantic_Differences.cfm.

crediting and credit for avoided deforestation. In the current environment, both have the capacity to inject a large number of credits into the system and may represent too much supply.[23] In the longer term, however, I believe they represent two of three useful directions. First, there needs to be a willingness to encourage developing country policy reforms at the sectoral level. Whether we are talking about efficiency standards, energy market reform, or other carbon-saving initiatives in developing countries, there should be incentives on the table. This might be a package of sector-based credits, or it might be linked progress in other areas of national interest (e.g., trade or technology). Second, I believe there needs to be a more flexible approach to project crediting that moves away from ton-for-ton accounting. Credit for avoided deforestation is one idea, but the broader approach would be to standardize projects that are desirable, ideally (but not necessarily) keeping the right incentive at the margin. For new technologies where there are likely to be learning spillovers, or for projects with other co-benefits, the incentive could be higher. The finicky approach to baselines in the CDM needs to be replaced with a more streamlined, though perhaps not as environmentally pure, approach.

Finally, Victor (2006) also makes the point that even more than projects and sectoral policies, major infrastructure deals have the potential to alter emission trajectories dramatically. If Russia, for example, could be encouraged to pipe gas to China, the potential emission reductions from less coal use in China could match the reductions attributable to the entire EU ETS over the next decade. Such deals are unlikely to happen under a purely climate-focused initiative, but approaching major developing countries about such choices, and looking for ways to move them along, ought to be a key element of an effort to engage developing countries.

The last part of a global climate agreement needs to recognize the undeniable role of technology policy – not as a substitute for mitigation policy, as in the stylized technologist's perspective, but as a complement. The motivation for this is the earlier observation that, for a variety of reasons, prices and/or price expectations may continue to be

[23] That is, it may drive allowance prices to zero. Whether this is too much supply from an economics perspective is really about whether the system has sufficient confidence to bank excess credits for future use – otherwise, there will be a loss of intertemporal efficiency. From an environmental perspective there may also be a desire to maintain a minimum price – perhaps to spur technology.

lower than the risk of climate change warrants. Further, there are market failures in the area of innovative activity more generally. Two directions for including technology in an international agreement emerge. One would be a cooperative technology agreement where countries agree to develop new technologies jointly and share in the benefits. As several authors have noted, this is likely to be a more successful cooperative effort, versus mitigation, because there are clear benefits to be shared and a loss associated with defection from the group. But even if countries are not interested in explicit cooperation, there should be an explicit recognition that technology development is one of the measures against which national action will be evaluated. While it cannot substitute completely for a mandatory mitigation program, a technology development and deployment program should be part of each country's mix and, at the margin, be allowed to trade off for some mitigation effort when scrutinized under an international agreement.

The key suggestions in this chapter – flexible commitments, a focus on technology and developing country engagement, and a clear process for evaluating national actions – could be implemented in a number of equally effective ways. For purposes of completeness, I will briefly outline one way.

First, nations sign an initial agreement that includes specific policy commitments in key countries (major economies and major emitters). The commitments are nonbinding and can be of any form – emission targets (absolute or intensity), prices, standards, technology development, or projects in other countries. Countries would presumably use their own commitments to leverage commitments from other countries, but there would be no minimum commitment.

Second, the agreement creates a common authority for generating project-based credits in regions of the world (or segments of economic activity) where mandatory policies are not currently in effect – e.g., CDM/JI. The agreement would streamline the current approach to encourage desired activities (e.g., key technology diffusion, avoided deforestation) without necessarily emphasizing ton-for-ton accounting. It would allow developing countries to receive credits for major emission-saving policy reforms.

Third, nations would be free to choose when and whether to link domestic policies. All nations would be encouraged to make the common credits (described above) fungible with any domestic requirement.

Fourth, the agreement would specify rolling, five-year reviews of national policies, modeled after the Organisation of Economic Co-operation and Development (OECD) country review process (each year, one-fifth of participating countries would be subject to the five-year reviews while all countries would receive a brief, updating review). The reviews would focus on three areas: mitigation and mandatory policies to incentivize emission reductions; technology development and programs to speed the development and deployment of climate-friendly technologies; (for industrialized countries) developing country engagement to encourage linked "deals" that simultaneously reduce emissions and achieve other goals (security, growth, or other environmental benefits):

- Mitigation policies would be evaluated using a variety of metrics, including emission levels, emission rates, emission prices, expenditures, and efficiency. Emissions relative to historical levels would be only one part of the measure.
- Technology policies would be evaluated on the basis of both spending and tangible progress in development and deployment. Measures would focus on absolute levels, rates, trends, and efficiency.
- Developing country engagement would be evaluated in terms of spending and tangible progress, including policy reforms, infrastructure projects, and technology transfer.

Fifth, countries would meet annually to discuss the reviews, provide feedback, and adjust their commitments. Every five years, there would be a major round of reviewing progress and re-setting commitments. Countries would be held accountable for their progress or lack of progress. However, rather than attempting explicitly to punish countries who fall behind, the consequences would be implicit in their lack of leverage in the negotiations, the weaker reciprocity from other countries, and perhaps the public shame provided by the discussion of their status.

Sixth, participation would be based on a willingness to make a threshold commitment to climate policy. Escalating commitments would be driven by bilateral and multilateral pressure for reciprocity, as well as linking climate to other issues, particularly in developing countries. Decisions to explicitly strengthen the obligations would be addressed at a later time.

It is worth noting that while this description could quite easily accommodate the status quo in terms of existing national policies, it includes one key difference: A regular and persistent mechanism to prod countries toward stronger domestic policies through periodic evaluation and commitment exercises. Rather than the current UNFCCC/Kyoto process, where annual negotiations are an opportunity to disagree over international policies, the proposed process would have annual meetings that allow scrutiny of *national* policies.

Conclusions

The focus of this paper has been to motivate and describe a practical global climate agreement. There are a number of arguments for tidier arrangements: the economist's benefit–cost analysis, the environmentalist's concern over safe concentration levels, or a technologist's technology-driven view of the world. While each of these leads to certain conclusions about the desired form of a global agreement, they also rest on assumptions that, at least currently, are challenged by real-world observations.

In particular, we see developing countries unwilling and perhaps unable to join in market-based programs. We see real obstacles to pricing emissions at a level that will properly spur technological development. And experience suggests that technology policy alone has a very hard time getting people to actually use the technologies that may be developed.

Against this backdrop of what is unlikely to work, we also see countries taking action in the absence of commitments (and countries with commitments failing to take action). We see a rough harmonization of prices among many programs even in the absence of a formal harmonizing mechanism. This suggests that a very basic assumption motivating the Kyoto approach and other suggestions for a strong, centrally directed regime – i.e., that a strong international structure is necessary, if not sufficient, to motivate domestic action – is perhaps not quite right. This paper explores how to translate these observations into a practical agreement.

In particular, it suggests we create an international agreement that provides opportunities to challenge nations to increased action as well as collectively evaluate that action, rather than attempting to incentivize that action in industrialized countries with compliance regimes. For the moment, it suggests we worry less about formal mechanisms to equalize marginal costs, and encourage national action that fits national circumstance. For developing countries, we need to explore every avenue to engage and encourage action – providing credits fungible in other trading programs, linking climate activities to other issues of concern (security and growth), and exploring larger deals that could alter emission trajectories. In the arena of technology, we need to recognize that mitigation policy alone will not be enough, to encourage technology cooperation, and to recognize technology policy efforts.

Why this approach? In addition to arguably being consistent with what might be implementable over the next decade, these steps also seem likely to yield useful building blocks for a future agreement with more cohesion. For one thing, virtually any group of market-based programs can eventually be sewn together. Permits can be traded against tax obligations. Tax rates can gradually be harmonized; safety valves can be escalated out of existence or relevance. Intensity or performance-based trading programs can be linked to absolute programs.[24] Perhaps most importantly, developing country credit programs can be morphed into developing country caps – when those countries are ready to sign on, such caps offer a far easier way to sell emission reductions to other countries.

Finally, nothing in this paper should be viewed as a lack of support for the longer-term vision of where an international agreement ought to go – and if it turns out to be achievable now, that would be great. Rather, this paper should be viewed as suggesting that, in the quest for that vision, we should not let the perfect be the enemy of the good and should not allow opportunities for improved global outcomes to go untapped in the meantime.

References

Aguilar, Soledad, Alexis Conrad, María Gutiérrez, Kati Kulovesi, Miquel Muñoz, and Chris Spence (2005). "Summary of the Eleventh Conference of the Parties to the UN Framework Convention on Climate Change and First Conference of the Parties serving as the Meeting of the Parties to the Kyoto Protocol: 28 November–10 December 2005," *Earth Negotiations Bulletin*.

Aldy, Joseph E. (2004). "Saving the Planet Cost-Effectively," in R. Lutter and J. F. Shogren (eds.), *Painting the White House Green*, Washington, DC: Resources for the Future, pp. 89–118.

[24] See special issue of *Climate Policy* 2004, edited by A. Marcu and W. Pizer, on linking systems.

Ambrose, Rona (2006). "Mandate to clean up air, water, soil," *Toronto Star*, March 1.

Babiker, Mustafa, Henry D. Jacoby, John M. Reilly, and David M. Reiner (2002). "The Evolution of a Climate Regime: Kyoto to Marrakech and Beyond," *Environmental Science and Policy* 5: 195–206.

Barrett, Scott (2001). "Towards a Better Climate Treaty," Policy Matters 01-29, AEI-Brookings, Joint Center for Regulatory Studies, Washington, DC.

Bell, Ruth Greenspan, and Clifford S. Russell (2002). "Environmental Policy for Developing Countries," *Issues in Science and Technology* 18(3): 63–70.

Bodansky, Daniel, Elliot Diringer, Jonathan Pershing, and Xueman Wang (2004). *Strawman Elements: Possible Approaches to Advancing International Climate Change Efforts*, Arlington, VA: Pew Center on Global Climate Change.

Bohringer, Christoph (2002). "Climate Politics from Kyoto to Bonn: From Little to Nothing?" *Energy Journal* 23(2): 51–72.

Bovenberg, A. Lans, and Lawrence H. Goulder (2001). "Neutralizing the Adverse Industry Impacts of CO_2 Abatement Policies: What Does It Cost?" in C. Carraro and G. Metcalf (eds.), *Behavioural and Distributional Effects of Environmental Policy*. Chicago, IL: University of Chicago Press, pp. 45–85.

Bradford, David (2002). "Improving on Kyoto: Greenhouse Gas Control as the Purchase of a Global Public Good," Dept. of Economics Working Paper, Princeton University.

Carraro, Carlo (1999). "The Structure of International Environmental Agreements," in C. Carraro (ed.), *International Environmental Agreements on Climate Change*, Dordrecht: Kluwer, pp. 9–26.

Carraro, Carlo, and Barbara Buchner (2004). "Economic and Environmental Effectiveness of a Technology-based Climate Protocol," FEEM Working Paper No. 61.

Cline, William (1992). *The Economics of Climate Change*, Washington, DC: Institute for International Economics.

Copeland, Brian R., and M. Scott Taylor (2004). "Trade, Growth, and the Environment," *Journal of Economic Literature* 42(1): 7–71.

Edenhofer, Ottmar, Kai Lessmann, Claudia Kemfert, Michael Grubb, and Jonathan Koehler (2006). "Technological Change: Exploring its Implication for the Economics of Atmospheric Stabilisation," *Energy Journal* (special issue, Endogenous Technological Change and the Economics of Atmospheric Stabilization): 57–107.

Fankhauser, S. (1995). *Valuing Climate Change: The Economics of the Greenhouse Gas Effect*, London: Earthscan.

Fischer, Carolyn, and Richard G. Newell (2005). "Environmental and Technology Policies for Climate Change and Renewable Energy," RFF Discussion Paper 04-05-Rev, Washington, DC.

Gagnier, Daniel (2005). *Emissions Trading: Industry Experience*, Paris: IEA.

Grubb, Michael, and James Wilde (2004). *European Emissions Trading Scheme: Implications for Industrial Competitiveness*, London: Carbon Trust.

Hayhoe, K., A. Jain, H. Pitcher, C. MacCracken, M. Gibbs, D. Wuebbles, R. Harvey, and D. Kruger (1999). "Costs of Multi-Greenhouse Gas Reduction Targets for the USA," *Science* 286: 905–906.

Hodgson, Pete (2002). *Climate Change: The New Zealand Response*. NZ Government. Available at www.beehive.govt.nz/.

(2005). *Carbon Tax Speech*. NZ Government. Available at www.beehive.govt.nz/.

Hyman, R. C., J. M. Reilly, M. H. Babiker, and A. De Masin (2003). "Modeling Non-CO_2 Greenhouse Gas Abatement," *Environmental Modeling and Assessment* 8: 175–186.

ICF Consulting (2005). *RGGI Electricity Sector Modeling Results*. RGGI.

IPCC (2001). *Climate Change 2001 Mitigation*, Cambridge: Cambridge University Press.

Jacoby, H. D., and A. D. Ellerman (2004). "The Safety Valve and Climate Policy," *Energy Policy* 32(4): 481–491.

Jaffe, A. B., R. G. Newell, and R. N. Stavins (2005). "A Tale of Two Market Failures: Technology and Environmental Policy," *Ecological Economics* 52(2–3): 164–174.

Kopp, Raymond, Richard Morgenstern, and William Pizer (2006). "Something for Everyone: A Climate Proposal that Both Environmentalists and Industry Can Live With," *Weathervane*, Resources for the Future, Washington, DC, September, 29. Available from www.rff.org/rff/Publications/weathervane/.

Kremer, Michael (2001). "Creating Markets for New Vaccine," In A. B. Jaffe, J. Lerner, and S. Stern (eds.), *Innovation Policy and the Economy*, Cambridge, MA: MIT Press, pp. 35–72.

McKibbin, Warwick J., M. Ross, R. Shackleton, and P. Wilcoxen (1999). "Emissions Trading, Capital Flows and the Kyoto Protocol," *Energy Journal* (Special Issue): 287–334.

McKibbin, Warwick J., and Peter J. Wilcoxen (1997). "A Better Way to Slow Global Climate Change," *Brookings Policy Brief*, No. 17 Brookings Institution, Washington, DC.

Mendelsohn, Robert, William D. Nordhaus, and Daigee Shaw (1994). "The Impact of Global Warming on Agriculture: A Ricardian Analysis," *American Economic Review* 84 (4): 753–771.

Montgomery, W. David, and Anne E. Smith (forthcoming). "Price, Quantity, and Technology Strategies for Climate Change Policy," in M. Schlesinger, H. Kheshgi, J. Smith, F. de la Chesnaye, J. Reilly, C. Kolstad, and T. Wilson (eds.), *Human-Induced Climate Change: An Interdisciplinary Assessment*, New York: Cambridge University Press.

National Commission on Energy Policy (2004). "Ending the Energy Stalemate," NCEP, Washington, DC.

Newell, Richard, and William Pizer (2003). "Regulating Stock Externalities Under Uncertainty," *Journal of Environmental Economics and Management* 45: 416–432.

Nordhaus, William D. (1994). *Managing the Global Commons: The Economics of Climate Change*. Cambridge, MA: MIT Press.

Nordhaus, William, and Joseph Boyer (1999). "Requiem for Kyoto: An Economic Analysis," *Energy Journal* (Special Issue): 93–130.

Pizer, William A. (2002). "Combining Price and Quantity Controls to Mitigate Global Climate Change," *Journal of Public Economics* 85(3): 409–434.

Pizer, William, and Kentaro Tamura (2005). "Climate Policy in the US and Japan," Resources for the Future Discussion Paper 05-28, Washington, DC.

Pope, C. Arden (2000). "Particulate Matter-Mortality Exposure-Response Relations and Threshold," *American Journal of Epidemiology* 152(5): 407–412.

Reilly, J., R. Prinn, J. Harnisch, J. Fitzmaurice, H. Jacoby, D. Kicklighter, J. Melillo, P. Stone, A. Sokolov, and C. Wang (1999). "Multi-Gas Assessment of the Kyoto Protocol," *Nature* 401: 549–555.

Repetto, Robert, and Duncan Austin (1997). "The Costs of Climate Protection: A Guide for the Perplexed," Washington, DC: World Resources Institute.

Ross, Heather (2001). "Clean Air: Is the Sky the Limit?" *Resources* (143): 13–16.

Smith, Joel B., and Dennis A. Tirpak (1989). "The Potential Effects of Global Climate Change on the United States: Report to Congress," US Environmental Protection Agency, Washington, DC.

Tol, Richard S. J. (2005). "The Marginal Damage Costs of Carbon Dioxide Emissions: An Assessment of the Uncertainties," *Energy Policy* 33: 2064–2074.

US Department of State (2006). *Work Plan for the Asia-Pacific Partnership on Clean Development and Climate*. US Department of State. Available at www.state.gov/g/oes/rls/ or/2006/59161.htm.

US EPA (1999). *The Benefits and Costs of the Clean Air Act, 1990 to 2010*, US Environmental Protection Agency, Washington, DC.

US EPA (2004a). "Acid Rain Program: 2003 Progress Report," US Environmental Protection Agency, Washington, DC.

(2004b). "Progressive Flow Control in the OTC NO_X Budget Program: Issues to Consider at the Close of the 1999 to 2002 Period," US Environmental Protection Agency, Washington, DC.

US Office of Management and Budget (2000). "Report to Congress on the Costs and Benefits of Regulation," Washington, DC.

United States Senate (1997). *A Resolution Expressing the Sense of the Senate Regarding the Conditions for the United States Becoming a Signatory to any International Agreement on Greenhouse Gas Emissions under the United Nations Framework Convention on Climate Change*. 105, Senate Resolution 98.

(2005). *Climate and Economy Insurance Act of 2005*. 109, Senate Amendment 868.

UNFCCC (2006). *CDM Scoreboard*. UNFCCC.

Victor, David G. (2006). "Engaging Developing Countries: Beyond Kyoto," paper presented at The Economics of Climate Change: Understanding Transatlantic Differences Conference, Resources for the Future, Washington, DC, March 3.

Weyant, John P., and Jennifer Hill (1999). "The Costs of the Kyoto Protocol: A Multi-Model Evaluation, Introduction and Overview," *Energy Journal* (Special Issue): *vii–xliv*.

White House (2002). *President Announces Clear Skies & Global Climate Change Initiatives*. White House, Washington, DC. Available at www.whitehouse.gov/news/releases/2002/02/20020214-5.html.

Wigley, T. M. L., R. Richels, and J. A. Edmonds (1996). "Economic and Environmental Choices in the Stabilization of Atmospheric CO_2 Concentrations," *Nature* 379: 240–243.

Commentaries on Pizer

7.1 *Is "practical global climate policy" sufficient?*

JAMES K. HAMMITT

William Pizer sets himself the task of evaluating the actions nations have and are taking to mitigate climate change to help determine what actions they are likely to be able to take in the future. He contrasts this approach with past discussions of what nations should do, which he describes in terms of the types of policies that might be favored by caricatures of three communities engaged in the debate: economists, environmentalists, and technologists. Pizer concludes that, while each of these communities brings useful perspective to the climate policy question, each view relies on assumptions that experience shows to be invalid, and so none of the recommended policies are likely to succeed. Based on his analysis of what nations can do, Pizer outlines an alternative architecture for international agreements that is consistent with past actions and hence arguably more likely to be followed than the Kyoto Protocol.

Global climate change presents a formidable challenge to society in the twenty-first century. Under current understanding of the science, continued anthropogenic increases in atmospheric concentrations of carbon dioxide and other greenhouse gases are likely to lead to changes in temperature, precipitation, sea level, and other characteristics of world climate that are unprecedented in human history. The consequences for unmanaged ecosystems, climate-dependent economic sectors including agriculture, forestry, and fisheries, and for land use and availability are difficult to predict but likely to be huge. Maintaining global climate in a state that does not require widespread adaptation of climate-sensitive industries, port-city infrastructure, and other aspects of human living patterns is likely to require limiting net anthropogenic emissions of greenhouse gases to well below current levels. Such a limitation marks a dramatic departure from the anticipated continuing rapid growth in emissions as developing countries such as China and India adopt more industrialized and energy-intensive living patterns, including major increases in their use of fossil fuels. Variation in greenhouse gas emissions on an annual and even decadal scale are of little importance because of factors that impart great inertia to climate change, notably the century-scale atmospheric residence times of carbon dioxide and most other greenhouse gases and the century-to-millennium-scale thermal response time of the global oceans. The level of sustainable emissions and time period over which it must be achieved are uncertain, but holding atmospheric concentration of carbon dioxide to levels that may avoid major disruptions to climate-sensitive activities is likely to require bringing emissions well below current levels by the end of the twenty-first century (e.g., Wigley, Richels, and Edmonds 1996).

If the basic science is correct, preventing substantial climate change requires a massive shift in the global use of fossil fuels and other activities that emit greenhouse gases away from their anticipated growth over the next decades. The challenge of achieving this shift appears to be so daunting that the most important attribute of any proposed international agreement to limit climate is the likelihood that it will catalyze the requisite shift. From this perspective, the Kyoto Protocol's focus on emission levels in the 2008–2012 period is of little concern except to the extent that it provides the beginnings of an incentive for development and adoption of the technologies, policies, and perhaps changes in social attitudes necessary to achieve the shift.

Pizer emphasizes that for an international agreement to attract participation and compliance, it must be perceived as in the national interests to potential signatories. Two comparisons are relevant. First, for a nation to accede to an international agreement, it must perceive that its interests are better served with the agreement than without. For the nation to abide by the agreement, it must perceive that its interests are better served by compliance than by free riding. The voluntary nature of international agreements necessarily limits the demands that can be placed on potential parties, since these cannot be more burdensome than the benefits from participation. The voluntary nature also has implications for the efficacy of different mechanisms (Wiener 1999).

In evaluating the extent to which participation and compliance with an international agreement further a nation's interests, an important question is how those interests are defined and perceived. Pizer discusses how national interests in economic development, specific technological choices, and policy mechanisms may differ across countries. A difference he does not discuss is the extent to which the damages from climate change, and hence the benefits of slowing or reducing its magnitude, may vary. Some high-latitude countries may experience increases in agricultural or forest productivity that may at least partially offset damages to other sectors. Tropical countries (which tend to have low income) are likely to suffer more from climate change than temperate countries because their climates are already warmer than what is apparently optimal for economic productivity (Mendelsohn, Dinar, and Williams 2006; Nordhaus 2006). Countries that span a range of climatic zones may adapt more readily to climate change by shifting activities within their borders. Other countries, particularly low-lying island or coastal states such as Bangladesh, may suffer grave damages from sea-level rise with little or no offsetting benefit. Holding other factors constant, countries that anticipate larger damages from climate change may be willing to accept higher burdens from a proposed international agreement.

Global climate change is often considered a global externality or commons problem for which the benefits of mitigation are not appropriable, so that no nation has much incentive to undertake unilateral mitigation. However, some nations (or strong coalitions such as the European Union) may be large enough that they account for a significant share of the benefits of mitigation. If so, it may be in the interests of such nations to mitigate climate change even in the absence of international agreement. Integrated assessment models suggest that a coalition of the industrialized countries may benefit enough to undertake substantial abatement efforts, even if the developing countries do not participate (e.g., Peck and Teisberg 1995; Hammitt and Adams 1996) and if no individual nation would do so unilaterally (Nordhaus and Yang 1996). These results suggest that coalitions consisting of a modest number of large countries, such as the Gleaneagles Dialog of the G8 + 5 and the L20 proposed by Canadian Prime Minister Paul Martin (discussed by Victor in this volume) may be willing to make substantial efforts toward abatement. With fewer members, such coalitions might also reach agreement more easily than coalitions with near-universal participation.

National interests relevant to participation in an international climate agreement are not limited to issues of climate damages and compliance costs. Nations interact with respect to a wide range of topics, including trade and defense, and a state may join and comply with an agreement to mitigate climate change that is not in its narrow national interest if there are sufficient compensating benefits in other aspects of its international relationships.

Given the decadal-to-century scale that a climate mitigation policy entails, the ability of a government to commit to future action is critical. A current government's limited ability to commit successor governments to the terms of an international agreement, or to other policies to mitigate climate change, may significantly constrain the set of feasible policies. Pizer discusses concerns about the possibility of using price instruments such as tradable carbon permits to stimulate the innovation and adoption of climate-friendly technologies. One issue, raised by Montgomery and Smith (forthcoming), is that governments cannot commit today to a future carbon price high enough to stimulate these activities because investors will fear that governments will subsequently reduce the carbon price to a level just sufficient to encourage adoption, greatly reducing the realized returns to innovators. Frankel (this volume) suggests that this commitment problem may be ameliorated by requiring a more stringent emission reduction in the near term than would be otherwise justified, although such an attempt could backfire (US withdrawal from the Kyoto Protocol may be an example of this). A better approach might be to adopt a policy with a predictable and automatic increase in carbon prices (such as the 5 percent annual increase in the safety-valve price in the Bingaman proposal that Pizer discusses). Such a policy might provide some assurance that the carbon price would not be dropped since it would develop constituents (e.g., producers of climate-friendly technologies) who would help resist rollback (much like retirees defend cost-of-living adjustments in social security programs). It is less clear what commitments a present-day government could make to assuage investors' fear that a future government would capture a substantial share of the returns to successful innovators through taxation.

More generally, limitations on governments' ability to commit to future policies may induce a rational-expectations effect through which the only credible policies are those that are consistent with current understanding of the climate issue and are amenable to modification as scientific understanding, technological development, and perceptions and preferences about climate change over time. Just as the Clinton Administration's agreement to the Kyoto Protocol was not credible in light of the Byrd–Hagel Resolution, so the current Bush Administration's unwillingness to consider any regulation of carbon dioxide emissions is not credible. Investors realize that successor administrations may have different views and must at least recognize the possibility that greenhouse gas emission regulations will be adopted in the future. Pizer notes that US electric utilities consider the possibility of future regulations in current decision making about the construction of new power plants.

Because of its effect on investment, uncertainty about future climate policy may be quite costly, yet because information relevant to climate policy will change, governments cannot credibly commit to specific regulations over many decades. What governments may be able to commit to is an adaptable policy with clear guidelines for the process by which amendments will be evaluated and adopted. In this case, investors could monitor developments in climate science, perceptions, and attitudes and would be able to forecast likely future policy developments with as much certainty as possible. In contrast, when a government states its commitment to a policy that is not credible, investors recognize that the policy is likely to change, but there is additional uncertainty about when and how the policy will be altered.

Pizer notes that economics has contributed more to the design of policy mechanisms for mitigating climate change (e.g., tradable permits, taxes, safety valves) than to setting the appropriate level of control. In climate change, as with many other areas of environmental policy, objectives have often been set in terms of "safe levels" rather than by equating marginal benefits and costs of more stringent control. Pizer suggests the reason that marginal analysis has had comparatively little effect on defining the appropriate stringency of climate measures is that it is "less engaging" to argue about the monetary value of damages than about how "clean, safe, or unspoiled we want the environment."

While it seems certain that advocates seeking greater environmental protection can mount a more effective political campaign by pointing to specific potential consequences (e.g., loss of species, increased severity of hurricanes), other factors may also contribute to the apparent preference for promoting safe-level approaches. One is the aversion many people feel to placing a monetary value on nonmarket attributes such as environmental quality and human health. This aversion may be associated with unwillingness to countenance explicit trade-offs between outcomes like these and market commodities. It may contribute to the view held by some advocates that economic-incentive mechanisms like emission taxes and tradable permits are objectionable because firms can treat the tax or permit fee as "just a cost of doing business." For some, human interference with the global climate system or climate change leading to widespread extinction of species may take the form of "protected values," which are absolute and cannot be traded for other values (Baron and Spranca 1997). Even for those who are willing to countenance explicit tradeoffs between the consequences of climate change and more pedestrian consequences, the scope, magnitude, effective irreversibility, and futurity of climate change consequences engender humility and great uncertainty. Many might feel that decisions about extinction of species should be based on some higher values than might be elicited in a typical valuation study.

A second likely reason that "safe levels" have been favored over marginal analysis is the concern that a marginal-damage-based approach, such as an emission tax, provides no assurance that environmental goals will be achieved. Indeed, when the costs of emission control are uncertain, quantity limits, like emission permits, guarantee the emissions level but leave the marginal and total control costs uncertain; price limits, like an emissions tax, guarantee the marginal cost but leave uncertain the emission level. As Pizer discusses, the price instrument is preferable for climate change because the marginal damages from emissions in a year are likely to be nearly constant. Because of the long atmospheric residence times of greenhouse gases, unexpectedly high emissions in one year can be offset by lower emissions in later years with very little or no effect on the environment.

Ironically, a marginal-damage-based approach may recommend more stringent control than a safe-level-based approach. A safe-level approach relies on setting some sort of bright line to distinguish "safe" from "unsafe," or in the terms of the United Nations Framework Convention on Climate Change "dangerous anthropogenic interference with the climate system" from interference that is not dangerous (abstaining from interference appears to be infeasible). If there were identifiable thresholds in the relationship between greenhouse gas emissions (or atmospheric concentrations) and damages to unmanaged ecosystems, economic activities, and other systems of concern, it might be possible to identify and defend such a bright line. However, it is equally plausible that the effects of climate change on many endpoints (e.g., sea level) are continuous rather than discrete, with few if any natural thresholds in the response. In this case, identifying a bright line requires defining what level on a continuum distinguishes dangerous from not dangerous with no clear standards. See Lorenzoni, Pidgeon, and O'Connor (2005) for a discussion of approaches to defining dangerous interference and Dowlatabadi (1999), Petschel-Held *et al.* (1999), and Yohe (1999) for discussion of "guardrails," "safe corridors," and "tolerable windows" for greenhouse gas concentrations and climate change.

Note that setting a bright line implies that all policies that do not violate the threshold have equivalent effects on climate. For example, if dangerous anthropogenic interference is defined as occurring if atmospheric concentrations of greenhouse gases rise to the equivalent of a doubling of preindustrial carbon dioxide levels, then policies that hold greenhouse gas concentrations to only a 50 percent increase above preindustrial levels are evaluated as no better than those that allow a 99 percent increase. The safe-level approach gives no credit, and provides no incentive, for policies that over-comply, i.e., policies that satisfy the threshold with a greater margin of safety. Any harm from allowing concentrations to increase from 50 percent to 99 percent above preindustrial levels is treated as negligible. Given uncertainty about the effects of greenhouse gas emissions on climate and about the consequences of climate change, a greater margin of safety is to be preferred.

If the safe level is associated with atmospheric concentrations, the least-costly compliance path is likely to be characterized by a delayed and relatively sharp reduction in greenhouse gas emissions (Wigley, Richels, and Edmonds 1996). In contrast, an emission path that balances marginal damages and marginal control costs is characterized by earlier and more gradual emission reductions (Hammitt 1999). The marginal-analysis-based path is more protective than the safe-level-based path, since it suppresses emissions and atmospheric concentrations in the relatively near term, providing some protection against the possibility of learning that climate change is more threatening than we currently estimate. Although the marginal-analysis-based path may ultimately lead to higher atmospheric concentrations than a safe-level-based path, this will only occur if the incremental damages from breaching the nominal safe level are smaller than the incremental costs of holding atmospheric concentrations below that level. In this case, it would be desirable to breach the safe level. Moreover, the atmospheric concentrations of greenhouse gases under a marginal-analysis-based path may not rise to the level of an otherwise similar safe-level-based path for many decades (Hammitt 1999), providing the opportunity to modify the marginal-analysis-based path to hold atmospheric concentrations below the safe level if better information suggests that is warranted. Note too that the marginal-analysis-based path is more credible than the safe-level-based path which requires future governments to enforce comparatively draconian emission reductions.

Pizer draws three lessons from his evaluation of the actions nations are taking to mitigate climate change: (1) binding international agreements are neither necessary nor sufficient for action, (2) the form of action taken by a nation is heavily influenced by domestic conditions, and (3) marginal control costs will be approximately equalized even without formal coordination. The first and second lessons are not surprising. As discussed above, individual nations may have an incentive to undertake some actions, perhaps because the nation captures a non-negligible share of the benefits of reduced climate change. Moreover, during this early stage of what may be a century-scale effort to control climate change, national policies may be strategic efforts to shape the eventual international response (e.g., by setting an example) or to gain first-mover advantages in developing new technology. The notion that nations would tailor their actions to domestic concerns requires little justification.

The third lesson, that marginal control costs may be reasonably similar across countries, even in the absence of emission trading or other formal methods of harmonization, merits comment. A first question is whether the statement is true. The evidence offered (that the proposed New Zealand emission tax and Canadian safety valve are comparable to the realized $15 and $25 per ton market price in the European Union Emission Trading Scheme) is not compelling, since neither the New Zealand nor Canadian system has been adopted and so the effective price of carbon in those countries may be closer to zero. Moreover, Pizer offers two counterexamples to his claim. The US proposal by Senator Bingaman specifies a safety valve of only $7 per ton (increasing 5 percent annually) and the Japanese considered a tax of $6 to $7 per ton. Victor (this volume) reports carbon prices in three other trading systems of less than $10 per ton. Second, this marginal control cost may reflect a focal point some nations are adopting in a setting of great uncertainty. This perspective is supported by New Zealand's reference to the EU emission trading system price as a guidepost. Third, the apparent convergence may simply reflect the modest level of effort directed toward reducing current greenhouse gas emissions, with many nations unlikely to meet their Kyoto Protocol targets (efforts to achieve those targets might have been much greater had the signatories' expectation that the United States would become a party to the Protocol been realized).

It is unclear from this evidence whether marginal control costs would be similar across nations (and hence gains from trading emission permits would be small) if nations were making more significant reductions in current emissions. One factor that would encourage equalization of marginal compliance costs would be international trade, since firms in a country facing high marginal compliance costs would be disadvantaged relative to international competitors facing lower marginal costs. But it is not clear how countries would equalize marginal compliance costs without explicit attention to it. In negotiating an international agreement, it seems intuitive that nations would evaluate their total compliance burden by comparison with other nations and it would not be surprising to observe some rough parity in total compliance burden (adjusted for differences in sensitivity to climate change, perhaps). It is not clear exactly how comparative burden would be assessed, but ratios of total compliance costs with economic output and with population like those described by Frankel (this volume) are possible. The main point is that however nations evaluate compliance burden, under a regime like the Kyoto Protocol which specifies national emission levels, harmonization of compliance burden is unlikely to simultaneously equalize marginal abatement costs unless either the abatement-cost functions are sufficiently similar across countries or compliance burden is defined by marginal abatement cost.

Based on his evaluation of what actions nations can be expected to take, Pizer proposes a "policies-and-measures" architecture for future climate agreements. In contrast to an agreement on outputs such as greenhouse gas emissions, the policies-and-measures approach is an agreement on inputs or efforts. Under this approach, each nation that accedes to the agreement will propose certain measures it will take to mitigate climate change. Implementation of the agreed measures and potentially their effectiveness would be evaluated later, perhaps as part of a five-year review cycle during which nations would propose new measures for subsequent periods. Pizer argues that a policies-and-measures approach is more likely to achieve real progress than a "targets-and-timetables" approach like that of the Kyoto Protocol, because it more easily accommodates differences among nations in preferred types of policies and because it postpones the hard work of evaluating and comparing national actions, rather than requiring *ex ante* agreement on specific emission targets. A further advantage is that the policies-and-measures approach can more easily include policies, such as research and development of climate-friendly technologies, that do not significantly affect near-term emissions (which, as discussed above, are probably not themselves of great importance) but that may prove invaluable for achieving the daunting changes in living patterns required to stabilize atmospheric concentrations at reasonable levels.

This policies-and-measures approach provides a method for encouraging national experimentation in the development of policies and institutions to mitigate climate change, with the potential to learn about which approaches work better and for nations to adopt the more successful measures. However, the efficacy of a nation's policies and measures will almost invariably be difficult to determine, as it rests on comparison of the observable outcome with the unobservable counterfactual situation of what would have occurred without these measures. In contrast, compliance with an agreed emission level is much easier to verify.

A key question is whether a policies-and-measures approach provides a mechanism through which nations can effectively pressure each other to make the kinds of large changes that will be required to constrain the buildup of greenhouse gases. In addition to difficulties in evaluating the efficacy of measures undertaken is the difficulty in even determining the extent to which promised measures were seriously attempted. Given the complexity of tax and regulatory codes in many nations, evaluators would need to determine whether a climate-friendly measure was effectively undermined by a compensating change in some other part of the tax or regulatory apparatus. Again, a targets-and-timetables approach like the Kyoto Protocol is easier to verify.

In summary, Pizer provides an insightful analysis of current efforts to reduce the threat of climate change and identifies important shortcomings in stylized perspectives of the economist, environmental advocate, and technologist. The policies-and-measures approach he proposes appears as if it might more easily gain the assent of a numerous and diverse set of states than a targets-and-timetables approach like Kyoto, but conditional on acceptance, it seems less certain than a Kyoto-like approach to achieve the desired environmental goals. A useful analogy may be drawn to the choice between quantity limits (e.g., tradable permits) and prices (e.g., emission taxes). A limit on output quantity (like Kyoto, but including all major emitters) is likely to achieve its environmental target, but risks excessive costs and with them, the possibility that nations will defect (or fail to join). An agreement on inputs and efforts, like the policies-and-measures approach Pizer proposes, provides greater protection against excessive costs but risks failure to achieve substantial control of greenhouse gas emissions. Which approach would maximize the chances of catalyzing the huge changes in technology and living patterns required to prevent disruptive climate change is ultimately the most important question.

References

Baron, J., and M. Spranca (1997). "Protected Values," *Organizational Behavior and Human Decision Processes* 70: 1–16.

Dowlatabadi, H. (1999). "Climate Change Thresholds and Guardrails for Emissions," *Climatic Change* 41: 297–301.

Hammitt, J. K. (1999). "Evaluation Endpoints and Climate Policy: Atmospheric Stabilization, Benefit-Cost Analysis, and Near-Term Greenhouse-Gas Emissions," *Climatic Change* 41: 447–468.

Hammitt, J. K., and J. L. Adams (1996). "The Value of International Cooperation for Abating Global Climate Change," *Resource and Energy Economics* 18: 219–241.

Lorenzoni, I., N. F. Pidgeon, and R. E. O'Connor (2005). "Dangerous Climate Change: The Role for Risk Research" (introduction to special issue on defining dangerous climate change), *Risk Analysis* 25: 1387–1398.

Mendelsohn, R., A. Dinar, and L. Williams (2006). "The Distributional Impact of Climate Change on Rich and Poor Countries," *Environment and Development Economics* 11: 158-178.

Montgomery, W. D., and A. E. Smith (forthcoming). "Price, Quantity, and Technology Strategies for Climate Change Policy," in M. Schlesinger, H. Khesgi, J. Smith, F. de la Chesnaye, J. Reilly, C. Kolstad, and T. Wilson (eds.), *Human Induced Climate Change: An Interdisciplinary Assessment*, New York: Cambridge University Press.

Nordhaus, W. D. (2006). "Geography and Macroeconomics: New Data and New Findings," *Proceedings of the National Academy of Sciences* 103: 3510–3517.

Nordhaus, W. D., and Z. Yang (1996). "A Regional Dynamic General-Equilibrium Model of Alternative Climate-Change Strategies," *American Economic Review* 86: 741–765.

Peck, S. C., and T. J. Teisberg (1995). "International CO_2 Emissions Control: An Analysis Using CETA," *Energy Policy* 23: 297–308.

Petschel-Held, G., H-J. Schellnhuber, T. Bruckner, F. L. Toth, and K. Hasselmann (1999). "The Tolerable Windows Approach: Theoretical and Methodological Foundations," *Climatic Change* 41: 303–331.

Wiener, J. B. (1999). "Global Environmental Regulation: Instrument Choice in Legal Context," *Yale Law Journal* 108: 677–800.

Wigley, T. M. L., R. Richels, and J. A. Edmonds (1996). "Economic and Environmental Choices in the Stabilization of Atmospheric CO_2 Concentrations," *Nature* 379: 240–243.

Yohe, G. (1999). "The Tolerable Windows Approach: Lessons and Limitations," *Climatic Change* 41: 283–295.

7.2 *An auction mechanism in a climate policy architecture*

JUAN-PABLO MONTERO

Introduction

William Pizer's "Practical Global Climate Policy" represents an important effort in trying to delineate the type of climate policies that have some chance of practical implementation. This is a most central policy exercise because, if done correctly, it defines the space of institutions, regulatory instruments, and policy measures that are realistically available for tackling the problems associated with climate change. The main message of the paper is that a practical global climate policy needs to make national action, rather than international coordination, the centerpiece. Pizer reaches this conclusion on the basis of observed behavior, that is, of current climate actions taken by different countries/regions.

I organize my reactions to Pizer's proposal in two parts. In the first part I provide some comments on the paper. Then, I advance a proposal of mine that builds around an auction of carbon permits in which the total number of permits is not fixed *ex ante* but is endogenous to the bids submitted by the different parties. My proposal is not necessarily coming to replace some of the (more comprehensive) policy proposals discussed in this book; it can be taken as an instrument or mechanism to complement those proposals.

Comments to Pizer

The main message that I take from Pizer's paper is that we should give up the idea of having coordinated international efforts – like the ones advanced by Jeffrey Frankel elsewhere in this book (this volume) – and instead rely on unilateral efforts. I have three problems with this

I am most grateful to Joe Aldy and Rob Stavins for truly constructive comments. I would also like to thank the Department of Education and Science of Spain for financial support (SEJ2006-1239/ECON).

message. My first problem is a practical one. If I would really believe that unilateral actions are the only way forward I should perhaps stop thinking about climate change policy. The economics profession – in particular the field of environmental economics – has already developed the tools for guiding the efficient implementation of environmental policy at the level of a region (e.g., country, state) where polluting agents within its borders can be brought under regulation. The profession, however, is still debating on how to tackle a (long-term) commons problem where affected and contributing parties (i.e., countries) only participate on a voluntary basis. This book and Aldy, Barrett, and Stavins (2003) are the best evidence of that.

The second problem relates to the empirical evidence used by Pizer to support his claim. Given the long-lived nature of the problem I think that too little time has passed to form a good idea of the types of climate policies that will prevail in the near and distant future. Furthermore, the types of initiatives that we observe today involve relatively mild emission control efforts (assuming Russia is allowed to sell all its Kyoto hot air) that are not even close to the control efforts needed to stabilize carbon concentrations. It is not clear to me that uncoordinated national actions could remain after countries are confronted with much more significant control efforts.

The third problem is a more normative one. Pizer emphasizes at the beginning that his paper is not about the ideal but about the possible. But when I revisit the "empirical evidence" I get the idea that coordinated international efforts are not entirely impossible. As many others, I share the view that Kyoto is a poor agreement (for many of the reasons provided in Olmstead and Stavins 2006) but, quoting Frankel (this volume), I think it is a good first (international) stepping stone on a practical path if we are to address the climate change problem effectively. The global public-good nature of the problem forces us to insist on the need for coordinated international efforts whether they are quantity based (e.g., Frankel, this volume) or price based (e.g., Nordhaus 2006). The economics literature is very clear if we fail to do so: too little will be done (I mean, much less than what could have been achieved collectively).

Before moving on to my proposal, let me add that I agree with Pizer about the tremendous difficulties in reaching agreement among such heterogeneous parties in issues as fundamental as the setting of the environmental goal and the participation of developing countries. I will touch on those issues as I describe the elements of my proposal; something that I now turn to.

An alternative proposal: a global auction mechanism

Countries are the bidders in this global auction mechanism. They in turn can implement domestic auctions of the same type. To appreciate my proposal let me start by briefly explaining the ongoing debate as to whether an international agreement to address climate change should be based on prices (i.e., taxes) or quantities (i.e., tradable permits). Pizer as well as others in this book also touch on this issue. When the regulation sets a tax on emissions, we know that firms' marginal costs will never go above the tax level, but it leaves us uncertain about the aggregate level of emissions. Conversely, when the regulation establishes an aggregate limit on emissions, what is uncertain is the (marginal) cost of reducing emissions which can go unexpectedly high (or low). If regulators knew parties' abatement costs, they would be indifferent between prices and quantities because they would be able to anticipate either the quantity response to a price limit or the price response to a quantity limit.

Because regulators rarely have good information on polluting agents' costs, Weitzman (1974) was the first to study how this information asymmetry affects the regulator's choice between prices and quantities. He finds that for some cases a price regulation will be better than a quantity regulation while in others the opposite will hold true. In the particular context of climate change, Weitzman (1974) would recommend the use of prices (i.e., taxes). The reason is that damages of emissions are related to the stock of carbon and other greenhouse gases in the atmosphere, while the cost of reducing emissions are related to the flow of emissions. This implies that the marginal costs of reducing emissions are highly sensitive to the level of emission reductions, while the marginal damages of emissions are almost invariant to the level of emissions.

Despite the theoretical superiority of prices, I share Frankel's inclination for quantities. As explained in his article (this volume), a price scheme in the form of a global harmonized carbon tax is just as difficult to implement as a quantity scheme (in the form of tradable permits), if not more so, provided that Kyoto parties have already demonstrated their ability to coordinate on differentiated national quantity targets. The difficulty in agreeing on a harmonized tax lies on the existing array of environmental and energy policies that vary substantially across countries. Furthermore, price schemes are more likely to face opposition from environmental groups who want to see emission limits explicitly set.

My proposal builds around an auction mechanism that establishes quantity limits. It nevertheless differs from the more conventional quantity-based proposals (e.g., Frankel, this volume)[1] in that it fully dissipates the information asymmetry that creates the tension between prices and quantities described above. Consequently, it is technically superior to both price and quantity schemes (it does not present advantages over prices only when the marginal damage function is totally flat).

The auction mechanism consists of a uniform price sealed-bid auction of emission permits with two key ingredients. The first ingredient is to let the total number of permits be endogenous to the demand schedules submitted by the regulated parties. In other words, the total number of permits (together with the price) are part of the auction clearing process. Ideally, the permits supply curve used by the regulator to clear the auction should be the pollution marginal damage curve. But it does not need to be so. The supply curve may be the result of some political process that, for example, resolves to put a ceiling on the price of permits. I will come back to this issue below. In any case, letting the total number of permits be endogenous is the correct thing to do because the regulator is clueless about the efficient number of permits to be allocated before communicating with the polluting parties, that is, before knowing their demands for permits (i.e., marginal abatement costs).

The second ingredient of my auction scheme is the introduction of paybacks or rebates. Part of the auction revenues are returned to regulated parties not as lump-sum transfers but in a way that parties would have incentives to bid truthfully. It is not difficult to see that we want to avoid full rebates because that would be equivalent to a grandfather allocation inducing parties to over-report their demand for permits to the point at which they end up emitting as if there were no

[1] By more conventional I mean a two-step scheme. First, the overall emission quota is decided. Then, the overall quota is grandfathered to (or auctioned off among) individual parties.

regulation. On the other hand, we do not want to eliminate rebates completely because that would induce parties to under-report their demand for permits in an effort to reduce prices. This demand-reducing phenomenon was first recognized by Wilson (1979) in his pioneer "auctions of shares" article. He explains that even when there is a very large number of bidders, uniform price auctions can exhibit equilibria with prices far below the competitive price (the price that would prevail if all bidders submit their true demand curves). The reason for this is that uniform pricing creates strong incentives for bidders to unilaterally (i.e., non-cooperatively) reduce their demand schedules in order to depress the price they pay for their inframarginal units. Consequently, one can establish that there must be some level of rebates (between 0 and 100 percent) that would induce firms to bid their true demand for permits.

As formally demonstrated in Montero (forthcoming), the incorporation of these two key ingredients – endogenous supply and paybacks – into a uniform price auction is not only simple but remarkably effective in implementing the efficient allocation of pollution reduction.[2] Paybacks are structured in such a way that each party will find it optimal to tell the truth regardless of what other parties do. It should be emphasized that paybacks are not known prior to the auction but are endogenously determined as part of the auction clearing process. All parties pay the exact same price at the margin for permits but their paybacks may differ unless they have identical demand functions. Paybacks, however, will rapidly fall with the number of parties, and as the number of parties grow large the auction scheme converges to the Pigouvian solution for correcting externalities.[3]

The auction mechanism is not only efficient but surprisingly equitable in terms of bidders' payments. The net payment of each bidder (i.e., permits purchases minus rebates) is exactly equal to the residual (or additional) damage that the bidder's pollution exerts upon society. If there is, for example, a large number of parties such that each party

[2] I should add that the auction scheme works in implementing the efficient outcome even if all parties or a subset of them are acting collusively.

[3] In Montero (forthcoming) I illustrate how paybacks change with the number of firms for linear demand and supply curves. For example, when there is only one party/bidder to be regulated the payback fraction that induces the bidder to tell the truth is 50 percent. For three (symmetric) bidders this number falls to 10 percent for ten bidders to 3 percent and for 100 bidders to less than 0.3 percent.

barely affects aggregate variables, the residual damage of each party would be (almost) equal to the marginal damage of pollution times the number of permits allocated to that bidder.

Application to climate change

Let me now attempt the exercise of discussing how this auction scheme could be applied to climate change and how it could complement (or become part of) other proposals. The auction scheme in Montero (forthcoming) is absolutely general; it presents the basic auction design but does not go into any particular application. In that sense it is silent to many aspects relevant to climate change policy. How is the environmental target to be decided? What is the time-horizon of the target? How often is the target to be revised? What are the bidders/parties in the auction? Countries? Firms? What happens to the auction revenues? How are the revenues recycled back to the economies? What are the dynamic properties of the auction scheme? Does the auction scheme lead to a time-consistent path of investment in new technologies? How does the scheme evolve over time? How does the scheme handle the participation of developing countries?

The implementation of my quantity-based approach should not be different from Frankel's (this volume) in many aspects. I will omit those here and focus on key differences: the process for setting the overall target, the allocation of permits to industrialized countries, the dynamic incentives for investments in new technologies, and the allocation of permits to developing countries. I will leave the last issue to the very end, so in the discussion that follows I momentarily restrict myself to an industrialized world willing to formulate and implement coordinated actions.

Setting the target

I agree with Pizer and many others in this book and elsewhere that for both technical and political reasons it is unrealistic to think that the setting of the target will follow benefit-cost maximization principles. Furthermore, due to the great uncertainties and possibilities of learning it makes no economic sense to choose the entire path of emission targets (or price targets for that matter) at once. In any case, I also see the setting of the target as a the result of a political process

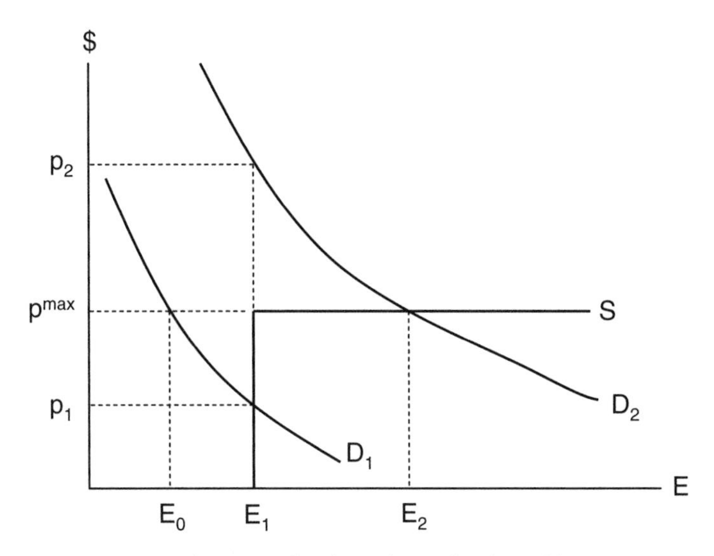

Figure 7.2.1. Setting the target for the industrialized world

involving all participating countries and evolving over time. In setting the target, parties may agree, for example, that for the next commitment period, say, for the five-year period following Kyoto, the "supply curve" of permits is the curve S in Figure 7.2.1. Ideally, this supply curve should reflect the emissions marginal damage curve but in this case is summarizing the "political consensus" that total emissions during the commitment period be capped at E_1 provided that the marginal costs of reducing emissions (i.e., price of permits) does not go above p^{max}.

Many readers will argue why not just use the well-known hybrid approach of allocating E_1 permits among affected parties (for example, according to Frankel's allocation rules) and letting these parties purchase additional permits at price p^{max} if they wish to do so. Provided that negotiating parties do not know the aggregate demand for permits (although each individual party is assumed to know its individual demand curve),[4] this hybrid approach is clearly superior to a pure tax p^{max} or a pure quantity limit E_1. If the aggregate demand for permits

[4] It is certainly an important issue whether countries know their true demand curves or not. I would argue that national governments can implement domestic auctions of the same type to obtain good estimates of their demand curves. I will come back to this later.

happens to be lower than expected, say D_1 in Figure 7.2.1, the quantity regime will lead to the correct level of emissions and equilibrium price of permits, respectively E_1 and p_1, while the price regime will lead to inefficiently low emissions, E_0. Conversely, if the aggregate demand for permits happens to be larger than expected, say D_2 in Figure 7.2.1, the price regime will lead to the correct level of emissions, E_2, while the quantity regime will lead to inefficiently low emissions and too high a price, respectively E_1 and p_2.

For the supply curve S, the hybrid approach solves the tension between pure prices and pure quantities, leading to the correct level of emissions regardless of whether the demand for permits happens to be low or high. My auction scheme implements the same outcome as the hybrid approach because parties have incentives to bid truthfully at the auction. If the "true" demand curve is D_1, parties will, on aggregate, bid D_1 and the auction will clear at price p_1 allocating a total of E_1 emission permits to the bidding parties. If, on the other hand, the true demand curve is D_2, the auction will clear at price p^{max} allocating E_2 permits to the bidding parties.

It is natural to wonder why we need the auction mechanism if the hybrid approach does the job of implementing the desired target (understood as the intersection of the negotiated supply curve and the true demand curve for permits). I have at least five reasons supporting the auction mechanism. The first reason may be less relevant for climate change than for other applications. If the supply curve of permits is not as depicted in S but exhibit ranges in which it is strictly increasing, the hybrid approach will fail to implement the desired target. The auction mechanism implements the efficient outcome for any supply curve. The second reason is that by construction the auction mechanism eliminates all incentives for the exercise of market power regardless of the size of the country. In a conventional emission trading scheme – where permits are allocated free or through an auction of fixed supply – the presence of large players may result in the exercise of market power (e.g., Olmstead and Stavins 2006).

The third reason is that the auction mechanism creates strong incentives for national governments to implement (cost-effective) approaches domestically that can reveal to them the true cost of abating emissions within their borders. They need this information when bidding at the international auction. The best way to obtain reliable cost information is by running internal auctions that share the structure of my auction mechanism.[5] These auctions must take place prior to the international auction and the supply of permits in those domestic auctions is not fixed but contingent on the country's results in the international auction.[6] Following Olmstead and Stavins (2006), my suggestion is for domestic auctions aimed at upstream sources. The fourth reason deals with the allocation of permits among the different countries, and the fifth with dynamic issues. I comment on these last two reasons separately.

Allocating permits to industrialized countries

Frankel (this volume) proposes a set of formulas for allocating the overall quota of permits among the different participating countries. These formulas change over time as they have to accommodate for changes in population, emissions, income, etc. They may also depend on invariants such as 1990 emissions. I find the use of these formulas an unnecessary complication to the negotiation process. As I explained in the previous section, the auction mechanism not only establishes the correct aggregate level of emissions but also makes sure that the individual allocations are efficient and equitable. All parties pay the same price for permits at the margin and their total payments (including rebates) are exactly equal to the residual damage they exert upon society. It is hard to argue against such (initial) allocation of permits.

There is one remaining issue however: What do we do with the auction revenues that remain after rebates have been completed? One possibility – following Bradford's (2002) suggestion – is to use part of these remaining revenues to buy reductions from the developing world. I will return to this point at the end of the note. But what if parties decide that all remaining revenues must be reimbursed to participating countries as a way to mimic a grandfathering allocation (at the country level) such as the one in Kyoto?

[5] Note that it is not sufficient to ask domestic sources to report their cost to the national government if that information will not be used later to allocate permits to these sources. Because a source does not internalize the cost of its deception on other domestic sources it would have incentives to over-report its costs (i.e., demand for permits) to some extent.

[6] For more details see the section on collusion in Montero (forthcoming).

Like any other mechanism that induces privately informed parties to report truthfully regardless of what other parties do,[7] the auction mechanism suffers from a budget-balance problem. The problem is that we cannot exactly reimburse all remaining revenues back to the parties (i.e., balance the budget) in a lump-sum fashion, that is, without distorting parties' incentives to report truthfully at the auction. Fortunately, the problem greatly diminishes as we increase the number of parties at the auction. In Montero (forthcoming), I explain how to fashion a lump-sum reimbursement system that assures perfect budget-balancing in the limiting case of a large number of parties. Using linear demand and supply curves, I show that if we have 10 (symmetric) parties at the auction, for example, 90 percent of the remaining revenues can be refunded without creating distortions. That figure jumps to 99 percent if we have 100 parties. Let me conclude this section by noticing that budget-balancing is less of a problem for the case of domestic auctions because national governments have different lump-sum ways in which they can recycle the remaining revenues back to their economies without distorting bidding behavior at the auction.

Dynamic properties of the auction mechanism

As explained by Pizer, Frankel (this volume) and others in this book, any practical climate change policy should be necessarily based on short-run negotiated targets (for a five-year period as in Kyoto, for example) that change over time. Given the large uncertainties we cannot expect parties to commit today to long-term targets. Furthermore, it would be economically inefficient do so in the presence of learning about damages and new technologies. I see, for example, parties negotiating by 2011 the target for the next commitment period, say 2013–2017; by 2016 the target for 2018–2022; and so on (the "evidence" shows that five-year targets are preferred over longer-period targets). Obviously, in negotiating the target, or more precisely the supply curve S, for the next commitment period, parties will necessarily take into account current information and future expectations

[7] These mechanisms are generally known as Vickrey–Clarke–Groves (VCG) mechanisms. Dasgupta, Hammond, and Maskin (1980) adapt the VCG mechanism for the case of pollution control. My auction mechanism has a completely different structure from theirs but still can be viewed as a VCG mechanism in that it makes each party pay only for the externality it exerts on society.

about the evolution of the global ecosystem and the global economy (including the development of new technologies) with all the irreversibilities that may involve (there is little I can say on whether parties will put a proper weight on future events).

In this long-term dynamic setting, it is important that firms face the proper incentives for the adoption and development of new and cleaner technologies. (By proper incentives I mean socially efficient ones). Since we cannot be using 1990 emissions as baseline year indefinitely, the problem that I see with other quantity-based approaches (e.g., Bradford 2002; Frankel, this volume) is that future permits allocations will be at some point linked to past behavior. This may affect firms' incentives to invest in new technologies in undesirable ways. The auction mechanism does not have such a problem. It provides firms with the socially efficient incentives to invest in new technologies. This is because the auction mechanism makes each party indirectly "solve" the social planner's problem at each point in time.

Participation of developing countries

There is no question that any successful global effort to address global warming must involve the broad participation of developing countries. The Clean Development Mechanism (CDM) of the Kyoto Protocol is one possible road, but I believe, as do many others, that the problems of setting baselines and proving additionality greatly erode its potential. Real-world evidence of the severity of these problems can be found in the voluntary provisions of the sulfur emissions trading scheme of the US Acid Rain Program (Montero 1999).

I agree with Frankel (this volume), Bradford (2002), and many others that developing countries should be (voluntary) brought as such to the international agreements. I am not proposing anything new here; I am just implementing Bradford's (2002) proposal of having an international agency buying reductions from all countries into the context of the global auction mechanism described above. The basic idea is to provide developing countries with a quota of permits approximately equal to their business-as-usual (BAU) emissions and let them come to the auction as sellers of emission reductions. Their emissions will remain unregulated if they decide not to come to the auction.

The idea is illustrated in Figure 7.2.2. As before, S is the (negotiated) supply curve of permits and D is the demand curve for permits from the

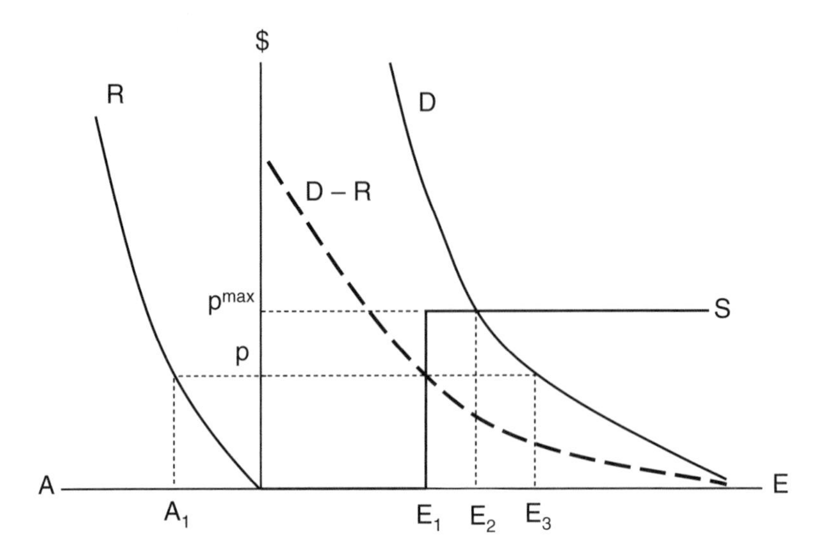

Figure 7.2.2. Adding the developing world

industrialized world. In the absence of developing countries, the auction will clear at price p^{max} with a total of E_2 permits. Let's now add the developing world. For a given BAU quota of emission, we can compute the aggregate supply curve of emission reductions coming from the participating developing countries. Let's denote this supply reduction curve by R (to avoid confusion in the notation A denotes emission abatement). Note that the curve D – R is simply the demand curve D from the industrialized world minus the supply curve R from the developing world. What are the new auction clearing price and quantity? Since the structure of the auction is such that each participating developing country has also incentives to bid its true marginal cost of reducing emissions, the auction will clear at price p and will allocate a total of E_3 permits to industrialized countries. Developing countries, on the other hand, will commit to a total abatement of A_1 below its BAU quota. Note that $E_3 - E_1$ is exactly equal to A_1. Because of the rebates needed to induce industrialized countries to bid truthfully, the revenues from the these countries is a little bit less than p^*E_3. These remaining revenues will be used to pay the A_1 emission reductions of developing countries. The total payment to developing countries is a little bit above p^*A_1 so as to induce them to bid truthfully as well (the larger the number of participating countries the closer the total payment to p^*A_1).

There are two remaining questions that I have little to add beyond and above what is in other proposals: (i) when should a developing country move to the right of Figure 7.2.2 and join the industrialized world?, and (ii) how will the BAU emissions of developing countries be estimated for later commitment periods provided that they have already engaged in emission reductions in the past?

References

Aldy, J. E. S. Barrett, and R. N. Stavins (2003). "Thirteen Plus One: A Comparison of Global Climate Policy Architectures," *Climate Policy* 3: 373–397.

Bradford, D. (2002). "Improving on Kyoto: Greenhouse Gas Control as the Purchase of a Global Public Good," Dept. of Economics Working Paper, Princeton University.

Dasgupta, P., P. Hammond, and E. Maskin (1980). "On Imperfect Information and Optimal Pollution Control," *Review of Economic Studies* 47: 857–860.

Montero, J.-P. (1999). "Voluntary Compliance with Market-Based Environmental Policy: Evidence from the US Acid Rain Program," *Journal of Political Economy* 107: 998–1033.

(forthcoming). "A Simple Auction Mechanism for the Optimal Allocation of the Commons," *American Economic Review*.

Nordhaus, W. D. (2006). "After Kyoto: Alternative Mechanisms to Control Global Warming," *American Economic Review, Papers and Proceedings* 96(2): 31–34.

Olmstead, S. M., and R. N. Stavins (2006). "An International Policy Architecture for the Post-Kyoto Era, *American Economic Review, Papers and Proceedings* 96(2): 35–38.

Weitzman, M. (1974). "Prices vs. Quantities," *Review of Economic Studies* 41: 477–491.

Wilson, R. (1979). "Auctions of Shares," *Quarterly Journal of Economics* 93: 675–689.

PART IV

Synthesis and conclusion

8 Epilogue: Architectures for agreement

THOMAS SCHELLING

THE participants took the title seriously – "architecture." Most authors are architects, some are architectural critics. The architects are of two kinds – two philosophies, two methodologies, two approaches to international institutions. One kind – the "true" architects – design complete integral systems, whole systems, leaving nothing out. They need no supplementary structures.

By "whole systems" I mean structures to cover all gases, all nations, all industries, all uses, and institutions that endogenize all incentives. The residual role of government is to enforce the rules; and the rules specify emission quotas and markets for purchase and sale of unused quotas. All the parts hang together. They are intellectually satisfying; they leave nothing out. There is a potential vulnerability here: if part of the system fails the rest may fail with it. But it all fits together economically and even aesthetically.

The second "architectural" approach, which I would not characterize as architecture but the authors do and I yield to their terminology, provides a set of substantially independent principles. The principles fit together with some completeness, but what I find to be their most attractive character is that the individual principles can stand alone, they have "separability"; if one fails it doesn't collapse a structure. The argument for each principle – the virtue of each principle – is independent of the arguments for the others.

Most of the complete architectures – the holistic, integral architectures – are dual, a domestic regime and an international. But the two display identical principles. The international regime entails national emission quotas enforced by some international mechanism, with trading of unused quotas – trading between national governments, not between individual emitters. The domestic regime is usually a tradable quota regime, sometimes a tax based on carbon (or carbon equivalents for other gases). The monitoring mechanism is usually not specified for the domestic regime; presumably for carbon it is fuel inputs, calibrated by carbon content, rather than CO_2 outputs, and for methane it usually goes unmentioned. For the international regime the measure of emissions is presumably fuel inputs, for which most nations may have decent data, at least for imported oil, gas, and coal, not CO_2 outputs. Methane, with the possible exception of pipeline leaks, is probably not reliably measurable. (Methane's residence time in the atmosphere, on the order of a decade, probably should excuse it from control for a few decades.)

Not considered in these six proposals is a carbon tax administered universally, i.e., on domestic producers or users, either levied by an international body that collects the proceeds (presumably from national governments), or mandated for domestic levies on greenhouse emissions, the proceeds kept internally. Thank goodness.

Missing from the holistic proposals is any reference to how the quotas should be enforced. Here there are two issues, participation and compliance. How are nations to be induced, coerced, or persuasively invited to participate in the scheme, not just Australia and Ukraine and Chile but Malaysia and Honduras and Namibia? There have been proposals – fortunately not among those represented here – that lavish quotas saleable for cash that might induce scores of developing nations to join and could be construed as "development aid." None of the holistic proposals included in this volume is specific about where to draw the line between those nations whose participation is essential to the success of the regime, and those that can safely be ignored or left till later. Allocating quotas by a democratic negotiation among close to two hundred governments would surely be difficult and probably time-consuming. Whether to impose sanctions, military, diplomatic, commercial – e.g., boycotts, embargoes, restrictions on finance or travel – and with what international recognition is a topic avoided in most discussion. And that issue arises equally in the enforcement of compliance.

"Compliance" refers to abiding by, conforming to, the rules, i.e., to the quotas established by the "participation" regime together with the rules for purchasing and selling unused quotas. Purely voluntary compliance is not taken for granted in the more holistic regimes: enforcement of some sort is assumed. I'm not sure it should be.

I believe there are many national governments that incur international commitments only with every intention of abiding by them and accordingly make serious efforts to conform to them. The United States is surely one. I believe it will be the case that many nations, having been allotted carbon (or carbon-equivalent greenhouse gas) quotas, and having made conscientious efforts to stay within them, will fail to stay within quota, will not consider themselves delinquent, and will not readily accept financial or other penalty for the failure. If they had willingly accepted the quotas at the outset, they will have erroneously estimated the requirements for fulfilling them. They will likely demand retroactive adjustment of the quota that could not be met rather than submit to penalty. If a "fine" is demanded and the nation does not pay it, who will enforce it, and by what means? If an embargo is declared, who will break off commercial relations? Is military enforcement the slightest possibility?

If quotas are very close to appropriate in nearly all cases, I can imagine nations that fail to meet them paying very modest penalties, perhaps disguised as "purchases" of others' unused quotas. But if quotas are set for periods of a decade or two, most governments will be poor at estimating what internal measures will actually bring emissions within quota. Shortfalls will be large and frequent and often not deserving a verdict of "guilty." Enforcement via severe penalty of any kind on an unwilling major nation I consider totally implausible; enforcement on some small nation that hadn't the authority or the resources or the capability to control emissions will be unpopular.

I am not certain that some kind of enforcement may not ultimately be devised, but it is certainly a huge obstacle to any multination negotiation involving major efforts over many decades among a hundred or more countries. At least, any proposal for "enforceable" quotas needs to be explicit about the means and the institutions for enforcement. Maybe more attention to purely "voluntary" compliance will prove necessary.

The other thing that I miss in the holistic architectures, which envision an overarching domestic regime of quotas and "trading" mechanisms over all carbon-emitting or fuel-producing activities, is some recognition of how innovative and unprecedented such a regime would be. Efforts over some decades to bring retirement, disability, unemployment, child welfare, poverty relief, and health insurance under one or more comprehensive regimes have gotten almost nowhere. Neither health insurance nor social security, even separated from poverty and unemployment, has gotten anywhere. Efforts to bring illicit drugs and abuse of prescription drugs under a comprehensive regime of policing, incarceration, treatment, prevention, domestic eradication, and border control have gotten nowhere. Every effort to "rationalize" and simplify federal taxes, even just the income tax, has foundered on the need for every interest, every lobby, every congressional district, every economic sector, to demand exceptions, exemptions, and special treatment.

The National Defense Reorganization of 1947 successfully restructured the armed forces quite substantially. It is hard to find any other large sector of the US economy or social structure that has been so radically revised so successfully in one large legal-political effort. And that, of course, restructured an institution that was fully under the jurisdiction of the US Government.

The domestic regimes proposed for the energy sector of the US economy have an advantage and a disadvantage. The advantage is that a very few simple (?) principles enacted into law would restructure the way fuel is marketed throughout the US economy – ranging from home heating to commuting to electricity production and distribution to airline travel to building construction, agricultural production, and aluminum containers. The disadvantage is the same: the farmers, the coal miners, the home builders, the retired, the people who drive long distances, the people who need air conditioning, all the people whose jobs will be in jeopardy, will need exceptions, exemptions, special consideration; every congressional district will have constituents who are threatened by the "impersonal," "noninterventionist," optimizing regime.

I find three necessary elements missing from the more holistic proposals – also from the less-integrated, more-independent principles, but that may matter somewhat less. One is how the quotas may be arrived at, procedurally and substantively. That is, what kind of diplomacy or international rationing procedure will assign a hundred or more quotas, and what criteria or standards will be applied. A second is no less daunting: how will agreement or acquiescence be arrived at on what the overall target should be, either in global annual emissions over several decades, or in the ultimate "ceiling" on concentrations, the "level that would prevent dangerous anthropogenic interference with the climate system," in the immortal words of the United Nations Framework Convention on Climate Change.

A third missing element is any attention to what kinds of costs should be accepted as bearable, or how a regime should adapt to evolving experience with emerging costs. One of the important differences between regimes that impose taxes and regimes that impose quantitative limits is that the former require explicit consideration of imposed costs (and may give less attention to averted damages), while the latter, though perhaps implicitly taking costs into account, focus on damages to be averted rather than on costs to be borne.

It is worth noting that the critical quantitative issues, involving what concentration of greenhouse gases will cause how much damage and what it is worth to avert it, is here as everywhere else left substantially open. The wide range of uncertainty on this issue – the "official" uncertainty entailing the traditional threefold uncertainty in the climate sensitivity plus the uncertainty of the damage function plus the uncertainty of abatement costs – makes any international agreement on the ultimate target so daunting as to be, up until now, unaddressed by any official body.

The less-integrated proposals, the "looser" architectures as I might characterize them, bring up a number of issues that may be assessed on their individual merits (and demerits) either as components of a more complete system or as ideas or principles that stand alone. Some of them are proposed as alternatives, some as new or neglected ideas. An example of the latter is the role of research and development (R & D), especially but not only in pursuit of mitigation technologies, a role that, as one paper points out, "should have been the centerpiece of the global climate regime, but it is barely mentioned in either of the two climate treaties" (Barrett, this volume, p. 244). Whether or not R & D should be explicitly the business of an international treaty, it surely should be central to any greenhouse gas program, especially as it is likely to be more popular than gasoline taxes or building code regulations.

Among the critical questions raised I highlight just a few. The most critical, because it is the foundation of any regime, is whether a greenhouse gas regime must be universal, or nearly so, or should "discriminate" (as one paper characterizes it) among nations essential to a regime's effectiveness and those that can be deferred or left out altogether. There are at least two issues here: what combination of nations may be most able to work together, and what nations are critical quantitatively to the success of a regime. (A third issue that I would raise is whether some critical nations, perhaps China and India, should be excused until the developed nations demonstrate that they are serious.)

A second issue is whether "binding" targets, or a pretense of them, should be preferred to looser diplomatic understandings. Another is whether trading arrangements need to be "universal" even among all the (limited) participants.

Two suggestions resonate with me especially. One, with which I have been identified, is that direct assistance to development should compete with greenhouse gas abatement. Many of the worst consequences of climate change will be felt in the poorest nations. Tropical diseases like malaria are cited: malaria may not only become more prevalent and more virulent over the coming decades, with warming, but is serious now and may be better combated with anti-malarial programs and accelerated development than with greenhouse gas abatement.

The other is the poor orphan that almost nobody wants to mention: directly intervening in the atmosphere to alter the earth's albedo, "geo-engineering" as it has been called. It is good that this volume helps to bring the subject out of the closet where it has been languishing except for occasional mention by the National Academy of Sciences or at a session of the American Association for the Advancement of Science (AAAS). I think it way premature to propose it for consideration in any formal treaty, maybe premature to propose it for consideration by the Intergovernmental Panel on Climate Change (IPCC), but long overdue for consideration in research, development, and experimentation.

The critical importance of poor-country development in avoiding the consequences of climate change, and the (barely?) possible eventual need for geoengineering, tend to remain below the purview of economists and others who focus on mitigation strategies. It is good to see them included in these papers.

Is there any alternative approach to reduction of carbon emissions that has a better chance of (a) doing the job and (b) getting enacted? And does it help or hinder that the privations and expenses, under either a holistic or a piecemeal regime, are perceived to be imposed by some international body or treaty?

It is easy to be a skeptic. And that is because these are hard issues. We have no historical model for the international regime that will be needed, nothing "enforceable" on the scale of a climate regime, not among sovereign nations. (The EU, certainly as formal and cohesive a regime as any that might be imagined for climate control in the foreseeable future, imposed severe penalties for any participating nation that incurred budget deficits exceeding 3 percent of GDP for three years running. As France and Germany approached delinquency in 2004, commentators expected nothing to happen; when the deadline passed nothing did happen.) Maybe we need to contemplate some kind of progress without a "regime" – something more opportunistic, more piecemeal, more purely diplomatic. (My model is NATO, an outstanding success on a huge scale with "commitments" unenforceable, related to actions rather than consequences.)

Domestically – and here I, like the members of this symposium, focus on the United States – we may need to think about alternatives to comprehensive regimes, even about experiments, accepting that politics will distort any plan, that what economists call "efficiency" – optimum restraints across the board on all uses of fossil fuels – will be an occasionally welcome outcome, not the guiding principle.

As I said, it is easy to be skeptical. I credit this volume with a wider array of proposals, suggestions, criteria, and considerations than any I have seen before.

9 *Architectures for an international global climate change agreement: lessons for the international policy community*

JOSEPH E. ALDY AND ROBERT N. STAVINS

GLOBAL climate change poses serious threats to the people and nations of the world, and the long-term, global nature of the problem has important implications for the design and implementation of effective policy, in particular for the design of a global climate policy architecture that is scientifically sound, economically rational, and politically pragmatic. The costs of abating greenhouse gas emissions occur locally and in the present, but the benefits accrue globally and over multiple generations.

The most important greenhouse gas that policy can affect, carbon dioxide, is a byproduct of the combustion of fossil fuels that serve as the primary source of energy for industrialized, emerging, and developing economies around the world. Relatively wealthy, developed countries are responsible for a majority of the anthropogenic greenhouse gases that have already accumulated in the atmosphere, but poor, developing countries will emit more greenhouse gases over this century than those currently industrialized nations if no efforts are taken to change their course of development. The problem of climate change with its global, long-term environmental, economic, energy, and development consequences necessitates the design of a sound international climate change policy architecture.

The Kyoto Protocol frequently has been described as a "first step" to address this awesome long-term, global problem. It is the first agreement to have imposed binding legal commitments on greenhouse gas emissions. In addition to representing the first step to mitigate greenhouse gas emissions, the Protocol represents the first step in the ultimate design of an effective, long-term international climate change policy architecture. It builds directly on the foundation set by the United Nations Framework Convention on Climate Change (UNFCCC).

The challenge for the international policy community lies in deciding on the next step, in terms of both climate-related goals and the design of policies to implement those goals. The Kyoto Protocol will inevitably be part of the foundation for all succeeding steps. Complementing the Protocol will be an array of other activities: efforts to pursue regional emission trading within key nations, such as the United States and Australia; international cooperation on energy research and development (R & D); climate- and energy-related development policies; and the implementation of domestic climate change policies under the Protocol itself. All of these will add to the foundation for moving forward.

In examining the Kyoto Protocol as well as international policy architectures that might serve to follow it in the second commitment period, we seek an architecture that is scientifically sound, economically rational, and politically pragmatic. More specifically, we have been guided by six criteria (Aldy, Barrett, and Stavins 2003) for evaluating climate policy architectures: (1) environmental outcome; (2) dynamic efficiency; (3) dynamic cost-effectiveness; (4) distributional equity; (5) flexibility in the presence of new information; and (6) participation and compliance.

Environmental outcome refers to a policy's time path of emissions or concentrations of greenhouse gases, or the impacts of climate change. A dynamically efficient policy maximizes the aggregate present value of net benefits of taking actions to mitigate climate change impacts. The criterion of dynamic cost-effectiveness refers to the identification of the least costly way to achieve a given environmental outcome. Distributional equity refers to the distribution of both benefits and costs across populations within a generation and across generations, and can account for responsibility for climate change, ability to pay to reduce climate change risks, and other notions of equity. Given the significant uncertainties that characterize climate science, economics, and technology, and the potential for learning in the future, a flexible policy infrastructure built on a sequential decision-making approach that incorporates new information may be preferred to more rigid policy designs. Finally, incentives for participation and compliance are important, since a climate policy architecture that cannot promote participation and compliance will not satisfactorily address the climate change problem.

Drawing upon lessons from designing and implementing the Kyoto Protocol, experience with complementary efforts, and insights from scholarly disciplines – including economics, international relations, political science, and law, among others – the contributors to this volume have set forth a range of ideas about how best to construct a future international climate change policy architecture. The targets and timetables approach embodied in the Kyoto agreement serve as the cornerstone of proposals advanced by Jeffrey Frankel and Axel Michaelowa. In their commentaries and in other writings, Daniel Esty, Joyeeta Gupta, Juan-Pablo Montero, Sheila Olmstead, and Jonathan Wiener discuss the design of climate change policy predicated on quantitative emissions targets with international emission trading. In contrast to such centralized approaches of setting emission commitments, David Victor, Warwick McKibbin, and Peter Wilcoxen propose policy architectures that aim to harmonize policies to address climate change among national and regional institutions. In a related vein, Richard Cooper suggests that countries should employ harmonized carbon taxes. So-called "bottom-up" approaches to climate change policy, through a mix of coordinated and unilateral country-level policies, are incorporated in the proposals by Scott Barrett and William Pizer. This notion of a mix of national policies evolving over time, perhaps through a system of pledge and review, also receives attention in commentaries by Carlo Carraro, Henry Jacoby, and Richard Morgenstern, and in Thomas Schelling's epilogue.

These ideas for a successor to the Kyoto Protocol cover nearly the entire spectrum of serious climate policy architectures. In this chapter, we synthesize and evaluate the key elements of the proposed architectures. The next section addresses the question of how to set climate change policy goals – an issue that fundamentally addresses the environmental outcome and the efficiency of policy. We then describe the near-consensus on the need for market-based implementation of climate change policies, noting a few important caveats. After focusing on cost-effectiveness, we turn to the related issues of equity and participation – how can policy architectures advance equitable outcomes and promote more participation by developing countries? We conclude with a discussion of the value of a broader suite of policies and the prospects for transitioning to the post-Kyoto policy architecture.

Setting climate policy goals

Setting climate change policy goals presents perhaps the greatest challenges to the international policy community. The emission commitments for the Kyoto Protocol's first commitment period (2008–2012) were among the most contentious elements negotiated at the Kyoto conference. The UNFCCC enshrines a qualitative long-term goal of atmospheric stabilization to prevent dangerous anthropogenic interference with the global climate, but agreement on a quantitative interpretation of this objective has not been reached. The setting of climate policy goals presumably reflects the desired environmental outcome (or benefits) of a climate change policy and the acceptable costs to society of undertaking the changes necessary to meet those goals. A variety of decision rules for informing the setting of such goals are available to policymakers, including the precautionary principle and economic efficiency (benefit-cost analysis). Alternatively, the setting of goals in the near term can be relegated to the politically feasible, as policymakers focus their efforts on the design of policy infrastructure.

Advocates of the use of the precautionary principle have argued that it can ensure that climate change does not cross a threshold beyond which the probability of substantial, dangerous impacts becomes unacceptably great. In this view, addressing a highly uncertain experiment with the only planet we have through precaution is a reasonable basis for action. Some analysts and policymakers have advocated a long-term stabilization goal of 550 parts per million (ppm) of CO_2 (about double the pre-industrial concentration) and 2°C above pre-industrial temperatures, reflecting the position taken by the European Union (Michaelowa, this volume). More ambitious goals, such as stabilization at 450 ppm may not be politically feasible, because the costs of achieving a 450 ppm target are much greater than those for a 550 ppm goal.

The difficulty in employing the precautionary principle is twofold: what is the appropriate threshold, and what are the countervailing risks of slower economic growth or inattention to other risks to society entailed by a precautionary approach? Climate change is already under way, so policymakers do not have the option of preventing climate change from occurring. With a problem imposing impacts over a continuum of atmospheric concentrations and temperatures, it may be difficult to identify the threshold that society wishes not to cross. Climate science suggests that nonlinear or threshold events – such as destruction of major coral reefs, melting of the West Antarctic ice sheet, or collapse of the thermohaline circulation in the Atlantic Ocean – are likely to occur at different temperature increases (Barrett, this volume).

Which threshold event should the precautionary principle target? Given uncertainties in the science, especially in relating atmospheric concentrations to specific temperature increases, being "safe" may require such low concentrations that they imply technological infeasibility in the short term and tremendous resource expenditures in the longer term.

A key question is whether setting a "safe level" is actually safer than choosing a goal through reliance on a criterion such as economic efficiency, that is use of benefit–cost analysis. Setting a threshold, say a concentration level, above which climate change is "dangerous" and below which climate change is "safe" is arbitrary in a world with climate change impacts at all possible concentration increases. Choosing a "safe" level may provide little incentive for activities that limit concentrations to a level below the threshold, even though these would deliver significant benefits.

Setting an emission path based on benefit–cost analysis – by accounting for damages at all concentration levels – would likely result in lower near-term emissions than a least-cost path to a concentration stabilization goal (Hammitt, this volume). By suppressing emissions more in the near-term and keeping the option open of very ambitious cumulative emission abatement in the long term, the benefit–cost approach could actually prove safer than setting a "safe level" if we learn over time that climate change impacts will be more damaging than previously thought.

Can benefit–cost analysis be used to help define a long-term emission pathway (or a pathway subject to change with refinements in benefit and cost estimates over time)? Some analysts – including economists – who would favor in theory employing benefit–cost analysis to set global quantitative greenhouse gas emission pathways have serious reservations about the feasibility of doing so. The central concern is that the world's political leaders face a daunting problem to constrain the decisions of future's leaders (Frankel, this volume; Aldy, Orszag, and Stiglitz 2001). It may simply not be feasible to impose modest emission abatement now, but much more ambitious and costly abatement several decades in the future (the cost-effective time-path endorsed by Olmstead in this volume and Olmstead and Stavins 2006, for example) – the future's leaders may reject such a deal when they come into power. Therefore, the argument goes, near-term emission targets should be more ambitious than suggested by benefit–cost analysis to reduce the incentive for the future's leaders to renege on deals made today. But such an approach may not solve the time inconsistency problem and may result in less-efficient policies than those informed by benefit–cost analysis (Wiener, this volume).

Rather than considering the normative question of how climate policy goals *should* be set, a more politically interesting question – at least in the short term – may be how *will* such goals be set? Historically, economic perspectives have had vastly greater effects on the means of environmental policies (such as cap-and-trade programs, which we discuss later) than on the identification of policy goals. Practical, political considerations – essentially domestic political interests – will dictate the actions countries are willing to undertake; these may deviate significantly from the prescriptions of either the precautionary principle or benefit-cost analysis (Pizer, this volume). Therefore, it may make sense to expend less effort quantifying the UNFCCC's ultimate objective, and dedicate greater effort to the design of policies, such as policies to promote technological development and deployment (Barrett, this volume). Most proposals for a successor to the Kyoto Protocol have not recommended specific emissions, concentration, or temperature goals (refer to the proposals in this volume, Aldy, Barrett, and Stavins 2003, and Bodansky 2004); rather, their focus has been on frameworks for setting and implementing goals.

This focus on institutions instead of emissions echoes the advice of Richard Schmalensee in his seminal assessment of climate policy architecture: "When time is measured in centuries, the creation of durable institutions and frameworks seems both logically prior to and more important than choice of a particular policy program that will almost surely be viewed as too strong or too weak within a decade" (Schmalensee 1998: 141). Clearly, goals will need to be set to guide long-term efforts to combat the problem of global climate change, but in the short-term, greater returns may be generated by developing the infrastructure for identifying and implementing goals.

Market-based implementation of climate policy

Few would argue that society ought to spend more than necessary to address the risks posed by global climate change. Market-based instruments, such as emission trading or emission taxes, can serve as the means for achieving climate policy goals at relatively low cost. In a world with scarce resources, the more cost-effective the means, the more feasible is an ambitious goal. In addition, lower costs of implementation can facilitate greater participation and compliance with climate goals.

The international policy community has embraced market-based implementation under the Kyoto Protocol. The European Union, formerly opposed to international emission trading during the Kyoto conference, launched the world's largest carbon dioxide cap-and-trade program in 2005, and efforts are under way in developing countries in a range of projects covered by the Protocol's Clean Development Mechanism. In addition, the emission trading concept has received support at the state and regional level in the United States and Australia, despite the fact that these countries have not ratified the Kyoto Protocol.

International emission trading serves as the cornerstone of several proposed international policy architectures (Frankel, and Michaelowa, this volume). In addition to yielding least-cost emission mitigation, such international emission trading can serve as a vehicle for transferring funds to developing countries. This brings two benefits. First, it provides the compensation that may be necessary to secure participation by developing countries, since developing countries could gain by becoming net exporters of emission permits. Second, a world of firm-based international emission trading would require transfers via the private sector, which are much less transparent than public sector appropriations. For example, instead of the private sector in developed countries transferring billions of dollars to oil exporters for crude oil, they could transfer (presumably fewer) dollars to emission permit exporters for their climate-friendly development. Such transfers may, nonetheless, suffer from critical political scrutiny (Summers, this volume).

National sovereignty means that countries cannot be legally coerced to take actions against their self-interest. Hence, the integrity and stability of an international emission trading regime may be in doubt. What happens if a low-cost supplier of emission permits decides to drop out of the regime? Or if a country with high abatement costs, and thus significant demand for permits, decides to drop out? Under Article 27 of the Kyoto Protocol, as in virtually all international treaties, a country can leave the agreement simply by giving notice of withdrawal, which becomes legally effective one year hence, although it is – in practical terms – effective immediately.

Such concerns motivate one proposal for coordinated but insulated national permit markets (McKibbin and Wilcoxen, this volume). Various regional emission trading regimes, such as the European Union Emission Trading Scheme (EU ETS), can and will evolve. Countries with common interests, legal systems, and financial relationships may develop regional trading regimes, but they may not engage with others who are dissimilar (Victor, this volume). Hence, it has been suggested that a set of national-level cap-and-trade programs may circumvent the potential problems of linking countries through international emissions trading.

To promote global cost-effective abatement, countries could coordinate their domestic programs by harmonizing the price of permits offered by governments. This implicit agreement on a safety-valve price would serve as the basis for equating costs across countries. Such a set of national cap-and-trade programs could comprise a system of pledge and review commitments by proponents of bottom-up policy design (Schelling, Pizer, Barrett, this volume).

Even those who are highly skeptical of international emission trading still support some form of cost-effective implementation. Some have called for an internationally harmonized carbon tax, which should provide similar incentives for abatement as an emission trading regime (Cooper, this volume; Nordhaus 1998). A tax may have considerably less appeal, because it eliminates the potential for an implementation mechanism to transfer resources to low-income countries that may otherwise be reluctant to undertake emission mitigation activities. Furthermore, countries could circumvent the effect of a carbon tax through "fiscal cushioning" – for example, a country could impose a carbon tax on industry while simultaneously reducing the excise taxes on fossil fuels (Wiener, this volume; 1999).

The immediate gains from linking national-level cap-and-trade programs may be modest, because permit prices of existing emission trading programs are similar (Pizer, this volume; c.f. Victor, Hammitt, and Montero, this volume). Even if this holds in the long-term, however, the value of the cap-and-trade approach remains. First, domestic and regional cap-and-trade programs would exploit the low-cost abatement opportunities within their markets; cost-effective implementation would characterize domestic actions. Second, policymakers who set domestic cap-and-trade program targets in a manner that impose comparable marginal costs with other countries' programs are effectively coordinating on the fairly transparent focal point of a carbon price. This would be similar to a system of national-level cap-and-trade programs with coordination about a safety-valve price (for annual emission permits that firms can buy from their governments) (McKibbin and Wilcoxen, this volume). Further, it is not very different from a system of harmonized domestic carbon taxes. A set of domestic and regional cap-and-trade programs with similar permit prices but little or no inter-program linkage is similar to coordinated but insulated national programs, and similar to harmonized carbon taxes. In each case, firms and households across participating countries would bear similar costs for emitting greenhouse gases, and modest, if any, international transfers would be involved.

The current system of market-based implementation under the Kyoto Protocol suffers from flaws that undermine its cost-effectiveness. The first phase of the EU ETS was characterized by substantial permit price volatility, and concerns about impacts on energy prices and competitiveness. In addition, it covers only half of carbon dioxide emissions in EU member states. The Clean Development Mechanism has channeled substantial resources to projects to abate synthetic greenhouse gases that likely would have been mitigated anyway (Victor, this volume). This project-based approach is a high-transaction cost and inefficient way of securing emission abatement in the developing world.

Meeting the challenge of addressing climate change will not be inexpensive. In the long term, it will require a substantial reworking of the energy foundation of industrial economies – either by changing the fuels that power economic activity or by developing technologies to capture the byproducts of burning these fuels. A successful climate policy architecture should promote cost-effective climate change mitigation; and market-based approaches, such as emission trading and emission taxes, are the best means to that end. The unresolved question is whether such systems can be imposed from the top-down, as in the Kyoto Protocol, or whether a more viable framework would evolve organically from a variety of national and regional emission trading regimes.

Designing a fair climate policy

The current climate policy architecture, as embodied in the UNFCCC and the Kyoto Protocol, calls for climate change mitigation efforts consistent with the principle of "common but differentiated responsibilities," which has been translated in practice into a set of specific, quantitative emission commitments for industrialized countries and no emission mitigation obligations for developing countries. While some interpretations of equity may indicate that it is appropriate for the relatively wealthy, industrialized countries, who are responsible for the lion's share of anthropogenic greenhouse gases which have already accumulated in the atmosphere, to take the first steps, there is little doubt that developing countries will – at some point – have to limit the growth of their emissions under any effective, long-term climate policy. The Kyoto Protocol provides no option for developing countries to take on quantitative targets, but future climate policy architecture will need to establish a fair process for securing developing country participation.

Several analysts have proposed rules for "graduation" into a system of quantitative emission commitments (Michaelowa, Gupta, this volume; Nordhaus 1998). In contrast to the UNFCCC and the Kyoto Protocol's bi-level regime of countries with quantitative emission commitments (Annex I/Annex B) and countries without commitments (Non-Annex I/Non-Annex B), the graduation criteria typically employ per capita income or per capita emissions as the basis for determining when individual developing countries should be obligated to take on commitments. Further, the stringency of commitments could vary by per capita emissions and income. Indeed, an explicit rule that accounts for income, population, historical emissions, and other factors could serve as the basis for graduation.

This notion of progressivity is implicit in the Kyoto Protocol commitments. The stringency of the Kyoto Protocol targets, measured as percentage reductions in carbon dioxide emissions from business-as-usual forecast, increases with per capita income. Although this progressive outcome resulted from an ad hoc negotiating process, future agreements could be based on formulas that maintain such progressivity (Frankel, this volume). Over many decades, the formula could place more weight on population, implying a per capita allocation in the distant future. For that matter, progressive commitments can also be employed with less stringent goals and more generous permit allocations for developing countries in a system of coordinated national-level cap-and-trade programs, or in pledge and review approaches.

The notion of a fair climate policy can also be viewed through the lens of international development policy. In other words, climate benefits could be leveraged from the energy and development efforts that developing countries want and need to pursue. Sustainable development policies and measures could incorporate climate change efforts into development priorities (Pershing, this volume). Revenues from industrialized countries, for example, through emission permit auctions, could finance such measures. Increased natural gas use in China and nuclear power in India have the potential to lower greenhouse gas emissions (Victor, this volume). Such efforts would simultaneously address local concerns about air quality and utility sector fuel diversification, and greenhouse gas emissions.

Complementing efforts to focus on development as a part of fair climate change policies is the need to facilitate adaptation. Developing countries may be more vulnerable and less resilient to the impacts of global climate change (Adger 2006). They have fewer resources, less human capital, less technological capacity, inadequate public health infrastructure, and weaker governance institutions to adapt to climate change than industrialized countries (Miranda *et al.* 2007). Developing countries also tend to have a larger share of their economies in climate-sensitive sectors, such as agriculture, forestry, and fisheries, and will likely experience larger percentage reductions in economic output than industrialized countries as the climate continues to change (Mendelsohn, Dinar, and Williams 2006). Promoting developing country adaptation – through financial transfers and R & D on technologies appropriate for developing country environments – could help satisfy notions of fairness in climate policy.

Promoting participation in climate policy

A policy architecture that cannot secure broad participation cannot in the long run deliver environmental benefits in a cost-effective or equitable manner. Promoting participation may be the greatest challenge for the design of climate policy architecture. No policy architecture can be successful without the United States, Russia, China, and India taking meaningful actions to slow their greenhouse gas emission growth and eventually reduce their emissions.

The characteristics of the climate change problem illustrate the difficulties of securing the support of world leaders to participate in serious efforts to mitigate climate change risks. First, the benefits of mitigation will accrue to future generations, while the costs will fall on the current generation. Second, the benefits of emissions mitigation accrue to the entire world, while the costs fall on those who undertake mitigation efforts.

One approach is to engage domestic constituencies. For example, the free allocation of permits in a cap-and-trade program can create vested interests in maintaining national emission programs. Giving emission permits to regulated firms represents a substantial transfer of assets. If a cap-and-trade program were weakened by the addition of more permits or eliminated entirely, this would reduce the asset value of these firms. On the other hand, it could be prohibitively expensive for governments to buy back permits if climate science commends a more ambitious climate policy (Pershing, this volume). Other ways of creating domestic constituencies include using revenues from cap-and-trade programs to finance shortfalls in social insurance programs.

Implementation policies can be tailored to provide incentives for participation. For example, progressive quantitative emission targets can attract participation by developing countries by imposing a relatively modest constraint on their emissions. Giving "headroom emissions" allocations to developing countries – granting them permits to more than cover their forecast business-as-usual emissions – can reduce downside risk of high compliance costs (Wiener, this volume; Stewart and Wiener 2003). Combined with a system of international emission trading, these less stringent targets can allow developing countries to become net exporters of emission permits. The trading regime becomes the mechanism for the side payments necessary to secure developing country participation.

Other approaches could be pursued to minimize downside risk to developing countries. Emission targets can be indexed to economic output to ensure that greenhouse gas limits do not constrain economic growth. An indexed target would be higher if the economy grew faster than expected but might increase less than proportionately with growth. The target would be lower if the economy grew slower than expected. This approach reduces the likelihood that developing country participation would relax the aggregate cap on participating countries' emissions via international emission trading (Aldy, Baron, and Tubiana 2003). This notion of an indexed quantitative emission target has some real-world experience in the case of the Government of Argentina's 1999 proposal and the US Government's 2002 intensity goal (Aldy 2004).

The problem with such approaches to promoting participation is that they assume that there are some countries willing to take on more stringent targets and make side payments. For example, if the United States does not want to take on an ambitious emission target, there will be much less demand and thus lower prices for the permits developing countries would aspire to export. Creating the incentive for some countries to participate should not simultaneously create the disincentive for others to do so. Moreover, developing countries who are opposed to taking on commitments for noneconomic reasons may not be particularly swayed by such economic incentives (Bodansky, this volume).

Alternatively, the approach taken by the Montreal Protocol of promoting developing country participation by direct payments could be pursued. Indeed, paying developing countries to use new, more climate-friendly technologies may be an effective first step (Esty, this volume). Engaging developing countries on energy and development issues will require more substantial investments by developed countries (Victor, this volume). The challenge therefore lies in whether governments of developed countries would be willing to finance substantial transfers to developing countries. The UNFCCC and the Kyoto Protocol include provisions for technology transfer financed by developed countries, but only limited funds have been directed to such efforts.

Instead of directing government funds for the deployment of climate-friendly technologies in developing countries, countries could fund directly the development of technologies through an R & D protocol (Barrett, this volume). This could draw on experience with other large, international R & D efforts that have involved substantial cost-sharing among participating countries. In theory, collaborative climate-friendly R & D efforts could then yield a new set of commercially viable technologies that would be deployed in a number of major countries. Such deployment could occur through domestic incentives (emission taxes, emissions cap-and-trade, or subsidies for climate-friendly technologies) or perhaps through a standards protocol, which would mandate the use of technologies in participating countries (Barrett, this volume). As long as a sufficient number of large countries employ the new technologies, they would become the de facto world standard. The challenge for this proposal is two-fold: securing sufficient funds for collaborative R & D, and developing a streamlined process for negotiating international technology standards. A bottom-up approach with domestic policies that generate revenues, such as a cap-and-trade program with a partial auction and a safety valve, could finance R & D efforts (Morgenstern, this volume).

Reflecting the concern that a top-down multinational cap-and-trade program is simply not viable in a world with national sovereignty, a number of analysts have focused on bottom-up, pledge and review approaches to climate policy architecture. The Marshall Plan and the formation of the North Atlantic Treaty Organisation serve as examples of how pledge and review produced successful and durable institutions. Climate change policy could evolve from national actions through small groups of like-minded countries (Victor, Pizer, this volume). These groups could eventually coordinate efforts in the long term as part of a broader climate policy architecture.

This bottom-up, pledge and review approach may be broad in terms of participation, but very shallow in terms of actions (Hammitt, this volume). Whether such an approach will yield more than unilateral, uncoordinated efforts depends upon whether it will evolve into a cohesive process. It has been argued that climate change policy focused on regional and fragmented programs could evolve in a way analogous to regional trade blocs (Carraro, this volume). Such regional trade blocs permitted countries to understand and inform the development of the world trade regime. They also provided an opportunity for countries to develop trust in one another. Such an approach on climate policy could complement the existing UNFCCC process.

Decentralized, pledge and review efforts may provide one more benefit beyond securing broader participation. The heterogeneity in policy approaches taken at national and regional levels can serve as a large set of case studies on policy design and implementation (Hahn 1998). By reducing uncertainties about the effectiveness and costs of various policy approaches, these tests can inform industrialized and developing countries alike about what does and does not work. This may be especially valuable for those leaders who are particularly risk averse about taking on climate change policies.

Despite existing concerns that climate change negotiations are already excessively complicated, it has been suggested that it may be possible to promote more participation by enlarging the issues under consideration. Integrating climate change measures in the development agenda and the trade agenda may serve as opportunities for effective issue linkage. Some have suggested that countries without emission commitments should have their exports subjected to border tax adjustments based on the carbon content of their goods. The Montreal Protocol allowed for trade sanctions, and climate change policies could also incorporate such penalties (Esty, this volume; Nordhaus 1998; and Aldy, Orszag, and Stiglitz 2001). In a related vein, trade policy and ratification of the Kyoto Protocol were linked by Russia when it negotiated with the European Union on its World Trade Organization accession. Russia agreed to ratify the Protocol when the EU dropped its objections to Russian pricing of natural gas. Expanding the negotiating game to include development and trade may allow for countries to make necessary trade-offs.

In contrast, it has also been suggested that negotiations should be reformed to a more manageable number of countries. The UNFCCC process, with some 190 member countries, is slow and bureaucratic, perhaps unnecessary. A smaller group of the most important industrialized and developing countries could agree on a new climate change policy architecture through a simpler, more efficient negotiating process (Bodansky, this volume). Perhaps the L20 process – representing leaders of twenty major countries – could provide a more effective venue for reaching agreement on climate policy. The approach initiated by the United Kingdom for a G8+5 process on climate change (where the five developing countries are Brazil, China, India, Mexico, and South Africa joining the G8 industrialized countries) is similar in approach.

Envisioning the next step for climate policy

The next step to address global climate change needs to be broader than the Kyoto Protocol, both in terms of the number of countries with obligations and perhaps the suite of policies to be employed. An important challenge is determining whether a bottom-up approach is superior to a top-down approach. If negotiations over a comprehensive, top-down approach hinder the development of domestic cap-and-trade programs or risk derailing the EU ETS for the post-2012 period, then the international community should consider transitioning to a pledge and review of mitigation efforts. An effective policy architecture is one that facilitates action, not imposes paralysis through a slow process.

Second, a climate policy architecture should support adaptation efforts. Climate change is occurring and will continue to occur to some extent regardless of greenhouse gas emission mitigation efforts.

Countries have more incentives to invest in adaptation, since the benefits of adaptation are more localized than the benefits of emission mitigation. A climate policy architecture might identify needs for those countries that lack the capacity and resources to adapt adequately to the changing climate. The architecture could also focus on leveraging R & D efforts to focus on the needs of those countries that are most vulnerable and lack the technology and human capital to invest in efforts to reduce their exposure to climate change.

Third, continued investment in R & D is necessary. R & D efforts should focus on emission mitigation, adaptation technologies, and geoengineering. Stabilizing long-term atmospheric greenhouse gas concentrations will require the development and deployment of low-carbon- and eventually zero-carbon-emitting technologies on an extensive scale. Continued R & D efforts can ensure that these technologies are both feasible and commercially viable at modest costs in the future. Such emission mitigation efforts should focus on zero-carbon energy sources, improvements in energy efficiency, and carbon capture and storage technologies.

Investments in adaptation R & D can deliver an array of benefits in mitigating the risks of climate change. Some of these efforts could focus on protecting developed coastal areas against sea-level rise and storm surges under climate change. Developing countries may benefit from improved technologies for coastal protection, irrespective of climate change, given the current risks posed by typhoon and monsoon seasons. R & D focused on the human health impacts of climate change, such as for a malaria vaccine, could reduce the risks of climate change and improve public health in low-income countries in a way that could facilitate economic development. Incorporating adaptation R & D efforts in a broad set of climate change policies could promote effective integration of climate and development policies and provide incentives for developing countries to participate more fully in the climate change policy architecture.

The potential for geoengineering to help address the risks of climate change merits consideration in the R & D portfolio (Schelling, Summers, Barrett, this volume). Geoengineering worries some participants in climate policy debates because they abhor the prospect of attempting a new experiment with the planet to correct an ongoing inadvertent experiment. But geoengineering need not be considered as a substitute for emission mitigation. Rather, it might be considered an "insurance policy for the insurance policy." If two, three, or four decades from now, we learn that the state of knowledge about climate change in 2007 was wrong – that climate change is much, much *worse* than we ever thought – then we would regret not undertaking the relevant research about the effectiveness and side effects of geoengineering options in the interim. Some have suggested that the direct costs of geoengineering solutions are quite low (Barrett, this volume), but clearly research on side effects and unintended impacts needs to be undertaken (Jacoby, this volume).

Finally, an effective climate policy architecture should be flexible enough to adapt to new information about climate science, as well as development, and other economic and technological factors. It should allow for informed updating of objectives in response to new information. Continued research on climate science and economics can inform the stringency of policies, and learning about the effectiveness and costs of various policy mechanisms can inform subsequent policy design. This flexible policy framework might also provide the opportunity for fruitful linkages with trade, development, and other policy agendas.

The world's first step to address global climate change, in the Kyoto Protocol, was not perfect. The next step does not need to be perfect either, but it ought to be an improvement. A next step needs to be taken, and it should reflect what has been learned through the experience thus far in the design and implementation of international climate change policy. There is no simple, universally accepted way for the world to move forward on this exceptionally difficult set of challenges. The climate policy architecture built for the second commitment period and beyond needs to provide the basis for continued efforts to address the problem, as well as the flexibility to adapt to new information. A policy architecture with these characteristics may be one that can secure sufficient international political support to move forward.

References

Adger, W. N. (2006). "Vulnerability," *Global Environmental Change* 16: 268–281.

Aldy, Joseph E. (2004). "Saving the Planet Cost-Effectively: The Role of Economic Analysis in Climate Change Mitigation Policy," in R. Lutter and J. F. Shogren (eds.), *Painting the White House Green: Rationalizing Environmental Policy Inside the Executive Office of the President*, Washington, DC: Resources for the Future Press, pp. 89–118.

Aldy, Joseph E., Richard Baron, and Laurence Tubiana (2003). "Addressing Cost: The Political Economy of Climate Change," in *Beyond Kyoto: Advancing the International Effort Against Climate Change*, Arlington, VA: Pew Center on Global Climate Change, pp. 85–110.
Aldy, Joseph E., Scott Barrett, and Robert N. Stavins (2003). "Thirteen Plus One: A Comparison of Global Climate Policy Architectures," *Climate Policy* 3(4): 373–397.
Aldy, Joseph E., Peter R. Orszag, and Joseph E. Stiglitz (2001). "Climate Change: An Agenda for Global Collective Action," paper presented at workshop on the Timing of Climate Change Policies, Pew Center on Global Climate Change, Washington, DC, October 2001, AEI-Brookings Joint Center on Regulatory Affairs Related Publication.
Bodansky, Daniel (2004). *International Climate Efforts Beyond 2012: A Survey of Approaches*. Arlington, VA: Pew Center on Global Climate Change.
Hahn, Robert W. (1998). *The Economics and Politics of Climate Change*, Washington, DC: American Enterprise Institute Press.
Mendelsohn, Robert, Ariel Dinar, and Larry Williams (2006). "The Distributional Impact of Climate Change on Rich and Poor Countries," *Environment and Development Economics* 11: 158–178.
Miranda, Marie Lynn, Joseph E. Aldy, Anna E. Bauer, and William H. Schlesinger (2007). "The Justice Dimensions of Global Climate Change," Nicholas School of the Environment and Earth Sciences Working Paper, Durham, NC: Duke University.
Nordhaus, William D. (1998). "Is the Kyoto Protocol a Dead Duck? Are There Any Live Ducks Around? Comparison of Alternative Global Tradable Emissions Regimes," Working Paper, Department of Economics, Yale University, New Haven, CT, 31 July.
Olmstead, Sheila M. and Robert N. Stavins (2006). "An International Policy Architecture for the Post-Kyoto Era," *American Economic Review, Papers and Proceedings* 96(2): 35–38.
Schmalensee, Richard (1998). "Greenhouse Policy Architectures and Institutions," in William D. Nordhaus (ed.), *Economics and Policy Issues in Climate Change*, Washington, DC: Resources for the Future Press, pp. 137–158.
Stewart, Richard B., and Jonathan B. Wiener (2003). *Reconstructing Climate Policy: Beyond Kyoto*, Washington, DC: American Enterprise Institute Press.
Wiener, Jonathan B. (1999). "Global Environmental Regulation: Instrument Choice in Legal Context," *Yale Law Journal* 108: 677–800.

Glossary and abbreviations

Annex B	The list of countries taking on legally binding commitments along with a listing of their actual commitments as defined in the Kyoto Protocol.
Annex I	Annex I Parties consist of countries belonging to the OECD and those with economies in transition. Annex I countries pledged to stabilize their greenhouse gas emissions at their 1990 levels, starting in the year 2000 per Article 4.2 of the UNFCCC.
AR4	Fourth Assessment Report of the IPCC, issued in 2007.
AAU	Assigned Amount Unit. An Annex B country's legally binding commitment under the Kyoto Protocol is referred to as its assigned amount. AAUs serve as the currency for international emission trading under Article 17 of the Protocol.
banking	Saving emission permits for future use in anticipation that these will accrue value over time.
basket	The six types of greenhouse gases of carbon dioxide (CO_2), methane (CH_4), nitrous oxide (N_2O), hydrofluorocarbons (HFCs), perfluorocarbons (PFCs), and sulfur hexafluoride (SF_6) form a basket in which the Kyoto commitments are denominated.
BAU	Business-as-usual. This refers to the projected level of greenhouse gas emissions expected without emission mitigation policies.
benchmark	A measurable variable used as a reference in evaluating the performance of projects or actions.
BTU	British Thermal Unit is a standard measure of the energy content of fuels.
bubble	The idea that emission reductions anywhere within a specific area count toward compliance.

The possibility of forming a "bubble" represents one of the flexible mechanisms included in the Kyoto Protocol.

cap Absolute emission limit.

cap-and-trade A policy that sets an aggregate emission cap, establishes emission permits that sum to the cap, allocates the permits to private firms, and allows firms to buy and sell emission allowances.

carbon dioxide equivalent The amount of CO_2 that would cause the same amount of radiative forcing as the given mixture of CO_2 and other greenhouse gases.

carbon sequestration The uptake and storage of carbon. Trees and plants, for example, absorb carbon dioxide, release the oxygen, and store the carbon.

carbon sink Any reservoir that takes up carbon released from some other part of the carbon cycle. For example, the atmosphere, oceans, and forests are major carbon sinks because much of the CO_2 produced elsewhere on the Earth ends up in these bodies.

CDM Clean Development Mechanism: In Article 12 of the Kyoto Protocol, the parties established the CDM for the purposes of assisting developing countries in achieving sustainable development and helping Annex I parties meet their emission targets; carbon currency: Certified Emission Reduction units (CERs).

CEA Council of Economic Advisers. The Council serves the President of the United States and provides the President with analysis of economic conditions and policies.

CER Certified Emission Reduction. A CER corresponds to a specific amount of emission reduction generated through a Clean Development Mechanism project.

CH_4 Methane. A greenhouse gas whose emission sources include landfills, rice paddies, livestock, coal mines, and natural gas systems.

CO_2 Carbon dioxide. CO_2 is the primary greenhouse gas affected directly by human activities, and its emission sources include fossil fuel combustion, land use change, and cement production.

commitment period The Kyoto Protocol commitment covers a five-year period from 2008 through 2012.

COP Conference of the Parties. The supreme body of the UNFCCC, comprising member countries to the Convention, that meets annually for negotiations.

economies in transition The industrialized countries listed in Annex I or Annex B that are undergoing the process of transition to a market economy. These include some former Soviet republics, including Russia, and several central and eastern European countries.

emission leakage A concept referring to the problem that emission abatement achieved in one location may be offset by increased emissions in unregulated locations.

emission permit In general a tradable entitlement to emit a specified amount of a substance. In the context of the EU ETS, the operators of covered installations need to hold a "greenhouse gas emissions permit" issued by a competent authority in accordance with Articles 5 and 6 of the EU ETS Directive.

EU ETS The European Union Emission Trading Scheme, specified by the Directive 2003/87/EC and launched in January 2005.

EMU European Monetary Union.

ERU Emission Reduction Unit. An ERU corresponds to a specific amount of emission reduction generated through a Joint Implementation project.

GATT General Agreement on Tariffs and Trade. The GATT was the multilateral agreement for international trade policy that was succeeded by the World Trade Organization.

GWP Global Warming Potential. GWPs measure the radiative forcing of a greenhouse gas over a specific period of time relative to that of CO_2. The Kyoto Protocol uses 100-year time horizon GWPs for comparing and aggregating greenhouse gas emissions under the Annex B commitments.

GHG	Greenhouse gas. Any trace gas that does not absorb incoming solar radiation but does absorb long-wavelength radiation emitted or reflected from the Earth's surface. The most important greenhouse gases are water vapor, carbon dioxide, nitrous oxide, methane, chlorofluorocarbons (CFCs), hydrofluorocarbons (HFCs), perfluorocarbons (PFCs), and sulfur hexafluoride (SF_6).
G8	Group of Eight. The G8 includes Canada, France, Germany, Italy, Japan, Russia, the United Kingdom, and the United States.
G20	Group of Twenty. The G20 includes the G8 members and major emerging market countries and provides a forum for finance ministers to address international finance issues.
HFCs	Hydrofluorocarbons. A class of greenhouse gases whose emission sources include refrigeration and insulating foam.
hot air	Hot air corresponds to the amount by which some eastern European countries' Kyoto Protocol emission commitments exceed their expected emissions over 2008–2012 without any abatement changes.
ICAO	International Civil Aviation Organisation.
IDA	The World Bank's International Development Association.
IEA	International Energy Agency.
IET	International Emission Trading, established by Article 17 of the Kyoto Protocol. Countries with Annex B commitments can participate in IET.
IIASA	International Institute for Allied Systems Analysis.
IMF	International Monetary Fund.
IMO	International Maritime Organization.
IPCC	Intergovernmental Panel on Climate Change. The IPCC was created in 1988 by the United Nations Environment Programme and the World Health Organization to advise the international policy community on the latest scientific research on global climate change.

ITER International Thermonuclear Experimental Reactor.

JI Joint Implementation. JI refers to emission mitigation projects between industrialized countries as defined in Article 6 of the Kyoto Protocol.

Kyoto Mechanisms Generic term for the flexible mechanisms of the Kyoto Protocol: bubbles, JI, CDM and international emission trading.

L20 An analog to the G20 whose membership includes the leaders of the G20 member countries. The L20, envisioned by former Canadian Prime Minister Paul Martin, would address a variety of multinational policy issues.

Mercosur The southern common market that aims to promote trade among Argentina, Brazil, Paraguay, Uruguay, and Venezuela. Several other South American countries hold associate member status.

MOP Meeting of the Parties. The supreme body of the Kyoto Protocol, comprising member countries to the Protocol, that meets annually for negotiations.

NAP National Allocation Plan. Under the EU ETS, national governments propose plans for allocating emission permits that require approval by the European Commission.

N_2O Nitrous oxide. A greenhouse gas whose emission sources include fossil fuel combustion, fertilizer manufacture, and agricultural production.

Non-Annex I country All countries that do not belong to Annex I of the UNFCCC, i.e. the developing countries and some countries with economies in transition.

OECD Organisation of Economic Co-operation and Development.

OPEC Organization of the Petroleum Exporting Countries.

PAM Policies and measures. Under the UNFCCC, Annex I countries should undertake policies and measures to demonstrate leadership in addressing global climate change.

PFCs Perfluorocarbons. A class of greenhouse gases whose emission sources include aluminium smelting and semiconductor manufacture.

ppm Parts per million.
ppp Purchasing power parity.
RGGI Regional Greenhouse Gas Initiative.
RTA Regional trade agreement.
SAR Second Assessment Report of the IPCC, issued in 1995.
SF_6 Sulfur hexafluoride. A greenhouse gas whose emission sources include high-voltage equipment insulation and other heavy industry uses.
TAR Third Assessment Report of the IPCC, issued in 2001.
targets and timetables Targets refer to the emission caps for countries and timetables refer to the timing of the commitment period.
US EIA US Energy Information Administration.
US EPA US Environmental Protection Agency.
WTO World Trade Organization.
UNFCCC United Nations Framework Convention on Climate Change. The multilateral agreement that provides the foundation for international climate negotiations.

Index

Page numbers followed by *f* indicate figures; those followed by *n* indicate footnotes; and those followed by *t* indicate tables.